LOGIC
for
Mathematics
and
Computer Science

Stanley N. Burris

Department of Pure Mathematics
University of Waterloo

Prentice Hall
Upper Saddle River, New Jersey 07458

Library of Congress Cataloging–in–Publication Data

Burris, Stanley N.
Logic for mathematics and computer science / Stanley N. Burris.
p. cm.
Includes bibliographical references and index.
ISBN 0–13–285974–2
1. Logic, Symbolic and mathematical. I. Title.
QA9.B86 1998
511.3– –dc21
97–15438
CIP

Acquisitions Editor: George Lobell
Editorial Assistant: Gale Epps
Editorial Director: Tim Bozik
Editor–in–chief: Jerome Grant
Assistant Vice President of Production and Manufacturing: David W. Riccardi
Editorial/Production Supervision: Robert C. Walters
Managing Editor: Linda Mihatov Behrens
Executive Managing Editor: Kathleen Schiaparelli
Manufacturing Buyer: Alan Fischer
Manufacturing Manager: Trudy Pisciotti
Marketing Manager: Melody Marcus
Marketing Assistant: Jennifer Pan
Creative Director: Paula Maylahn
Art Director: Jayne Conte
Cover Designer: Bruce Kenselaar
Cover photo: Orange and black geometric design. Photo © by Susan Meller/ Design Library, NY.

Printed in the United States of America
10 9 8 7 6 5 4 3 2 1

ISBN 0–13–285974–2

Prentice–Hall International (U.K.) Limited, *London*
Prentice–Hall of Australia Pty. Limited, *Sydney*
Prentice–Hall Canada Inc., *Toronto*
Prentice–Hall Hispanoamericana, S.A., *Mexico City*
Prentice–Hall of India Private Limited, *New Delhi*
Prentice–Hall of Japan, Inc., *Tokyo*
Simon & Schuster Asia Pte. Ltd., *Singapore*
Editora Prentice–Hall do Brasil, Ltda., *Rio de Janeiro*

To my son, Kurosh

Contents

Preface

Mathematical logic has been, in good part, developed and pursued with the hope of providing practical algorithmic tools for doing reasoning, both in everyday life and in mathematics, first by hand or mechanical means, and later by electronic computers. In this text elementary traditional logic is presented side by side with its algorithmic aspects, i.e., the syntax and semantics of first–order logic up to completeness and compactness, and developments in theorem proving that were inspired by the possibilities of using computers. Here we are referring to Robinson's resolution theorem proving, and to the Knuth–Bendix procedure to obtain term rewrite systems, both using the key idea of (most general) unification.

Thus we find the choice of topics important as well as accessible to a wide range of students in the mathematical sciences. These topics are rich in basic algorithms, giving the students a desirable hands–on experience. *No background in abstract algebra or analysis is assumed,* yet the material is definitely *mathematical* logic, logic for mathematics and computer science that is developed and analyzed using mathematical methods.

One simple theme is that there are really several self–contained proof systems of interest to mathematical logic, some more suitable than others for particular kinds of questions. Not counting Aristotle's syllogisms, Chapters 1–5 correspond to the five logical systems we consider, and Chapter 6 is a continuation of Chapter 5.

Chapter 1 briefly reviews Aristotle's *syllogisms*, which dominated logic for two millennia, and then presents Boole's work on the *algebra of logic*. Boole's work inaugurated modern mathematical logic. Boole said that he had reduced logical reasoning to reasoning about equations concerning classes of objects, and for the latter he employed methods analogous to well–known methods of working with equations in elementary algebra. The algebra of logic as applied in the *calculus of classes* was the main form of mathematical logic for half a century and attracted such well–known persons as Lewis Carroll, the author of *Alice's Adventures in Wonderland.*

Chapter 2 examines *propositional logic* in considerable detail. Propositional logic is similar to the calculus of classes, but here the focus is on reasoning about propositions, not classes of objects. In modern treatments of mathematical logic it has completely replaced the calculus of classes, in good part because it provides a convenient stepping stone to the study of first–order logic (where you are allowed to quantify over elements).[1]

After studying the notion of truth and truth–equivalence in the propositional logic, we look at PC, a Frege–Hilbert style proof system. PC is designed to be reasonably easy to learn and work with, and to provide lots of simple exercises for students. It is related to the system RS (*Rosser's System*) used by Copi in his text *Symbolic Logic*, a system that he derived from earlier work of Rosser. However PC is significantly different from RS as it is not based on natural deduction and has a restricted replacement axiom.

Next, we look at a *resolution* version of propositional logic because resolution theorem proving is a mainstay of modern automated theorem proving. Both the Davis–Putnam algorithm and general resolution theorem proving are explained. The chapter concludes with examples of sets of clauses that have been used to show limitations on the speed of general resolution theorem proving.

These notes are unusual in the detail in which they look at propositional logic. This choice was made because propositional logic is a good training ground for students and also because it is an area of interest in modern automated theorem proving, for two reasons. First, the work of Löwenheim, Skolem, and Herbrand showed that one can reduce any mathematical question to a question about the satisfiability of a sequence of propositional formulas that one can algorithmically generate from the mathematical question. We will see how to do this in Chapter 5. The second reason is that rather modest problems in propositional logic can be very difficult for computers. (This is directly connected with the outstanding open question of whether or not polynomial time is the same as nondeterministic polynomial time, i.e., does P = NP?)

[1]The calculus of classes more naturally leads to the calculus of relations. Then, by the addition of quantifiers over relations, it becomes essentially a second–order logic. Second–order logic is rather difficult to study, since it lacks a decent set of axioms and rules of inference, so we will have little to say about it.

Chapter 3 turns to *equational logic*. Reasoning with equations is one of the most basic skills we learn in mathematics, and, as already mentioned, Boole's calculus of classes was based on reasoning about equations. In this chapter we first show that the rules one learns in high school are all one needs to do proofs with equations. This was first rigorously proved by G. Birkhoff in the mid–1930s. However, from the computational point of view, this is not particularly helpful in finding a proof that an equation follows from some other equations. This leads us into the study of term rewrite systems, a proposal of Knuth to use "directed equations" to expedite the search for proofs in equational logic. In this section we learn one of the fundamental algorithms of automated theorem proving, namely, how to find the most general unifier of two terms. This algorithm is then used in the Knuth–Bendix procedure, a procedure to convert equations into directed equations.

Chapter 4 looks at resolution theorem proving for *predicate clause logic* as developed by Robinson in the mid–1960s. His goal was to short–circuit the method of Löwenheim, Skolem, and Herbrand by stopping at an intermediate step of their procedure (to reduce a mathematical statement to a sequence of propositional formulas), namely, at a step when one has clauses, and then apply resolution to the clauses to arrive at the desired proof. It was in this work that Robinson used most general unifiers, also employed by Knuth a few years later in his work on term rewrite systems. One of the best known logic programming languages, Prolog, incorporates resolution theorem proving.

Chapter 5 returns to traditional *first–order logic*. There is a leisurely introduction to expressing mathematical ideas in first–order logic in two familiar contexts, namely, that of the natural numbers and that of graphs. Then the general first–order logic is formulated, fundamental equivalences are studied, and the method of putting a formula into prenex form is given. Next follows the crucial technique of skolemization, a procedure to convert a formula into a closely related quantifier–free formula. Skolemization gives the final tool needed to formulate the Löwenheim, Skolem, and Herbrand method of reducing a first–order sentence to a set of clauses for resolution theorem proving, as well as to a sequence of propositional formulas. These reductions lead to the conclusion that any of these systems is *theoretically* adequate to tackle heavy–duty problems in mathematics.

In all the proof systems presented here we are interested in formulating a simple set of axioms and rules of inference for proving theorems, and we make a special point in each case of proving *soundness*, i.e., that the axioms are true and the rules lead to correct conclusions, and *completeness*, i.e., that we have enough axioms and rules to prove all the theorems in the logic under consideration. The completeness proofs tend to be quite different in the different proof systems, with Gödel's completeness theorem for first–order logic being the most difficult.

Chapter 6 gives a detailed proof of Gödel's completeness theorem for a particular first–order proof system, using Henkin's technique of adding witness formulas. For clarity each key step of the proof is allotted its own section.

Another common feature of these logics is that they have a *compactness theorem*. Compactness basically says that something holds in an infinite situation iff it holds in its finite parts, e.g., an infinite graph is 3-colorable iff each finite subgraph is 3-colorable. The compactness theorems follow easily from the completeness theorems. However, our approach also shows that one can simply bootstrap the compactness theorem from propositional logic to predicate clause logic to first–order logic. This sequence gives a rather simple proof of compactness for first–order logic.

There are four appendices. Appendix A gives a simple *timetable* for the development of mathematical logic and computing. Appendix B is a worksheet detailing the steps in a logical development of the natural numbers from *Peano's axioms*—I like for my students to be familiar with this classic work before graduating, and it is an excellent review of mathematical induction. I usually give a portion of this on the first problem assignment. Appendix C has explicit suggestions for writing up definitions and proofs *by induction over propositional formulas*. Appendix D is another worksheet, this one providing a detailed proof of the completeness of the elegant *propositional proof system* of Frege and Łukasiewicz.

There is one notational device used in this text that is not so common, namely, the use of two symbols for equality, $\approx$ and $=$. Starting with Chapter 3, Equational Logic, we use the former symbol when we are working with equations in a particular logical system, and the latter in any context to mean "is the same as." For example, the expression $F(x, y) = x \cdot y \approx y \cdot x$ is to be read: $F(x, y)$ is the equation $x \cdot y \approx y \cdot x$.

These notes have been designed to give the student an understanding of the most basic ideas from mathematical logic, namely an introduction to the syntax and semantics of formal proof systems, completeness and compactness, as well as significant connections with automated theorem proving. We hope that this introduction and the historical comments made throughout the notes will help the reader see the natural unfolding of the ideas.

The reader will find supplementary text and access to software at the World Wide Web site

http : //thoralf2.uwaterloo.ca

For correspondence my e–mail address is

snburris@thoralf2.uwaterloo.ca

Acknowledgments: For my initial involvement in the study of automated theorem proving I am much indebted to George McNulty, Nachum Dershowitz, and John Lawrence. This book evolved from a course given by the author on the foundations of formal theorem proving during the Special Year for Logic and Universal Algebra at the University of Illinois at Chicago in the Fall of 1991, at the invitation of Joel Berman and Wim Blok.

Other colleagues have played important roles by listening and providing feedback to portions of this work as it was being developed. In particular I want to mention Fernando Ferreira, Isabel Ferreirim, Ralph Freese, Jaroslav Ježek, Silvia LaFalce, William Lampe, J. B. Nation, James Raftery, Margarita Ramalho, Maria Saramago, Leslie Schwartzman, and Louxin Zhang. The historical studies were improved by Paweł Idziak who provided me with a copy of the first (1928) edition of Hilbert and Ackermann's book on logic. Joohee Jeong helped with development of software for use with the propositional logic.

For help in the latter stages of designing and proofreading this book the contributions of Joel Berman, Dejan Delic, William McCune and Ross Willard are sincerely appreciated. The secretaries in the Department of Pure Mathematics, Debbie Brown and Carolyn Jackson, provided enthusiasm and final proofreading assistance. George Lobell and the staff at Prentice Hall were most helpful in the timely production of this book.

This book was written in *AMS-Latex* using figures produced by *xfig*, and the final document was converted to a high resolution postscript file using *dvips*. Barbara Beeton (AMS Technical Support) provided assistance with the use of *AMS-Latex*.

Thanks go to the National Sciences and Engineering Research Council of Canada for supporting my investigations in universal algebra, logic, and computation, investigations that inspired this book—and to the Faculty of Mathematics in the University of Waterloo for providing excellent working conditions.

Stanley Burris
Waterloo, 1997

The Flow of Topics

Arrangement of the Text

Chapter	Traditional Logic	Universal Algebra	Computational Logic
1	Syllogisms §1.1–1.2		Algebra of logic §1.3–1.4
2	Basics §2.1–2.9		Resolution §2.10–2.13
3	Algebras, terms §3.1–3.3	Equational logic §3.4–3.11	Unification §3.12 TRSs §3.13–3.15
4	Structures, formulas §4.1–4.2		Resolution §4.3–4.9 Ground clauses §4.10
5	Semantics §5.1–5.14		Skolemization §5.12–5.13
6	Completeness §6.1–6.10		

The preceding diagram indicates the basic structure of the text, showing the topics that are normally covered in a logic course, topics from universal algebra, and topics from computer science.

Two fundamental aspects of logic, the foundational and the computational, are presented here. Traditional logic topics can be taught without going into computational aspects by staying with the first column of the diagram, and one can work through the algorithms of the computational side without being too concerned about the traditional theory. One can see possible ways of designing a course by considering the following dependency diagram, where a bold line from a higher topic to a lower topic means that the lower topic depends on the higher topic.

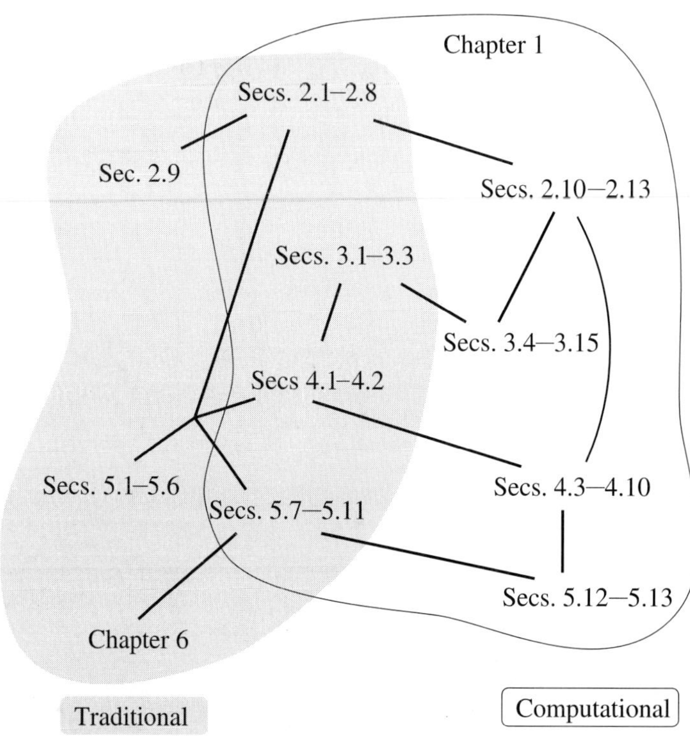

Dependency of topics

Thus one can design a **traditional course** based on the shaded portion. Note that the topic in Sec. 2.9 is not required for other topics. Thus one can, if desired, completely omit the study of propositional proof systems, with the idea that the interesting proof system to be covered will be the first–order case. Or one can replace Sec. 2.9 with any of a number of different proofs systems available. One particularly elegant choice is given in Appendix D. Likewise, one can omit the leisurely introduction to the first–order logic given in Secs. 5.1–5.6.

One can design a very **computational** course by choosing the topics in the loop. Note that Chapter 1 and Secs. 3.4–3.15 are not requisite for other topics, and thus could be optional, but all the other computational topics have important interdependencies.

Personally I prefer a mix of the two, with some variation from year to year. The University of Waterloo has a one–semester elective course in logic that is intended for third– and fourth–year students in *all* areas of the mathematical sciences. A substantial portion of the students in mathematics and computer science take this course.

Most recently I have taught the following topics in this course: Chapter 1 §1.1–1.2, Chapter 2 §2.1–2.13, Chapter 3 §3.1–3.11, Chapter 4 §4.1–4.2, and Chapter 5 §5.1–5.11.

Part I

Quantifier–Free Logics

Chapter 1

From Aristotle to Boole

1.1 SOPHISTRY

Around 800 B.C. Greek civilization emerged from a dark age, of nearly four centuries, that had started shortly after the fall of Troy. The Greeks (in Asia Minor) rediscovered writing, and about this time the epic poems of Homer were recorded. The extraordinary idea of mathematics as a deductive science was introduced by Thales of Miletus (624–545 B.C.). The followers of Pythagoras (566–497 B.C.) adopted the natural number as the fundamental notion from which to explain the universe.

During these developments in mathematics other Greeks were gaining fame for their skill at producing arguments in everyday life that confounded common sense. One such group, the sophists, traveled about the country, teaching their skills of arguing, and holding public competitions. When confronted with such arguments one had difficulty determining whether or not the reasoning was correct—perhaps this was what inspired Aristotle (384–322 B.C.) to single-handedly invent logic in the 4th century B.C. Here are four samples of sophistry that have been favorites over the years:

1. **The liar paradox by Epimenides (ca. 600 B.C.)** can be found in the writings of St. Paul (Titus 1:12–13): *One of themselves, a prophet of their own, said,* Cretans are always liars, wily beasts, lazy gluttons. *This testimony is true.*
2. **The Achilles and the tortoise paradox of Zeno** says that if the tortoise is given a head start in a race then Achilles can never catch the tortoise because he must always get to where the tortoise was.
3. **Protagoras and his student Euathlus in court:** Euathlus was studying the art of argumentation with the master Protagoras so that he could become a lawyer. He had paid Protagoras half the fees in advance and agreed to pay the

other half when he won his first case in court. Euathlus delayed taking up the law practice, so finally Protagoras took him to court in an effort to obtain the other half of his fees.

In court Protagoras argued to the court as follows: If I win the case, then Euathlus should pay me. If I lose the case, then by our agreement he should pay me.

Euathlus, a good student, responded as follows: If I win the case, then by the court's decision I should not pay. However if I lose the case, then by our agreement I should not pay.

4. **The judges and gallows.**[1] To cross a certain bridge of the island of Baratavia (governed by Sancho Panza) one had to answer the questions of four judges. The law was the following: *Anyone who crosses this river shall first take oath as to whither he is bound and why. If he swears to the truth he shall be permitted to pass, but if he tells a falsehood, he shall die without hope of pardon on the gallows that has been set up there.* One day a traveler came who swore that his destination was to die upon the gallows there . . . .

For a fuller discussion of the history of ancient logic, and the role of sophistry, the following classics are recommended: A. Church [**8**], H. Delong [**13**], and W. and M. Kneale [**30**].

Exercises

1.1.1 Suppose Lizalot is the name of the Cretan who said "Cretans are always liars, wily beasts, lazy gluttons." If Lizalot is telling the truth then, since he is a Cretan, he also always lies, i.e., if he is telling the truth he must be lying. So what does St. Paul mean when he says "this testimony is true"?

1.1.2 How can speedy Achilles catch the slow tortoise if every time he gets to where the tortoise was, the tortoise has moved on?

1.1.3 Suppose you are the judge in the Protagoras vs. Euathlus case. What is your verdict regarding Protagoras's suit for the other half of his fees?

1.1.4 If the traveler was telling the truth in the passage from *Don Quixote*, that his destination is to die on those gallows, then the judges must give him safe passage for telling the truth. But then he

[1] From *Don Quixote* by Cervantes (1547–1616).

is lying. If he lies, then the judges must send him to the gallows. But then what he said becomes the truth. What are the judges to do in this case?

1.2 THE CONTRIBUTIONS OF ARISTOTLE

Aristotle's approach to codifying correct modes of reasoning was to concentrate on arguments using the following four types of statements:[2]

A	All S is P.	universal affirmative
E	No S is P.	universal negative
I	Some S is P.	particular affirmative
O	Some S is not P.	particular negative

The word *some* is used as in mathematics, where it means *at least one*. The letters S and P are called the *terms*,[3] S being the *subject* term, and P the *predicate* term. S is also referred to as the *first* term, and P as the *second* term.

One mnemonic for the choice of letters A E I O is:

Aff**I**rmo n**E**g**O**

In the following we will use the symbol $\therefore$ for *therefore*. The simplest arguments are immediate inferences:

All S is P.
$\therefore$ All non–P is non–S.

Aristotle's main interest was the **syllogism** consisting of two *premisses*[4] and one *conclusion*:

Major premiss
Minor premiss
$\therefore$ Conclusion

He required that each of the three statements in a syllogism be in one of the forms A E I O.

---

[2] We are taking license in this section to talk about assertions with quantifiers, even though Part I is about quantifier–free logics.

[3] We will follow Boole's usage here, but starting in Chapter 3, we will give a new definition to the word *term*.

[4] This spelling of the word *premiss* follows that of C. S. Peirce and A. Church. See Church [**8**].

Let us agree that the conclusion has subject S and predicate P:

$$— S — P.$$

Aristotle required that the predicate P appear in exactly one premiss, the *major* premiss, the subject S appear in exactly one premiss, the *minor* premiss, and a third term M, called the *middle* term by Boole, must appear exactly once in each of the two premisses.

Thus, considering only the possible orderings of S, P, and M in the statements in a syllogism, we have the following four possible cases, called **figures:**

	1st Figure	2nd Figure	3rd Figure	4th Figure
Major	$— M — P$	$— P — M$	$— M — P$	$— P — M$
Minor	$— S — M$	$— S — M$	$— M — S$	$— M — S$
Conclusion	$— S — P$	$— S — P$	$— S — P$	$— S — P$

Figure 1.2.1 The four figures for syllogisms

Remark 1.2.2 Aristotle did not recognize the fourth figure. Also, he assumed that A implies I, and E implies O; hence he did not admit the empty class.

We can analyze the syllogisms using Venn diagrams, namely, we start with the following diagrams for the four kinds of statements:

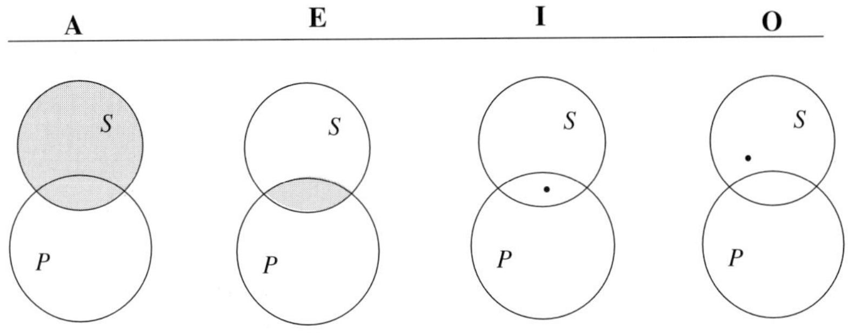

Figure 1.2.3 Venn diagrams for the A,E,I,O statements

Note that shading a region signifies that it is empty, and a • indicates there is an element in a region. A region that is not shaded and does not have a • in it could be empty, or it could have elements in it.

Example 1.2.4 Let us analyze the second figure AEE syllogism,

All P is M .
No S is M .
$\therefore$ No S is P .

We have the following diagram for the premisses:

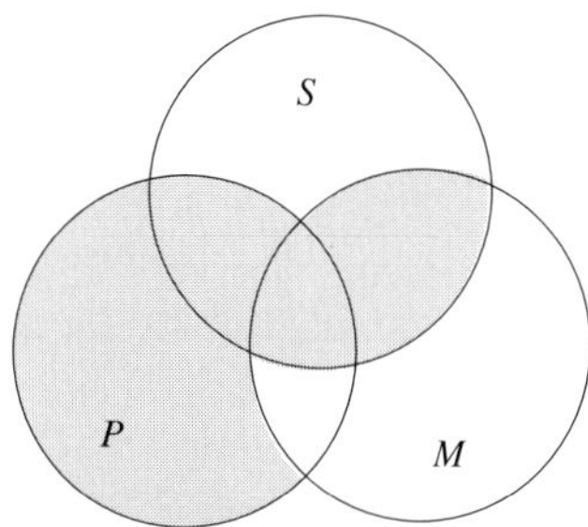

Figure 1.2.5 Venn diagram of premisses

In the diagram we see that S and P have no elements in common, i.e., their intersection has been completely shaded. Thus we can conclude: No S is P; hence the syllogism is valid.

Example 1.2.6 Now let us analyze the second figure AIO syllogism,

All P is M .
Some S is M .
$\therefore$ Some S is not P .

For the premisses we have:

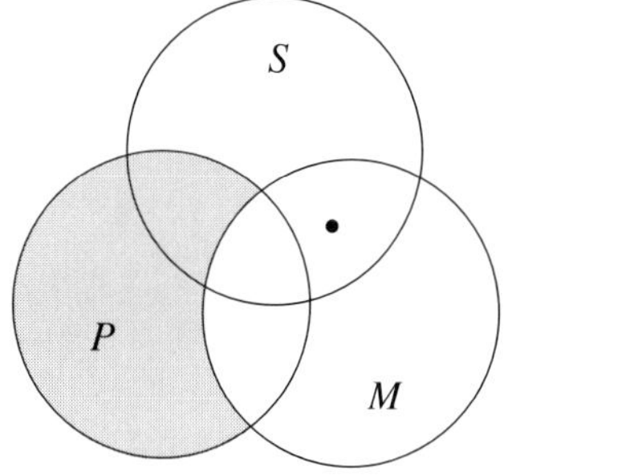

or

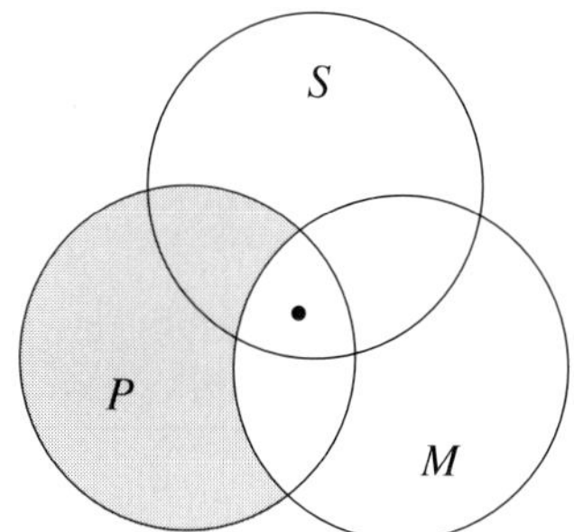

Figure 1.2.7 Venn diagrams of premisses

The second diagram gives a *counterexample* to the syllogism, namely, let $P = S = M = \{a\}$. (There are other counterexamples.)

Since there are four figures, and each of the three lines of a syllogism could be in any of the four forms A E I O, we have $4 \times 4^3 = 256$ syllogisms. Of these only 15 give valid forms of argument; if we assume $M, P, S \neq \emptyset$, then there are 24 valid forms (6 in each figure).

The following table[5] describes the 15 valid syllogisms, indicated by a •; the 9 additional syllogisms marked by a □ are those we add if we assume that the classes M, P, and S are not empty.

Major premiss	$\longrightarrow$	A	A	A	A	E	E	E	E	I	I	I	I	O	O	O	O
Minor premiss	$\longrightarrow$	A	E	I	O	A	E	I	O	A	E	I	O	A	E	I	O
	Conclusion $\downarrow$																
First figure	A	•															
	E					•											
	I	□		•													
	O					□		•									
Second figure	A																
	E		•			•											
	I																
	O		□		•	□		•									
Third figure	A																
	E																
	I	□		•						•							
	O					□		•						•			
Fourth figure	A																
	E		•														
	I	□								•							
	O		□			□		•									

Figure 1.2.8 The valid syllogisms

Example 1.2.9 (*Lewis Carroll*[6]) Consider the premisses:

No pigs can fly.
All pigs are greedy.

This is third–figure EA, so looking at the preceding table we see that no valid syllogism applies. However, if we add the extra (and reasonable) information "There are pigs," then we can apply EAO to conclude "Some greedy creatures cannot fly." (This is the conclusion Carroll reached.)

---

[5]Professor Howard DeLong has kindly permitted the inclusion of this modification of Table 3, p. 21, from his book [**13**] *A Profile of Mathematical Logic*.

[6]This is item 25 on p. 102 of [**7**].

Remark 1.2.10 Euclid (ca. 300 B.C.) was certainly influenced by Aristotle's clarity. Following Aristotle another great school developed in Greece, the stoics. Some of the ancients considered the stoic Chrysippus (280–207 B.C.) to be even greater than Aristotle; unfortunately, little of his work has survived. An enormous amount of writing on logic was done in the Middle Ages. In particular, various clever tricks and mnemonics were invented to help in remembering valid syllogisms. Venn diagrams were not invented until the late 1800s.

For further reading on logic before Boole the reader is encouraged to consult the texts [**8**], [**13**], and [**30**] mentioned at the end of the previous section.

Exercises

1.2.1 Classify the following syllogisms (e.g., as second figure OEA), and use the table on p. 8 to determine if they are valid. Consider the case that all classes are nonempty (Aristotle's assumption) as well as the case that permits classes to be empty (modern logic).

a. All pirates are thieves.
No thieves are reliable.
∴ No pirates are reliable.

b. Some pirates are not thieves.
No thieves are reliable.
∴ Some pirates are reliable.

c. Some thieves are not reliable.
Some thieves are not pirates.
∴ Some pirates are not reliable.

d. Reliable people are not thieves.
Some thieves are pirates.
∴ No pirate is reliable.

1.2.2 Use Venn diagrams to analyze the syllogisms in each of the following cases, where the conclusion $\Diamond$ can be any of the four types of Aristotelian statements, i.e., each question asks about a column of the table on p. 8. Consider the case that all classes are nonempty (Aristotle's assumption) as well as the case that permits classes to be empty (modern logic).

a. The first figure EA$\Diamond$.

b. The second figure AE$\Diamond$.

c. The third figure OA◊.
d. The fourth figure IA◊.

1.2.3 What valid conclusions can you draw, using syllogisms, from the following premisses?

a. All pirates are thieves.
No thieves are reliable.

b. Some pirates are not thieves.
No thieves are reliable.

c. Some thieves are not reliable.
Some thieves are not pirates.

d. Reliable people are not thieves.
Some thieves are pirates.

1.3 THE ALGEBRA OF LOGIC

SYMBOLS	$=$
	$\cup, \cap, ', 0, 1$
	variables

The logic of classes was introduced by Boole in 1847 in order to apply techniques of algebra to the process of reasoning. It is easy to see how classes come into consideration if we look at the syllogism

All pirates are thieves.
No thieves are reliable.
∴ No pirates are reliable.

First, we determine that three classes are involved:

$A =$ pirates
$B =$ thieves
$C =$ reliable people.

Then the argument becomes

All A are B.
No B is C.
∴ No A is C.

Since we are now talking about classes, we can express this symbolically by

$$\begin{aligned} A &\subseteq B \\ B \cap C &= 0 \\ \therefore A \cap C &= 0, \end{aligned}$$

where 0 denotes the empty class (and 1 the universe). Actually, Boole did not have a notation for *subclass* ($\subseteq$) but instead worked with *union* ($\cup$), *intersection* ($\cap$), and *complement* ($'$). So rather than $A \subseteq B$ he might have written

$$A \cap B = A,$$

or, even better, he would have expressed this as an equation with one side 0, much as is the practice in solving equations in high school, namely,

$$A \cap B' = 0.$$

Now the argument becomes

$$\begin{aligned} A \cap B' &= 0 \\ B \cap C &= 0 \\ \therefore A \cap C &= 0. \end{aligned}$$

Let us see how we can deduce the third line from the first two. Observe that from the first line, $A \cap B' = 0$, we have

$$A \cap B' \cap C = 0; \qquad (1)$$

and from the second line, $B \cap C = 0$, we have

$$A \cap B \cap C = 0. \qquad (2)$$

Taking the union of (1) and (2), we have (using the fact that $0 \cup 0 = 0$)

$$(A \cap B' \cap C) \cup (A \cap B \cap C) = 0,$$

which we can rewrite as

$$(A \cap C \cap B') \cup (A \cap C \cap B) = 0,$$

and then using a distributive law, we have

$$(A \cap C) \cap (B' \cup B) = 0,$$

or

$$(A \cap C) \cap 1 = 0.$$

But then

$$A \cap C = 0.$$

We have managed to verify the correctness of the original argument by using algebraic manipulations of equations about classes. Methods for manipulating equations about classes form the content of the *calculus of classes*. To be proficient at this you need the fundamental identities of the calculus of classes at your fingertips:

1.	$X \cup X$	$=$	X	idempotent
2.	$X \cap X$	$=$	X	idempotent
3.	$X \cup Y$	$=$	$Y \cup X$	commutative
4.	$X \cap Y$	$=$	$Y \cap X$	commutative
5.	$X \cup (Y \cup Z)$	$=$	$(X \cup Y) \cup Z$	associative
6.	$X \cap (Y \cap Z)$	$=$	$(X \cap Y) \cap Z$	associative
7.	$X \cap (X \cup Y)$	$=$	X	absorption
8.	$X \cup (X \cap Y)$	$=$	X	absorption
9.	$X \cap (Y \cup Z)$	$=$	$(X \cap Y) \cup (X \cap Z)$	distributive
10.	$X \cup (Y \cap Z)$	$=$	$(X \cup Y) \cap (X \cup Z)$	distributive
11.	$X \cup X'$	$=$	1	
12.	$X \cap X'$	$=$	0	
13.	X''	$=$	X	
14.	$X \cup 1$	$=$	1	
15.	$X \cap 1$	$=$	X	
16.	$X \cup 0$	$=$	X	
17.	$X \cap 0$	$=$	0	
18.	$(X \cup Y)'$	$=$	$X' \cap Y'$	DeMorgan's law
19.	$(X \cap Y)'$	$=$	$X' \cup Y'$	DeMorgan's law.

In this chapter we will be using the convention that an equality written with class variables X, Y, Z, etc., such as $X = X \cap (X \cup Y)$ will mean that we are dealing with an equation that holds for all classes.

The *dual* equation of a given equation is obtained by interchanging $\cup$ and $\cap$ as well as 0 and 1. Thus the dual of the distributive law $X \cap (Y \cup Z) = (X \cap Y) \cup (X \cap Z)$ is $X \cup (Y \cap Z) = (X \cup Y) \cap (X \cup Z)$. Indeed in the above table we see that the following pairs of equations are duals: (1,2), (3,4), (5,6), (7,8), (9,10), (11,12), (14,17), (15,16), and (18,19). Equation (13) is *self–dual*, i.e., it is its own dual. Since the dual of each of the fundamental equations is also a fundamental equation, it follows that if we can derive an equation from the listed identities, then we can also derive its dual.

If one looks at Aristotle's syllogisms, one starts with two assertions regarding three terms, and one wants to arrive at one assertion regarding two of the terms. Thus one is quite clearly interested in eliminating one of the terms, the middle term. Boole recognized that when working with equations there was no need to limit oneself to three classes in the premisses, nor was there any need to have just two premisses. The sides of the equations could be quite complicated as well.[7]

Thus Boole saw that it was straightforward to consider arguments with many assertions involving many terms, with several "middle terms" to be eliminated. When translated into the logic of classes the argument would look like

$$
\begin{aligned}
F_1(A_1,\ldots,A_m,B_1,\ldots,B_n) &= 0\\
&\bullet\\
&\bullet\\
&\bullet\\
F_k(A_1,\ldots,A_m,B_1,\ldots,B_n) &= 0\\
\therefore F(B_1,\ldots,B_n) &= 0.
\end{aligned}
$$

Figure 1.3.1 General arguments in the logic of classes

This is an argument in which the terms involved in the premisses are $A_1,\ldots,A_m,B_1,\ldots,B_n$. The middle terms $A_1,\ldots,A_m$ are eliminated, giving a conclusion involving the terms $B_1,\ldots,B_n$.

Let us do an example using a problem taken from Boole's *Laws*, pp. 118–120, noting first the following convention:

Notational Convention 1.3.2 The phrase *if and only if* is abbreviated as *iff*.

[7]Boole used the word *function* to describe the expressions $F(A,B,\cdots)$ he built up from names of classes, the empty class and the universe, and the basic class operations of union, intersection and complement. This use of the word *function* conflicts with our modern usage, so we will simply use the word *expression* in this chapter. In Chapter 3 we will be more precise about how to make equations.

Example 1.3.3 Determine if the following argument is valid, where $a,\ b,\ c,\ d$ are four properties of substances:

If a and b appear, so does precisely one of c or d.
If b and c appear, then both or neither of a, d appears.
If neither of a, b appears, then neither of c, d appears.
If neither of c, d appears, then neither of a, b appears.
$\therefore$ If both a and b do not appear, then neither does c.

Here we have an argument involving four terms and four premisses, and the conclusion eliminates one of the terms. First, we want to set this up as a problem in the logic of classes. Let A be the class of substances in which a appears, etc. Then the translation into equations about classes is as follows, where *now we adopt the notation* AB for $A \cap B$, i.e., we use juxtaposition for intersection (as Boole did), and we use his convention that *intersection precedes union*, i.e., $AB \cup C$ means $(A \cap B) \cup C$:

$$\begin{aligned} AB &\subseteq CD' \cup C'D \\ BC &\subseteq AD \cup A'D' \\ A'B' &\subseteq C'D' \\ C'D' &\subseteq A'B' \\ \therefore A'B' &\subseteq C'. \end{aligned}$$

We express this in the form Boole preferred by converting each line into an equation of the form $F(\cdots) = 0$. We will do the conversion for the first line in detail:

$$\begin{aligned} AB \subseteq CD' \cup C'D \quad &\text{iff} \quad AB(CD' \cup C'D)' = 0 \\ &\text{iff} \quad AB\big((CD')'(C'D)'\big) = 0 \\ &\text{iff} \quad AB\big((C' \cup D)(C \cup D')\big) = 0 \\ &\text{iff} \quad AB(CD \cup C'D') = 0. \end{aligned}$$

We leave it to the reader to check that the following equations use correct conversions of the various lines of the preceding argument:

$$\begin{aligned} AB(CD \cup C'D') &= 0 && (3) \\ BC(AD' \cup A'D) &= 0 && (4) \\ A'B'(C \cup D) &= 0 && (5) \\ C'D'(A \cup B) &= 0 && (6) \\ \therefore A'B'C &= 0. && (7) \end{aligned}$$

Actually this argument in the logic of classes is very easy to justify because we can simply look at equation (5) and apply the distributive law to obtain

$$A'B'C \cup A'B'D = 0,$$

and thus we immediately have

$$A'B'C = 0.$$

Here we have used the simple observation that $A \cup B = 0$ is equivalent to both $A = 0$ and $B = 0$, a powerful tool in the calculus of classes.

Let us make the problem a bit more challenging by changing the original conclusion to

$\therefore$ Not all of a, b, and c appear.

This translates into the equational conclusion

$$\therefore ABC = 0. \tag{8}$$

To see that this follows from premisses (3)–(6) we need to do a bit more work. From equation (3), using a distributive law, we have

$$ABCD \cup ABC'D' = 0,$$

so

$$ABCD = 0. \tag{9}$$

Next, using (4) and a distributive law we have

$$ABCD' \cup A'BCD = 0,$$

so

$$ABCD' = 0. \tag{10}$$

Combining (9) and (10) by taking the union, we have

$$ABCD \cup ABCD' = 0,$$

and applying a distributive law gives

$$ABC(D \cup D') = 0,$$

so

$$ABC \cap 1 = 0,$$

or

$$ABC = 0,$$

as desired.

We can take the union of our two conclusions and arrive at a stronger conclusion, namely,

$$A'B'C \cup ABC = 0. \tag{11}$$

Can we further strengthen this conclusion about the relationships among A,B,C? This was the issue Boole tackled, namely, how to obtain the strongest possible conclusion when you eliminate certain terms, i.e., classes. This problem, called the *Elimination Problem*, was destined to become one of the cornerstones of the algebra of logic. As it turns out for our example, equation (11) is the strongest conclusion one can draw, i.e., it is the solution to the elimination problem in this case. We will see why in the next section on the method of Boole.

Example 1.3.4 Now we turn to the most complicated example that we will express and analyze in the logic of classes.

1. Good–natured tenured mathematics professors are dynamic.
2. Grumpy student advisors play slot machines.
3. Smokers wearing Hawaiian shirts are phlegmatic.
4. Comical student advisors are mathematics professors.
5. Untenured faculty who smoke are nervous.
6. Phlegmatic tenured faculty members who wear Hawaiian shirts are comical.
7. Student advisors who are not stock market players are scholars.
8. Relaxed student advisors are creative.
9. Creative scholars who do not play slot machines wear Hawaiian shirts.
10. Nervous smokers play slot machines.
11. Student advisors who play slot machines are nonsmokers.
12. Creative stock market players who are good–natured wear Hawaiian shirts.
13. Therefore no student advisors are smokers.

If we take the universe of this discussion to be faculty members, then we have the following 13 classes and their complements involved in this argument, and the translation of the argument into equations:

Class	Defined by	Complement
A	good–natured	grumpy
B	tenured	untenured
C	mathematics professor	
D	dynamic	phlegmatic
E	wears Hawaiian shirts	
F	smoker	nonsmoker
G	comical	
H	relaxed	nervous
I	stock market player	
J	scholar	
K	creative	
L	plays slot machines	
M	student advisor	

	Equations
1.	$ABCD' = 0$
2.	$A'ML' = 0$
3.	$FED = 0$
4.	$GMC' = 0$
5.	$B'FH = 0$
6.	$D'BEG' = 0$
7.	$MI'J' = 0$
8.	$HMK' = 0$
9.	$KJL'E' = 0$
10.	$H'FL' = 0$
11.	$MLF = 0$
12.	$KIAE' = 0$
13.	$\therefore MF = 0.$

Figure 1.3.5 A complicated puzzle problem[8]

The next picture gives a favorite strategy[9] for tackling such puzzle problems by hand. We start with the conclusion that we want to draw, namely, $MF = 0$. For any class X we have $MF = MFX \cup MFX'$. Now, we look over the shorter premisses to see if there is a class X such that either $MFX = 0$ or $MFX' = 0$. Say we have $MFX' = 0$. Then to show $MF = 0$ it suffices to show $MFX = 0$. The strategy is to repeat this procedure with MFX instead of MF, etc.

What happens if no such X exists? Then we choose an X such that the fewest number of additional classes $Y_1, \ldots, Y_k$ are needed to have $MFXY_1 \cdots Y_k = 0$.

For this example we see that $MFL = 0$ by premiss 11. Thus we need only show $MFL' = 0$. Next, we observe that $MFL'A' = 0$ by premiss 2, and so forth.

---

[8]Lewis Carroll's most fascinating puzzle problems, from the end of the 19th century, were not published until 1977 (see [**3**]). This example was inspired by Problem 32 on p. 403 of [**3**].

[9]This method is a slight variation on the tree method developed and used by Lewis Carroll (see [**3**]).

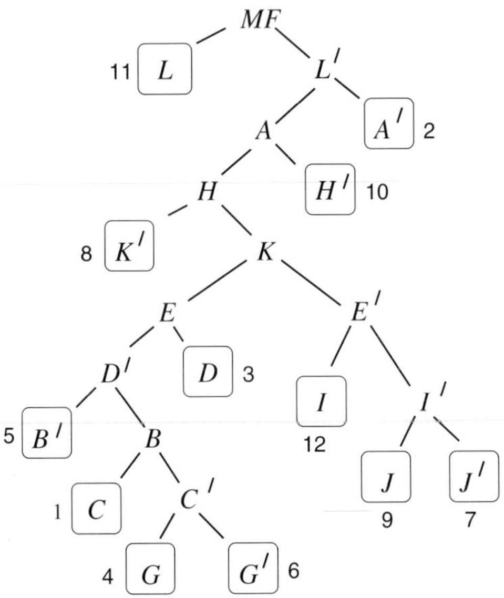

Figure 1.3.6 Showing the validity of the argument

Note that each branch starting at MF terminates in a box with a number beside it. The box indicates that we have enough information to guarantee that the intersection of the classes along the branch is 0, and the number attached to the box tells us the number of a premiss that guarantees that the product is 0. For example, looking at the branch that terminates with the attached number 8, we see that the product $MFL'AHK' = 0$ because premiss 8 says that $HMK' = 0$, and $MFL'AHK' = (HMK')FL'A = 0FL'A = 0$.

We have managed to construct a tree such that each branch terminates in a box, which means that the conclusion follows from the premisses. Looking at the numbers used in this proof we see that every premiss appears attached to some box, i.e., we used all the premisses to prove this result.

Exercises

1.3.1 Translate the following assertions as equations in the form $F = 0$ in the logic of classes, using the indicated universe in each case.

a. A professor who spends too much time at the student donut shop will be cheerful and overweight. (*Universe*: people)

b. A student training to be a computer scientist is either a mathematics major or an engineering major. (*Universe*: students)
c. Philosophers like to give advice to people who are happier than they are. (*Universe*: people)
d. Well begun is half done. (*Universe*: jobs)
e. A stitch in time saves nine. (*Universe*: people)
f. An ounce of prevention is worth a pound of cure. (*Universe*: people)
g. One who hesitates is lost, but one will be in trouble if one does not look before leaping. (*Universe*: people)
h. Practice makes perfect. (*Universe*: musicians)
i. An apple a day keeps the doctor away. (*Universe*: children)
j. A book can be either accurate or interesting, but not both. (*Universe*: books)
k. A single man looks twice, but a married man can look only once. (*Universe*: men)
l. A paper a day means promotion and pay, but a paper a year and you are out on your ear. (*Universe*: researchers)
m. Deans who like to smoke are either trying to lose weight or they are worried about the budget. (*Universe*: people)
n. Professors who like to smoke have to keep their office doors closed. (*Universe*: people)
o. An instrument that tickles the human ear has a bow made of hairs from a horse's tail and strings made from cat's entrails. (*Universe*: instruments)

1.3.2 Use the fundamental identities on p. 12 to show that the following arguments are correct.

a. $(A(B \cup C))' = 0$
$((C \cup A)C)' = 0$
$BC = 0$
$\therefore A' \cup B = 0$

b. $(BC' \cup B'C)' = 0$
$(A \cup B')' \cup B = 0$
$B \cup (AC \cup C) = 0$
$\therefore A'B \cup AB' = 0$

c. $(BC)' = 0$
$AC = 0$
$\therefore (A' \cup B)' = 0$

d. $(BC \cup B')' = 0$
$\therefore BC' = 0$

e. $(A \cup B \cup C)' = 0$
$AB' = 0$
$\therefore (B \cup C)' = 0$

f. $AC = 0$
$(B(A \cup B))' = 0$
$\therefore (A' \cup B)' = 0$

g. $AB = 0$
$A(B \cup C) = 0$
$\therefore AC = 0$

h. $AB = 0$
$(B \cup C)' = 0$
$\therefore AC' = 0$

i. $AB \cup BC = 0$
$(BC)'B = 0$
$\therefore B = 0$

j. $(A \cup D)(A \cup B \cup C) = 0$
$(C \cup D)C' = 0$
$\therefore D = 0$

1.4 THE METHOD OF BOOLE, AND VENN DIAGRAMS

The basic objective in the calculus of classes is to find methods to answer the following two questions, given some equations $F_1 = 0, \ldots, F_n = 0$:

1. Does a given equation $F = 0$ follow from them?
2. [ELIMINATION PROBLEM] What is the most general equation $F = 0$, involving a particular subset of the variables, that follows from them?

A key concept in Boole's work is that of a *constituent*, namely, an intersection of classes or their complements such that each class or its complement appears, but not both, e.g., $A' \cap B' \cap C$ is a constituent if we are working with the three classes A, B, C. There are eight constituents for the classes A, B, C, namely,

$$ABC, ABC', AB'C, AB'C', A'BC, A'BC', A'B'C, A'B'C'.$$

These correspond to the eight regions in the following Venn diagram for the classes A, B, C. Venn diagrams offer a simple and clear picture of what is meant by a constituent, but Boole did not use Venn diagrams—they were not invented until 1881.[10]

[10]Note that the region labeled $A'B'C'$ consists of all points outside the three circles, i.e, the points that are not inside any of the circles. Lewis Carroll objected to this unbounded region that seemed so different from the others and, to achieve harmony, designed a geometrical picture for the constituents such that all regions considered were "confined".

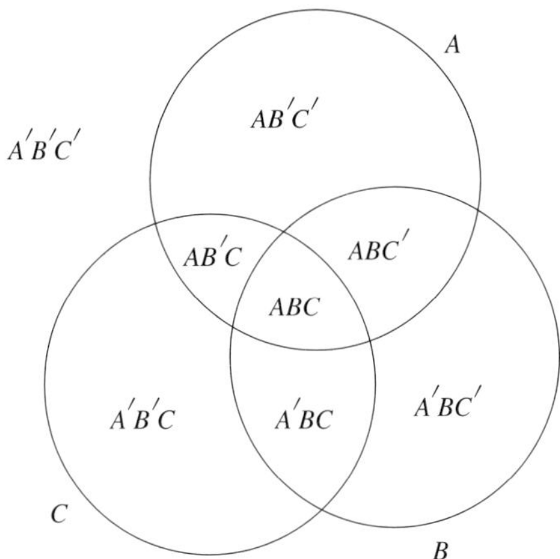

Figure 1.4.1 Constituents of a Venn diagram

What is the key property Venn wanted for his diagrams? It is simply that a Venn diagram for the classes $A_1, \ldots, A_n$ divides the plane into 2^n constituents, each of which is a connected region. For example the following diagram would not qualify for a Venn diagram of the two classes A, B. The constituent AB' is composed of two disconnected regions (the shaded regions).

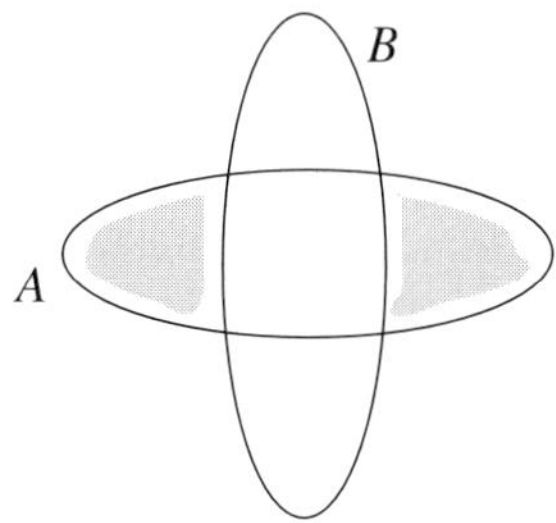

Figure 1.4.2 Not a Venn diagram

Here are (essentially) the diagrams that Venn used for $2 \leq n \leq 5$.

Figure 1.4.3 Venn's Venn diagrams

Remark 1.4.4 It is a bit challenging to draw such diagrams for $n \geq 4$. See how long it takes you to do one for four classes before studying this picture! And for the one with five classes, note that Venn has a region in the center with a hole cut out. He did not find it so easy to draw a diagram for five classes with simply connected classes. And beyond five classes he did not find these pictures helpful.

Since one cannot draw a Venn diagram for four classes with four circles, even with different radii, perhaps ellipses are the next most elegant choice. But it might be surprising to know that for *any* n one can put together a Venn diagram with n–ellipses, i.e., n–ellipses can provide all 2^n constituents. This is not obvious even for $n = 5$. See the article by Branko Grünbaum, "Venn diagrams and independent families of sets", *Mathematics Magazine*, Jan.–Feb. 1975, 12–23.

Boole observed that any $F(\cdots)$ arising in an equation can be expressed as a union of (disjoint) constituents.

Example 1.4.5 We have

$$(AB \cup C)' = A'BC' \cup AB'C' \cup A'B'C'. \tag{12}$$

This follows easily from the fundamental identities. First, we use DeMorgan's laws and the distributive laws as follows:

$$\begin{aligned} (AB \cup C)' &= (AB)'C' \\ &= (A' \cup B')C' \\ &= A'C' \cup B'C'. \end{aligned} \tag{13}$$

Now we have expressed $(AB \cup C)'$ as a union of intersections of the classes or their complements. All that remains is to obtain the constituents needed. First, $A'C'$ from (13) is missing the class B, so we intersect it with $B \cup B'$, which is just 1, to obtain

$$\begin{aligned} A'C' &= A'(B \cup B')C' \\ &= A'BC' \cup A'B'C'. \end{aligned} \tag{14}$$

Likewise, we intersect $B'C'$ from (13) with $A \cup A'$ to obtain

$$\begin{aligned} B'C' &= (A \cup A')B'C' \\ &= AB'C' \cup A'B'C'. \end{aligned} \tag{15}$$

Now, substituting (14) and (15) into (13) gives the desired result:

$$\begin{aligned} (AB \cup C)' &= A'C' \cup B'C' \\ &= A'BC' \cup A'B'C' \cup AB'C' \cup A'B'C' \\ &= A'BC' \cup A'B'C' \cup AB'C', \end{aligned}$$

which is the same as (12) except for the order of the constituents.

Given classes $\vec{A} = A_1, \ldots, A_m$ we say K is an $\vec{A}$ *constituent* if it is a constituent for the classes $A_1, \ldots, A_m$. Boole used the following theorem to *expand* a given expression $F(A_1, \ldots, A_m)$ into its constituents. It says basically that F is a union of $\vec{A}$–constituents.

The next result is Boole's expansion theorem (see his *Laws*, pp. 73–76).

Theorem 1.4.6 [THE EXPANSION THEOREM] Let $F(X_1, \ldots, X_n)$ be given. Then

$$\begin{aligned} F(X_1, \ldots, X_n) = F(1, 1 \ldots, 1)X_1X_2 \cdots X_n \\ \cup\ F(0, 1, \ldots, 1)X_1'X_2 \cdots X_n \ \cup\ F(1, 0, \ldots, 1)X_1X_2' \cdots X_n \\ \cup\ \cdots\ \cup\ F(0, 0, \ldots, 0)X_1'X_2' \cdots X_n'. \end{aligned}$$

Example 1.4.7 Let $F(A, B, C) = (AB \cup C)'$. Boole's expansion theorem for three variables says

$$\begin{aligned} F(A, B, C) = F(1, 1, 1)ABC \ \cup\ F(1, 1, 0)ABC' \\ \cup\ F(1, 0, 1)AB'C \ \cup\ F(1, 0, 0)AB'C' \ \cup\ F(0, 1, 1)A'BC \\ \cup\ F(0, 1, 0)A'BC' \ \cup\ F(0, 0, 1)A'B'C \ \cup\ F(0, 0, 0)A'B'C'. \end{aligned}$$

Now we have

$$
\begin{aligned}
F(1,1,1) &= (11\cup 1)' = 0\\
F(1,1,0) &= (11\cup 0)' = 0\\
F(1,0,1) &= (10\cup 1)' = 0\\
F(1,0,0) &= (10\cup 0)' = 1\\
F(0,1,1) &= (01\cup 1)' = 0\\
F(0,1,0) &= (01\cup 0)' = 1\\
F(0,0,1) &= (00\cup 1)' = 0\\
F(0,0,0) &= (00\cup 0)' = 1.
\end{aligned}
$$

Thus

$$
\begin{aligned}
F(A,B,C) &= F(1,0,0)AB'C' \ \cup\ F(0,1,0)A'BC' \ \cup\ F(0,0,0)A'B'C'\\
&= AB'C' \ \cup\ A'BC' \ \cup\ A'B'C'.
\end{aligned}
$$

For a generalization of the expansion theorem see Exercise 1.4.6 on p. 28.

1.4.1 Checking for Validity

The following uses Jevons' adaptation (from the 1860s) of Boole's constituents to solve the first of the two questions posed on p. 20.

Theorem 1.4.8 The argument

$$
\begin{aligned}
F_1(A_1,\ldots,A_m,B_1,\ldots,B_n) &= 0\\
&\bullet\\
&\bullet\\
&\bullet\\
F_k(A_1,\ldots,A_m,B_1,\ldots,B_n) &= 0\\
\therefore F(B_1,\ldots,B_n) &= 0
\end{aligned}
$$

in the logic of classes is valid (i.e., correct) iff each $\vec{A},\vec{B}$–constituent of F is an $\vec{A},\vec{B}$–constituent of at least one of the F_i's.

Proof. Throughout this brief proof all constituents are assumed to be $\vec{A},\vec{B}$–constituents. Each of the constituents of the F_i's must be 0, since each F_i is a union of its constituents, and we are given that each F_i is 0. Thus if each constituent of F is a constituent of some F_i then each constituent of F is 0. But then F must be 0.

Conversely, if some constituent $K(A_1,\ldots,A_m,B_1,\ldots,B_n)$ of F is not a constituent of any F_i, then choose classes A_i, B_j such that

$K(A_1, \ldots, A_m, B_1, \ldots, B_n)$ is not 0, but all other constituents are 0. This gives a counterexample to the argument, so the argument cannot be correct. □

Example 1.4.9 Let us apply this method to test conclusion (8), namely, $\therefore ABC = 0$ from Example 1.3.3 on p. 14, i.e., to the argument

$$\begin{aligned} AB(CD \cup C'D') &= 0 \\ BC(AD' \cup A'D) &= 0 \\ A'B'(C \cup D) &= 0 \\ C'D'(A \cup B) &= 0 \\ \therefore ABC &= 0. \end{aligned}$$

Since we have the four variables A, B, C, D in this argument, we want to expand both premisses and conclusion using these four variables. Doing so we obtain the following:

$$AB(CD \cup C'D') = ABCD \cup ABC'D' \tag{16}$$
$$BC(AD' \cup A'D) = ABCD' \cup A'BCD \tag{17}$$
$$A'B'(C \cup D) = A'B'CD \cup A'B'CD' \cup A'B'C'D \tag{18}$$
$$C'D'(A \cup B) = ABC'D' \cup AB'C'D' \cup A'BC'D' \tag{19}$$
$$ABC = ABCD \cup ABCD'. \tag{20}$$

Now, we observe that the first constituent $ABCD$ of the conclusion ABC occurs as a constituent in (16), and the second constituent $ABCD'$ is in (17). Thus this is a valid argument.

Let us see how we can analyze this argument by using a Venn diagram for four classes. In the following the shaded area is the part that is 0, or empty, according to the stated premisses:

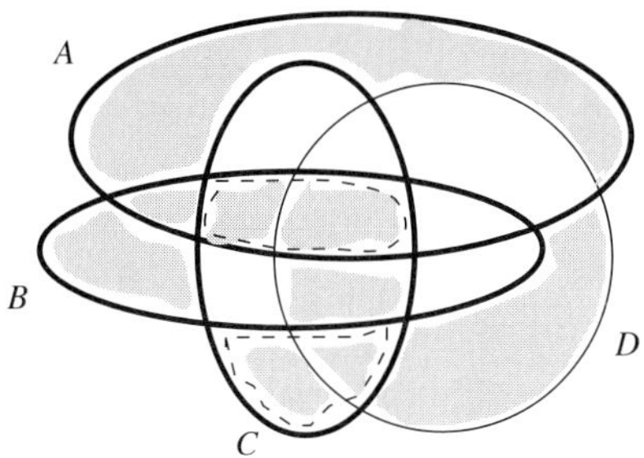

Figure 1.4.10 Venn diagram for the premisses

The result is that nine A, B, C, D–constituents are shaded. To find conclusions involving only A, B, C note that we have drawn the classes A, B, C with a heavier line. Look at the diagram and find the A, B, C–constituents of the heavy–lined regions that are completely shaded. They are the two regions indicated with a dashed line, namely, ABC above and $A'B'C$ below. (*Caution*: The three regions A, B, C do not form a Venn diagram, so you have to make sure all pieces of an A, B, C–constituent are shaded. Note that only one of the two pieces of ABC' is shaded.) Thus $ABC = 0$ and $A'B'C = 0$ are valid conclusions.

1.4.2 Finding the Most General Conclusion

Now let us turn to the second question on p. 20, the elimination problem. We are given the premisses

$$\begin{aligned} F_1(\vec{A}, \vec{B}) &= 0 \\ &\bullet \\ &\bullet \\ &\bullet \\ F_k(\vec{A}, \vec{B}) &= 0 \end{aligned}$$

and we want to find the most general conclusion $F(\vec{B}) = 0$. Boole's first step was to collapse the premisses into a single equation, say, $E(\vec{A}, \vec{B}) = 0$, by letting

$$E(\vec{A}, \vec{B}) = F_1(\vec{A}, \vec{B}) \cup \cdots \cup F_k(\vec{A}, \vec{B}).$$

Now we give Boole's main theorem, the elimination theorem (see his *Laws*, p. 103).

Theorem 1.4.11 [THE ELIMINATION THEOREM] The most general equation $F(\vec{B}) = 0$ that follows from $E(\vec{A}, \vec{B}) = 0$ is obtained by setting

$$F(\vec{B}) = E(1, 1, \cdots, 1, \vec{B})E(0, 1, \cdots, 1, \vec{B}) \cdots E(0, 0, \cdots, 0, \vec{B}),$$

or, more briefly,

$$F(\vec{B}) = \bigcap E(a_1, \dots, a_m, \vec{B}),$$

where each a_i can be either 0 or 1.

Example 1.4.12 Let us apply Boole's solution to the elimination problem to find the most general conclusion that we can regarding the classes A, B, C, given the premisses of Example 1.3.3 on p. 14, namely, we are given

$$\begin{aligned} AB(CD \cup C'D') &= 0 \\ BC(AD' \cup A'D) &= 0 \\ A'B'(C \cup D) &= 0 \\ C'D'(A \cup B) &= 0. \end{aligned}$$

First we put the equations together to obtain the single equation $E(A, B, C, D) = 0$, where $E(A, B, C, D)$ is

$$AB(CD \cup C'D') \ \cup \ BC(AD' \cup A'D) \ \cup \ A'B'(C \cup D) \ \cup \ C'D'(A \cup B).$$

Then we have

$$\begin{aligned} E(A, B, C, 1) &= AB(C \cup 0) \ \cup \ BC(0 \cup A') \ \cup \ A'B'(C \cup 1) \ \cup \ 0(A \cup B) \\ &= ABC \ \cup \ BCA' \ \cup \ A'B' \\ E(A, B, C, 0) &= AB(0 \cup C') \ \cup \ BC(A \cup 0) \ \cup \ A'B'(C \cup 0) \ \cup \ C'(A \cup B) \\ &= ABC' \ \cup \ BCA \ \cup \ A'B'C \ \cup C'A \ \cup \ C'B. \end{aligned}$$

Thus

$$\begin{aligned} F(A, B, C) &= (ABC \cup BCA' \cup A'B') \\ &\qquad (ABC' \cup BCA \cup A'B'C \cup C'A \cup C'B) \\ &= ABC \cup A'B'C, \end{aligned}$$

so the most general conclusion about A, B, C is

$$ABC \ \cup \ A'B'C \ = \ 0.$$

We can also check this using the Venn diagram on p. 25. The two A, B, C–constituents that are shaded are indicated by the dashed lines, namely, ABC and $A'B'C$. Thus $ABC \cup A'B'C$ gives the union of the shaded A, B, C–constituents, and thus $ABC \cup A'B'C = 0$ is the most general conclusion.

Exercises

1.4.1 Use the fundamental identities on p. 12, but not Theorem 1.4.6 on p. 23, to answer the following:

a. Find the A, B–constituents of $(AB' \cup A'B)'$.
b. Find the A, B, C–constituents of $(AB' \cup A'B)'$.
c. Find the A, B, C, D–constituents of $(AB' \cup A'B)'$.

1.4.2 Use Theorem 1.4.6 on p. 23 to answer the following:

a. Expand $F(A, B) = (AB \cup A'B')'$ into A, B–constituents.
b. Expand $F(A, B, C) = (AB \cup A'B')'$ into A, B, C–constituents.
c. Expand $(AB \cup A'B')'$ into A, B, C, D–constituents.

1.4.3

a. Expand a general $F(A, B)$ into A, B, C–constituents.
b. Expand a general $F(A, B, C)$ into A, B, C, D–constituents.

1.4.4

a. Prove that

$$F(A, \vec{B}) = F(1, \vec{B})A \cup F(0, \vec{B})A'.$$

This is called the *expansion of F about A*.

b. Expand $(AB_1)'(A \cup B_2)$ about A.

1.4.5

a. Prove that

$$F(A_1, A_2, \vec{B}) = F(1, 1, \vec{B})A_1A_2 \cup F(1, 0, \vec{B})A_1A_2' \\ \cup F(0, 1, \vec{B})A_1'A_2 \cup F(0, 0, \vec{B})A_1'A_2'.$$

This is called the *expansion of F about* A_1, A_2.

b. Expand $(A_2B_1)'(A_1 \cup B_2)$ about A_1, A_2.

1.4.6 [Boole, *Laws*, p. 73] For $\vec{A} = A_1, \ldots, A_m$ prove that

$$F(\vec{A}, \vec{B}) = \bigcup_K F(\sigma_K, \vec{B})K,$$

where K runs over the A–constituents, and σ_K is the unique sequence of 1's and 0's that makes K evaluate to 1. This is called the *expansion of F about* $\vec{A}$.

1.4.7 In the following use Venn diagrams to determine if the argument is correct.

a. $(AB)'A = 0$
$(AC)'B' = 0$
$\therefore C' = 0$

b. $(A \cup C')' = 0$
$(A'C)'(BC)' = 0$
$\therefore (BC')' = 0$

c. $(A' \cup B)' = 0$
$B' \cup C' = 0$
$\therefore A \cap B' = 0$

d. $(A' \cup B)' = 0$
$(A \cup C)' = 0$
$\therefore AB' = 0$

e. $(B' \cup C)' = 0$
$(A' \cup B)' = 0$
$\therefore (A \cup C)C = 0$

f. $(AB)'A = 0$
$(C \cup D')' = 0$
$(A \cup B \cup D)' = 0$

g. $(A' \cup B)' = 0$
$(B' \cup BC)' \cup AC' = 0$
$\therefore BD = 0$

h. $(AB \cup B')' = 0$
$(A'D \cup BCD)' = 0$
$\therefore BC' = 0$

i. $D \cup (A \cup B)' = 0$
$(B' \cup D)' = 0$
$\therefore AD \cup B = 0$

j. $(A \cup C \cup D)' = 0$
$CB = 0$
$BD = 0$
$\therefore (A' \cup B)' = 0$

1.4.8 For the arguments in Exercise 1.4.7 use Theorem 1.4.8 on p. 24 to determine the validity of the argument.

1.4.9 From each set of premisses in Exercise 1.4.7 use Venn diagrams to find the most general conclusion one can draw involving the classes A, B.

1.4.10 From each set of premisses in Exercise 1.4.7 use Boole's Elimination Theorem (Theorem 1.4.11, p. 26) to find the most general conclusions involving the classes A, B.

1.4.11 Repeat Exercise 1.4.9, parts f–j, this time using Venn diagrams to find the most general conclusions involving the classes A, B, C.

1.4.12 Repeat Exercise 1.4.10, parts f–j, this time using Theorem 1.4.11 on p. 26 to find the most general conclusions involving the classes A, B, C.

1.5 HISTORICAL REMARKS

George Boole (1815–1864)
1847 - *The Mathematical Analysis of Logic*
1854 - *An Investigation of the Laws of Thought*

Mathematical logic officially arrived in 1847 with the publication of Boole's *The Mathematical Analysis of Logic,* and in 1854 in an expanded form, *An Investigation of the Laws of Thought.* Boole said in the latter (p. 1):

> The design of the following treatise is to investigate the fundamental laws of the operations of the mind by which reasoning is performed; to give expression to them in the symbolical language of a calculus, and upon this foundation to establish the science of Logic and construct its method ....

One could write a substantial treatise analyzing Boole's work on logic. Certainly Boole was convinced that ordinary logic dealt with assertions that could be considered as assertions about *classes* of objects. These could be translated into his symbolic logic of classes as equations. Modern logic deals with more complicated assertions than these.

But there are more troubling aspects of Boole's writings on logic, four of which we will discuss here, namely, (1) his overly zealous attempts to make the algebra of logic behave like the algebra of numbers, (2) his restrictive definition of the sum (or union) of two classes, (3) his claim to be able to handle existential statements with equations, and (4) the use of an all–encompassing universe.

Item 1: One of the most striking features about Boole's work is the extent to which he found that the laws of thought agree with the laws of number. This led him to use the notation $+$ for union, $\times$ for intersection, and $1 - A$ for the complement of A. On the basis of observing that $x(y + z) = xy + xz$ and $xy = yx$ he jumped to the conclusion that the laws of thought are essentially the same as the laws of number, the main exception being $x^2 = x$. With this he proceeded to do strange things like applying Maclaurin expansions and dividing expressions in the logic of classes.

The modern viewpoint is that the extent to which the laws of thought agree with the laws of number is pure coincidence, and one is in no way justified in invoking the tools of analysis, etc. So why did Boole make this leap? At the time Boole was developing his logic of classes it was not a mathematical subject, and there were no axioms to describe what one could and could not do with classes. Rather than taking the time to develop this as an independent subject by finding a suitable set of axioms, i.e., what we would now call Boolean algebra, he simply borrowed methods from the traditional algebra and calculus.

Consequently, many of the proofs that Boole gave to justify his methods are best described as "complete nonsense". Here is a sample (starting on p. 101 of *An Investigation of the Laws of Thought*): from

$F(A) = 0$ show that $F(1)F(0) = 0$. First Boole correctly uses the expansion

$$F(A) = F(1)A + F(0)(1 - A) = 0.$$

Then he says

$$(F(1) - F(0))A + F(0) = 0,$$

a step that needs some caution. If $F(0)$ is not 0, then the statement is false with the interpretation of $+$ as union. One can remedy this by letting both $+$ and $-$ be *symmetric difference*. (See Item 2 for how Boole actually handled $+$, i.e., unions.)

The real fun begins with the next steps, namely,

$$A = \frac{F(0)}{F(0) - F(1)}$$

and thus

$$1 - A = -\left(\frac{F(1)}{F(0) - F(1)}\right).$$

Now he substitutes these fractional expressions for A and $1 - A$ into the fundamental equation $A(1 - A) = 0$ to obtain

$$-\left(\frac{F(0)F(1)}{(F(0) - F(1))^2}\right) = 0\,.$$

Looking at the numerator he derives

$$F(0)F(1) = 0.$$

And he gives two more "proofs".

Item 2: Boole said that he could see no way to interpret the expression $A+B$ if the classes A and B were not *disjoint*. Nonetheless he was quite willing to use the expression $A + B$ in the *intermediate* steps of a proof. His justification was that it was just like using imaginary numbers in the intermediate steps of ordinary algebra — they have no meaning, but they do no harm!

Item 3: Boole also thought that he could use equations to handle existential statements. To say that "some x is y" he would use $vx = vy$, where v is a symbol for an "indefinite class". Nearly four decades later Schröder would point out that this does not work if one is to consider x, y, v as classes (on an equal footing). However one does have the following correct translations into the language of the calculus of classes augmented by $\neq$.

A	All A's are B's	$A \cap B' = 0$
E	No A's are B's	$A \cap B = 0$
I	Some A's are B's	$A \cap B \neq 0$
O	Some A's are not B's	$A \cap B' \neq 0$

Figure 1.5.1 Expressing Aristotelian statements in the logic of classes

Item 4: Boole had only one universe, the class 1 of all things. This concept causes philosophical problems, i.e., is 1 in the class 1? It was soon replaced by the *universe of discourse*, that is, 1 could be any class that had as subclasses the classes in the argument being considered, e.g., 1 could be the class of university students.

Needless to say Boole's book is not used as an introduction to logic, except perhaps to study how so many good ideas could be mixed up with so much confusion. In spite of the turmoil in Boole's proofs of the methods he used, his methods for analyzing arguments were correct. The possibility of reasoning by using algebra caught on, and the weak points of Boole's work were soon cleared up. As we mentioned before, the fundamental identities listed on p. 12 were not the starting point for Boole. Instead he used a formula to expand expressions $F(\cdots)$ as unions of intersections of classes and their complements. The fundamental identities not mentioned by him were soon discovered by those who followed, and for a while they had the discoverer's name attached (Jevons, DeMorgan, Peirce, etc). Today, only the DeMorgan identities (1860) have a name attached to them.

Jevons was one of the most precise scholars to work on Boole's logic of classes. He avoided the nonsense, introduced the modern definition of union in 1864, and in 1870 wrote the textbook *Elementary Lessons in Logic*, which has gone through at least 35 reprintings. He also built a lovely small piano–like machine in 1869 for carrying out calculations in the calculus of classes for arguments involving 4 variables. You can still see the original machine at the Oxford Museum for the History of Science. He wanted to build one for 10 variables until he realized that it would fill the entire wall of his study.

A big practical problem that dogged the calculus of classes was that, in spite of its beauty, the computations could be overwhelming. To apply Boole's methods to the lengthy argument in Example 1.3.4 on p. 16 would require thousands of computations. This led to a major attempt to find a better method. One of the most popular ideas was the Venn diagram, introduced in 1881. As we mentioned earlier in the chapter, Venn used them for arguments of up to 5 variables. Lewis Carroll presented two–dimensional arrays in his book *The Game of Logic*, 1886, that he claimed are practical for arguments involving up to 10 variables. Martin Gardner, the popular mathematics writer for *Scientific American*, has an informative and entertaining book, *Logic Machines and Diagrams*, 1982, on the history of computational methods in logic.

The nineteenth century came to a close with the publication of Ernst Schröder's three volumes of *The Algebra of Logic*, a tribute to the efforts that had followed Boole's introduction of the calculus of classes. Volume I has a survey of the study of equations in the calculus of classes based on a foundation provided by C. S. Peirce. In the 1880s Peirce gave a simple axiomatic foundation for the logic of classes, starting with essentially what we would now call axioms for partially ordered sets with greatest lower bounds (glbs) and least upper bounds (lubs), i.e., axioms for lattices. From these axioms one can prove most of the desirable properties for $+$ and $\times$, where these are given by the lub and glb. Then Peirce said the distributive law easily followed. Proving this to be incorrect seems to have provided considerable stimulus for Schröder (see his Vol. I). Later, Peirce would explain his mistake by saying that he had the flu when he was writing up that paper.

Schröder briefly surveyed the progress on algorithms in the logic of classes, presenting procedures of MacColl, Jevons, Venn, and Peirce. He came to the conclusion that no one had really improved on the method of Boole (as presented in this chapter). He lamented that he could not hope to do justice to all the contributors to this problem as it would "require a thousand volumes". This gives a good indication of the enthusiasm with which the new algebra of logic was welcomed.

In Volume II Schröder points out that one cannot use equations to handle existential quantifiers, that one needs to introduce negated equations. Volume II goes on to analyze, in depth, systems of equations and negated equations in the logic of classes, especially the elimination problem for these systems. Also this volume has Schröder's treatment of the propositional calculus. He concluded that it is really just a restricted version of the calculus of classes.

Volume III looks at the *logic of relations*, as developed by Peirce. Peirce was following up on an idea of DeMorgan, that one had a richer algebraic structure with binary relations R than with classes A, since the relations were closed under *converse* ($\breve{R}$) and *composition* ($R \circ S$) as well as union, intersection, and complement.

The study of equations for the algebra of relations is called the *calculus of relatives*. This becomes very complicated. Schröder exhibited hundreds of expressions $F(R)$ in one binary relation variable R that are pairwise inequivalent. He gave up the search without realizing that there are really infinitely many such expressions $F(R)$ that are pairwise inequivalent. (In the logic of classes there are only four pairwise inequivalent expressions in one variable A, namely, $A, A', A \cup A', A \cap A'$.)

The third volume also has a thorough treatment of Peirce's use of the operations $\bigcup_R$ and $\bigcap_R$, the union and intersection over all relations R. These operations on relations, introduced in the early 1880s, give the expressive power of quantifiers over the relations. (He followed the tradition of Boole's arithmetical notation, using $\sum_R$ and $\prod_R$.)

Now let us try to summarize the work in the algebra of logic. The single main objective was to apply mathematical methods to logical reasoning, to turn logical reasoning, a branch of philosophy, into algebra. This algebra of logic was devoted to algorithmic aspects, to the rules of computing used to determine if an argument was correct, or to find the best conclusion from a given set of premisses. Hence we have the names *calculus of classes* and *calculus of relatives*. A passionate expression of this program was given by Schröder:

> getting a handle on the consequences of any premisses, or at least the fastest methods for obtaining these consequences, seems to me to be one of the noblest, if not the ultimate, goal of mathematics and logic. (Schröder, Vol. III, p. 241)

Two leading figures in mathematical logic would continue the work of Schröder into the twentieth century, namely, Löwenheim and Skolem.

However, another direction was gaining the upper hand at the beginning of the twentieth century, a project to devise a special logic in which one could develop all of mathematics, a logic with a nice notation and a few axioms. This was spearheaded by Whitehead and Russell in their three–volume project called *Principia Mathematica* (1910–1913) based on (1) Peano's notation and ambitious *Formulario* project (of writing up all the great theorems of mathematics) and (2) Frege's brilliant but unpopular work on mathematical foundations for the natural numbers. They thanked Peano for inventing new notation to free them from the "undue obsession" with ordinary algebra. And they adopted the logic of Frege, but not his ill–fated two–dimensional notation, as their logic, with modifications to avoid known contradictions.

Hilbert was fascinated by the *Principia Mathematica* project of Whitehead and Russell and gave lectures on the Principles of Mathematics at Göttingen, starting in 1917. But he was also mindful of what the algebra of logic, with its emphasis on algorithms, could mean for all of mathematics:

> Once logical formalism is established one can expect that a systematic, so–to–say computational, treatment of logical formulas is possible, which would somewhat correspond to the theory of equations in algebra. (Hilbert and Ackermann, 1928, p. 72).

So with Hilbert we have a synthesis of the two programs, the algorithmic and the foundational, in a notation and system following that of Whitehead and Russell. The slender 1928 book, *Foundations of Theoretical Logic*, by Hilbert and Ackermann, set the direction and standards for the development of modern mathematical logic.

By 1930 the logic of classes had virtually vanished from the scene. One consequence was that works written in the notation of Schröder would become difficult to read, and even ignored, whereas Hilbert and Ackermann is easily readable today, and *Principia* is reasonably readable. For example Skolem's elegant 1920 analysis of the decidability of the universal theory of lattices was not "discovered" until 1992, even though it was the second section of a paper whose first section had been combed over for many years, that is, the first section has Skolem's simple proof of Löwenheim's theorem (and hence it has the Löwenheim–Skolem theorem).

The calculus of relatives was revived after World War II, when there was a renewed interest in the algebraization of logic, a subject now called *algebraic logic*. This was initiated by Halmos (polyadic algebras) and Tarski (relation algebras, cylindric algebras).

Chapter 2

Propositional Logic

SYMBOLS	connectives ($\vee, \wedge, \rightarrow, \leftrightarrow, \neg, 0, 1$) variables

2.1 PROPOSITIONAL CONNECTIVES, PROPOSITIONAL FORMULAS, AND TRUTH TABLES

Propositional logic started to appear about twenty–five years after Boole's work with the calculus of classes. The idea was simply that one could work with basic units called *propositions*, each of which was *true* or *false*. However, this viewpoint did not become the dominant one, replacing the logic of classes, until Whitehead and Russell's *Principia Mathematica* appeared in the years 1910–1913.

We assume the reader is as comfortable with the concept of a proposition as with the concepts of points, lines, and natural numbers. Nonetheless, it seems worthwhile to repeat that *a proposition is a statement that is true or false*. We can modify or connect such propositions with connectives like *not, and, or, implies,* and *iff*[1] to form other propositions. Propositions are a key ingredient in correct arguments, in law as well as in mathematics. The object of propositional logic is to study general principles regarding the notion of correct argument.

To obtain general principles we work with *propositional variables* $P, Q, R, \cdots$ rather than with particular propositions. We will use the letter X for our *set of propositional variables*. To make the notation more concise we introduce the following symbols for the *standard*

[1]Following Convention 1.3.2, we use *iff* as an abbreviation for the phrase *if and only if*.

connectives, including the *standard constants*:

1 true				0 false
		¬ not		
∧ and				∨ or
→ implies				↔ iff

To say that $\neg$ is a connective is somewhat of an abuse of terminology, since we apply it to a single proposition, and the constants do not connect anything. But it is simpler to refer to the collection of seven symbols as the set of connectives.

In propositional logic the connectives *not*, *and*, *or*, etc., will be used as in mathematics, e.g., we say $P \vee Q$ holds if *at least one* of P, Q holds.

2.1.1 Defining Propositional Formulas

If we combine propositional variables in a meaningful way with these symbols we obtain propositional formulas, e.g.,

$$(P \vee Q) \rightarrow (P \wedge Q).$$

Although this description of propositional formulas by example may be quite clear to some readers, we still want to give a mathematically precise definition of what is meant by a propositional formula. Our goal is to give an *inductive definition*, that is, to initially describe some strings of symbols that we will call formulas and then to give a procedure to create new formulas from ones that we already have. All the strings of symbols that we can obtain in this manner, and *only* in this manner, will constitute our collection of propositional formulas. [This parallels the inductive description of the natural numbers, namely, starting with 0, and given any natural number n, its successor is also a natural number.]

The *propositional formulas* in the standard connectives are (inductively) defined by the following statements:

- Each propositional variable is a propositional formula.
- 0 and 1 are propositional formulas.
- If F is a propositional formula, then $(\neg\,\mathsf{F})$ is a propositional formula.
- If F and G are propositional formulas, then $(\mathsf{F} \vee \mathsf{G})$, $(\mathsf{F} \wedge \mathsf{G})$, $(\mathsf{F} \rightarrow \mathsf{G})$, and $(\mathsf{F} \leftrightarrow \mathsf{G})$ are propositional formulas.

The propositional formulas are precisely the strings of symbols that we can make using the listed statements.

Notational Convention 2.1.1 In practice we often like to drop the outer parentheses in a formula for readability, putting them back in when we want to use them as part of a larger formula, e.g., $P \to Q$, but $(P \to Q) \to R$.

Remark 2.1.2 We are using *infix* notation for the binary connectives. If we used *prefix* notation, e.g., $\vee PQ$ instead of $P \vee Q$, then we would be able to dispense with parentheses, e.g., $\vee P \wedge QR$ instead of $P \vee (Q \wedge R)$. If we choose to use prefix notation, then we need to convince ourselves that the parentheses are not needed, i.e., that we have *unique readability* of prefix notation.

Notational Convention 2.1.3 In algebra we know that $x + y \cdot z$ means $x + (y \cdot z)$ and not $(x + y) \cdot z$. This is because we have agreed on this convention to make algebraic formulas more readable. Likewise, to make propositional formulas more readable, we adopt *precedence* conventions for the standard connectives, namely:

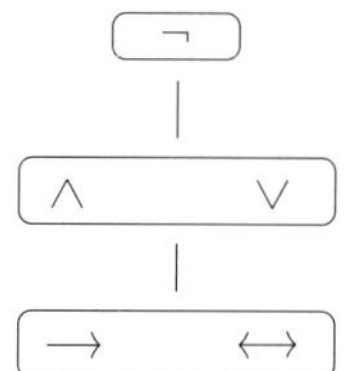

Figure 2.1.4 Precedence convention

Higher connectives in this figure have greater binding power, but we will need parentheses to determine the precedence for those on the same level. For example, $\neg P \to Q \vee R$ means $(\neg P) \to (Q \vee R)$, but $Q \vee R \wedge S$ is ambiguous. We need to know if $(Q \vee R) \wedge S$ or $Q \vee (R \wedge S)$ is correct.

When discussing propositional formulas we will, on occasion, refer to certain components called subformulas.

Definition 2.1.5 The *subformulas* of a formula F are defined inductively by the following conditions:

- The only subformula of a propositional variable P is P itself.
- The only subformula of a constant c is c itself (c is 0 or 1).
- The subformulas of $\neg \mathsf{F}$ are $\neg \mathsf{F}$, and all subformulas of F.

- The subformulas of G □ H are G □ H and all subformulas of G and all subformulas of H, where □ denotes any of the four binary connectives $\vee$, $\wedge$, $\rightarrow$, $\leftrightarrow$.

Example 2.1.6 The subformulas of $(P \rightarrow Q) \vee \neg (P \rightarrow Q)$ are $(P \rightarrow Q) \vee \neg (P \rightarrow Q)$, $\neg (P \rightarrow Q)$, $P \rightarrow Q$, P, Q.

Remark 2.1.7 A brief summary of how to write up inductive definitions over formulas is given in Appendix C.

2.1.2 Truth Tables

A formula by itself is not true or false—it becomes so only after the propositional variables are replaced with concrete propositions. To determine whether such a concrete instantiation is true or false one needs to know the impact of the connectives on truth values as determined by the standard truth tables:

P	$\neg P$
1	0
0	1

P	Q	$P \wedge Q$
1	1	1
1	0	0
0	1	0
0	0	0

P	Q	$P \vee Q$
1	1	1
1	0	1
0	1	1
0	0	0

P	Q	$P \rightarrow Q$
1	1	1
1	0	0
0	1	1
0	0	1

P	Q	$P \leftrightarrow Q$
1	1	1
1	0	0
0	1	0
0	0	1

One can easily justify the choice of truth values in all the tables but the one for "implies."

- $\neg P$ is true iff P is false.
- $P \wedge Q$ is true iff both P and Q are true.
- $P \vee Q$ is true iff at least one of P or Q is true.
- $P \leftrightarrow Q$ is true iff P and Q are both true, or both false.

The truth table for $P \rightarrow Q$ has been a longstanding source of doubt and irritation. If P is true, then of course $P \rightarrow Q$ is true iff Q is true. That takes care of two of the rows of the truth table, the ones with a 1 in the P column. But the dubious part is to say that $P \rightarrow Q$ is always true when P is false. A statement like "If the moon is made of yellow cheese, then I am the Prince of Denmark" would be *true* by this definition, a statement to which one is perhaps not so ready to assign a truth value.

One way out of this problem with the meaning of "implies" is to agree that $P \rightarrow Q$ expresses the same thing as $\neg P \vee Q$. Indeed, this is how we use the word "implies" in mathematics. Thus the statement above becomes "Either the moon is not made of yellow cheese or I am the Prince of Denmark," which is transparently true. With this agreement on the use of "implies" the truth table for $P \rightarrow Q$ is easily seen to be as shown.

Now, given any propositional formula F we have a *truth table* for F, e.g., for $(P \vee Q) \rightarrow (P \wedge Q)$ we have

P	Q	$(P \vee Q) \rightarrow (P \wedge Q)$
1	1	1
1	0	0
0	1	0
0	0	1

A longer version of the truth table includes the truth tables for the subformulas:

P	Q	$P \vee Q$	$P \wedge Q$	$(P \vee Q) \rightarrow (P \wedge Q)$
1	1	1	1	1
1	0	1	0	0
0	1	1	0	0
0	0	0	0	1

Definition 2.1.8 A *truth evaluation* $\mathsf{e} = (e_1, \ldots, e_n)$ for the list $P_1, \ldots, P_n$ of propositional variables is a sequence of n truth values. Given a formula $\mathsf{F}(P_1, \ldots, P_n)$ let $\mathsf{F}(\mathsf{e})$ denote the propositional formula $\mathsf{F}(e_1, \ldots, e_n)$ obtained by replacing each P_i by e_i, and let $\widehat{\mathsf{F}}(\mathsf{e})$ be the truth value of $\mathsf{F}(\mathsf{e})$ when the connectives are evaluated according to their truth tables.

Example 2.1.9 Let $\mathsf{F}(P, Q, R, S)$ be the formula

$$(\neg (P \leftrightarrow Q)) \rightarrow (R \wedge S),$$

and let e be the truth evaluation $(1, 0, 1, 1)$ for P, Q, R, S. Then $\mathsf{F}(\mathsf{e})$ is the propositional formula $(\neg (1 \leftrightarrow 0)) \rightarrow (1 \wedge 1)$, and by carrying out the truth table calculations on this expression, we see that $\widehat{\mathsf{F}}(\mathsf{e}) = 1$.

Notational Convention 2.1.10 If F is a propositional formula, then we use the notation $\mathsf{F}(P_1, \ldots, P_n)$ to indicate that the variables occurring in F are *among* the variables $P_1, \ldots, P_n$. It is *not* necessary for all these variables to occur in F.

Example 2.1.11 Let $F(P, Q, R)$ be the propositional formula $R \to (P \to R)$. Note that only the variables P and R actually occur in the formula.

If $\mathsf{F}(P_1, \ldots, P_n)$ is a propositional formula and e is a truth evaluation for $P_1, \ldots, P_n$, then

e_1	$\cdots$	e_n	$\widehat{\mathsf{F}}(\mathsf{e})$

gives a row of the truth table of F. Furthermore, each row of the truth table of F comes from some truth evaluation e. *Therefore, truth evaluations correspond to individual lines of truth tables.*

Exercises

2.1.1 Fill in the following truth tables:

a.

P	Q	$P \to Q$	$\neg Q \to \neg P$	$\neg P \vee Q$
1	1			
1	0			
0	1			
0	0			

b.

P	Q	$P \leftrightarrow Q$	$(P \to Q) \wedge (Q \to P)$	$(P \wedge Q) \vee (\neg P \wedge \neg Q)$
1	1			
1	0			
0	1			
0	0			

c.

P	Q	$P \wedge (P \vee Q)$	$P \vee (P \wedge Q)$
1	1		
1	0		
0	1		
0	0		

d.

P	Q	$\neg(P \vee Q)$	$\neg P \wedge \neg Q$	$\neg(P \wedge Q)$	$\neg P \vee \neg Q$
1	1				
1	0				
0	1				
0	0				

e.

P	Q	R	$P \wedge (Q \vee R)$	$(P \wedge Q) \vee (P \wedge R)$
1	1	1		
1	1	0		
1	0	1		
1	0	0		
0	1	1		
0	1	0		
0	0	1		
0	0	0		

f.

P	Q	R	$P \vee (Q \wedge R)$	$(P \vee Q) \wedge (P \vee R)$
1	1	1		
1	1	0		
1	0	1		
1	0	0		
0	1	1		
0	1	0		
0	0	1		
0	0	0		

2.2 EQUIVALENT FORMULAS, TAUTOLOGIES, AND CONTRADICTIONS

There are many different ways to say the same thing in our propositional logic. This fact motivates the notion of *equivalent* formulas.

2.2.1 Equivalent Formulas

Definition 2.2.1 We will say that two propositional formulas F and G are *(truth) equivalent*, written $\mathsf{F} \sim \mathsf{G}$, if $\widehat{\mathsf{F}}(\mathsf{e}) = \widehat{\mathsf{G}}(\mathsf{e})$ for every truth evaluation e for the variables, i.e, if they have the same truth tables (using tables with enough variables to include all the variables appearing in either formula).

Proposition 2.2.2

$$1 \sim P \to P \qquad\qquad P \vee Q \sim \neg P \to Q$$

$$0 \sim \neg(P \to P) \qquad\qquad P \wedge Q \sim \neg(P \to \neg Q)$$

$$P \leftrightarrow Q \sim \neg((P \to Q) \to \neg(Q \to P)).$$

Proof. Just look at the truth tables. □

We have just expressed the standard connectives in terms of $\neg, \rightarrow$ (insofar as truth equivalence is concerned). Thus as far as expressive power is concerned, the two connectives $\neg, \rightarrow$ are all we really need to build propositional formulas.

The following is a very useful list of equivalent formulas.

Theorem 2.2.3 [FUNDAMENTAL TRUTH EQUIVALENCES]

1.	$P \vee P$	$\sim$	P	idempotent
2.	$P \wedge P$	$\sim$	P	idempotent
3.	$P \vee Q$	$\sim$	$Q \vee P$	commutative
4.	$P \wedge Q$	$\sim$	$Q \wedge P$	commutative
5.	$P \vee (Q \vee R)$	$\sim$	$(P \vee Q) \vee R$	associative
6.	$P \wedge (Q \wedge R)$	$\sim$	$(P \wedge Q) \wedge R$	associative
7.	$P \wedge (P \vee Q)$	$\sim$	P	absorption
8.	$P \vee (P \wedge Q)$	$\sim$	P	absorption
9.	$P \wedge (Q \vee R)$	$\sim$	$(P \wedge Q) \vee (P \wedge R)$	distributive
10.	$P \vee (Q \wedge R)$	$\sim$	$(P \vee Q) \wedge (P \vee R)$	distributive
11.	$P \vee \neg P$	$\sim$	1	excluded middle
12.	$P \wedge \neg P$	$\sim$	0	
13.	$\neg \neg P$	$\sim$	P	
14.	$P \vee 1$	$\sim$	1	
15.	$P \wedge 1$	$\sim$	P	
16.	$P \vee 0$	$\sim$	P	
17.	$P \wedge 0$	$\sim$	0	
18.	$\neg (P \vee Q)$	$\sim$	$\neg P \wedge \neg Q$	DeMorgan's law
19.	$\neg (P \wedge Q)$	$\sim$	$\neg P \vee \neg Q$	DeMorgan's law
20.	$P \rightarrow Q$	$\sim$	$\neg P \vee Q$	
21.	$P \rightarrow Q$	$\sim$	$\neg Q \rightarrow \neg P$	
22.	$P \rightarrow (Q \rightarrow R)$	$\sim$	$(P \wedge Q) \rightarrow R$	
23.	$P \rightarrow (Q \rightarrow R)$	$\sim$	$(P \rightarrow Q) \rightarrow (P \rightarrow R)$	
24.	$P \leftrightarrow P$	$\sim$	1	
25.	$P \leftrightarrow Q$	$\sim$	$Q \leftrightarrow P$	
26.	$(P \leftrightarrow Q) \leftrightarrow R$	$\sim$	$P \leftrightarrow (Q \leftrightarrow R)$	
27.	$P \leftrightarrow \neg Q$	$\sim$	$\neg (P \leftrightarrow Q)$	
28.	$P \leftrightarrow (Q \leftrightarrow P)$	$\sim$	Q	
29.	$P \leftrightarrow Q$	$\sim$	$(P \rightarrow Q) \wedge (Q \rightarrow P)$	
30.	$P \leftrightarrow Q$	$\sim$	$(P \wedge Q) \vee (\neg P \wedge \neg Q)$	
31.	$P \leftrightarrow Q$	$\sim$	$(P \vee \neg Q) \wedge (\neg P \vee Q)$	

Proof. (Exercise.) □

Proposition 2.2.4 Truth equivalence is an equivalence relation, i.e., for all formulas F, G, H we have

- $\mathsf{F} \sim \mathsf{F}$
- $\mathsf{F} \sim \mathsf{G}$ implies $\mathsf{G} \sim \mathsf{F}$
- $\mathsf{F} \sim \mathsf{G}$ and $\mathsf{G} \sim \mathsf{H}$ implies $\mathsf{F} \sim \mathsf{H}$.

Proof. This follows from the definition of $\mathsf{F} \sim \mathsf{G}$, namely, that $\widehat{\mathsf{F}}(\mathsf{e}) = \widehat{\mathsf{G}}(\mathsf{e})$ for all truth evaluations e for the variables, and the fact that ordinary equality, =, is an equivalence relation. □

2.2.2 Tautologies

There is a close connection between formulas that are always true and truth equivalence.

Definition 2.2.5 A propositional formula F is called a *tautology* if $\widehat{\mathsf{F}}(\mathsf{e}) = 1$ for every truth evaluation e. We also say that F is *identically true*, or *always true*.

Theorem 2.2.6 Two propositional formulas F and G are truth equivalent iff the formula $\mathsf{F} \leftrightarrow \mathsf{G}$ is a tautology.

Proof. Let $P_1, \cdots, P_n$ be the variables occurring in $\mathsf{F} \leftrightarrow \mathsf{G}$. Then $\mathsf{F} \sim \mathsf{G}$ iff $\widehat{\mathsf{F}}(\mathsf{e}) = \widehat{\mathsf{G}}(\mathsf{e})$ for every truth evaluation e of the variables $P_1, \ldots, P_n$. The latter holds iff $(\widehat{\mathsf{F} \leftrightarrow \mathsf{G}})(\mathsf{e}) = 1$ for every truth evaluation e for the variables $P_1, \ldots, P_n$, and this holds iff $\mathsf{F} \leftrightarrow \mathsf{G}$ is a tautology. □

Note that to check a formula F to see if it is a tautology we need to consider only the evaluations e of the variables occurring in F.

2.2.3 Contradictions

Of course, there are formulas that are never true, e.g., $P \wedge \neg P$.

Definition 2.2.7 A formula F is a *contradiction* if $\widehat{\mathsf{F}}(\mathsf{e}) = 0$ for every truth evaluation e.

Obviously F is a tautology iff $\neg \mathsf{F}$ is a contradiction.

Exercises

2.2.1 Verify the truth equivalences stated in Theorem 2.2.3.

2.2.2 Which of the following propositional formulas are tautologies? contradictions? (Give reasons.)

a. $(\neg Q \vee (R \rightarrow (P \wedge S))) \vee (Q \rightarrow R)$
b. $((Q \rightarrow R) \leftrightarrow S) \rightarrow ((P \wedge S) \rightarrow S)$
c. $((R \wedge (S \vee P)) \rightarrow Q) \vee \neg Q$
d. $Q \wedge \neg((R \wedge P) \rightarrow Q)$
e. $\neg(\neg R \rightarrow ((S \wedge Q) \rightarrow S))$

2.2.3 Which pairs of the following propositional formulas are truth equivalent? (Give reasons.)

a. $R \vee (Q \wedge R)$
b. $\neg R \rightarrow (Q \wedge R)$
c. $((\neg R \vee P) \rightarrow P) \wedge P$
d. $\neg R \rightarrow ((S \wedge R) \wedge R)$
e. $\neg((Q \rightarrow (P \vee Q)) \leftrightarrow \neg P)$

2.2.4 Which pairs of the following propositional formulas are truth equivalent? (Give reasons.)

a. $\neg(P \leftrightarrow (R \leftrightarrow P))$
b. $P \vee ((P \leftrightarrow R) \vee P)$
c. $R \vee ((\neg Q \leftrightarrow P) \leftrightarrow Q)$
d. $(R \rightarrow (\neg P \rightarrow P)) \vee P$
e. $(R \leftrightarrow P) \vee ((P \vee (Q \vee R)) \rightarrow P)$

2.3 SUBSTITUTION

Given a formula, say,

$$\mathsf{F} = P \vee \neg(Q \rightarrow P),$$

if we want to *substitute* the formula $\mathsf{G} = P \wedge Q$ for P in F, then we need to put the formula G in place of *every* occurrence of P in F. This would give $\mathsf{G} \vee \neg(Q \rightarrow \mathsf{G})$, or, in full detail,

$$(P \wedge Q) \vee \neg(Q \rightarrow (P \wedge Q)).$$

The key necessary result about substitutions is that they preserve truth equivalence.

Theorem 2.3.1 [SUBSTITUTION THEOREM] If we are given an equivalence $\mathsf{F}(P_1, \ldots, P_n) \sim \mathsf{G}(P_1, \ldots, P_n)$, and $\mathsf{H}_1, \ldots, \mathsf{H}_n$ are formulas, then $\mathsf{F}(\mathsf{H}_1, \ldots, \mathsf{H}_n) \sim \mathsf{G}(\mathsf{H}_1, \ldots, \mathsf{H}_n)$.

Proof. $\mathsf{F} \sim \mathsf{G}$ means $\widehat{\mathsf{F}}(e_1, \ldots, e_n) = \widehat{\mathsf{G}}(e_1, \ldots, e_n)$ for any truth evaluation e for the variables $P_1, \ldots, P_n$. *In particular*, the equation is true whenever $e_1, \ldots, e_n$ are the values of $\mathsf{H}_1, \ldots, \mathsf{H}_n$ at some truth evaluation. Hence $\mathsf{F}(\mathsf{H}_1, \ldots, \mathsf{H}_n) \sim \mathsf{G}(\mathsf{H}_1, \ldots, \mathsf{H}_n)$. □

Example 2.3.2 From the distributive law

$$P \wedge (Q \vee R) \sim (P \wedge Q) \vee (P \wedge R)$$

we can conclude

$$\mathsf{F} \wedge (\mathsf{G} \vee \mathsf{H}) \sim (\mathsf{F} \wedge \mathsf{G}) \vee (\mathsf{F} \wedge \mathsf{H})$$

for any propositional formulas $\mathsf{F}, \mathsf{G}, \mathsf{H}$, e.g., we have

$$(P \vee \neg R) \wedge ((P \to S) \vee (Q \wedge S)) \sim ((P \vee \neg R) \wedge (P \to S)) \vee ((P \vee \neg R) \wedge (Q \wedge S)).$$

Using this same idea we can generalize all the fundamental truth equivalences in Theorem 2.2.3 (p. 44), replacing P, Q, R with any propositional formulas $\mathsf{F}, \mathsf{G}, \mathsf{H}$ to obtain the more general form of the fundamental equivalences (see p. 81).

Exercises

2.3.1 Determine which of the following propositional formulas are substitution instances of the formula $P \to (Q \to P)$. If a formula is indeed a substitution instance, give the formulas substituted for P, Q.

a. $R \to (R \to R)$
b. $R \to (\neg R \to R)$
c. $\neg R \to (R \to R)$
d. $(R \leftrightarrow T) \to ((R \leftrightarrow T) \to (R \leftrightarrow T))$
e. $(P \to (Q \to P)) \to (P \to (P \to (Q \to P)))$
f. $(P \to (Q \to P)) \to (P \to ((P \to P) \to (Q \to P)))$

2.3.2 Determine which of the following propositional formulas are substitution instances of the formula

$$(P \to (Q \to R)) \to ((P \to Q) \to (P \to R)).$$

If a formula is indeed a substitution instance, give the formulas substituted for P, Q, R.

a. $(Q \to (Q \to Q)) \to ((Q \to Q) \to (Q \to Q))$
b. $((P \to P) \to (P \to P)) \to$
$(((P \to P) \to P) \to ((P \to P) \to P))$

c. $((P \to P) \to (P \to P)) \to ((P \to P) \to ((P \to P) \to P))$
d. $(P \to ((Q \to \neg P) \to (P \to \neg Q)))$
$\quad \to ((P \to (Q \to \neg P)) \to (P \to (P \to \neg Q)))$
e. $(P \to ((Q \to \neg P) \to (P \to \neg Q)))$
$\quad \to ((P \to Q) \to (P \to (P \to \neg Q)))$
f. $((P \to P) \to ((P \to P) \to P))$
$\quad \to (((P \to P) \to (P \to P)) \to ((P \to P) \to P))$

2.4 REPLACEMENT

Another key operation on propositional formulas is to replace one occurrence of a subformula with another formula. We refer to *one* occurrence because a formula may have many occurrences of a given subformula.

Example 2.4.1 We noted in Example 2.1.6 (p. 40) that $(P \to Q) \vee \neg(P \to Q)$ has five subformulas. The entire formula, and $\neg(P \to Q)$, occur only once; $P \to Q$ occurs twice; and the variables occur twice each.

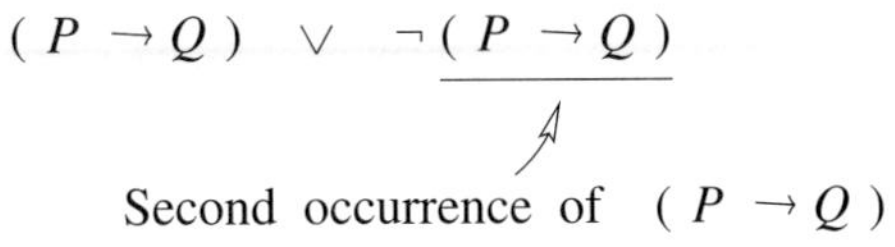

$$(P \to Q) \vee \neg\underline{(P \to Q)}$$

Second occurrence of $(P \to Q)$

Figure 2.4.2 An occurrence of a subformula

Example 2.4.3 If we replace the second occurrence of $P \to Q$ in $(P \to Q) \vee \neg(P \to Q)$ with the formula $\neg P \vee Q$, we obtain $(P \to Q) \vee \neg(\neg P \vee Q)$. We know that $P \to Q \sim \neg P \vee Q$, and it is easy to check that $(P \to Q) \vee \neg(P \to Q) \sim (P \to Q) \vee \neg(\neg P \vee Q)$. Thus we have an example in which replacement of an occurrence of a subformula of a formula with an equivalent formula produces a formula equivalent to the original.

Before we give the main result about replacement, namely, that it also preserves truth equivalence, we need to introduce a standard method of proof in logic courses: *induction on formulas*. This method has obvious similarities with *induction on numbers*.

2.4.1 Induction Proofs on Formulas

Suppose we have an assertion about formulas F, say $\mathsf{Assert}(\mathsf{F})$, that we want to prove is true for all formulas F. One of the most basic methods for doing this is called *induction on formulas*. The following outline gives the necessary steps:

Ground Cases:

- Prove that $\mathsf{Assert}(0)$ and $\mathsf{Assert}(1)$ are true.
- Prove that $\mathsf{Assert}(P)$ is true for P, a propositional variable.

Induction Cases:

- Assume $\mathsf{Assert}(\mathsf{F})$ is true (*induction hypothesis*) and prove $\mathsf{Assert}(\neg\,\mathsf{F})$ is true.
- Assume $\mathsf{Assert}(\mathsf{F})$ and $\mathsf{Assert}(\mathsf{G})$ are true (*induction hypothesis*) and prove $\mathsf{Assert}(\mathsf{F} \,\square\, \mathsf{G})$ is true for $\square$, where $\square$ is any of the four binary connectives $\vee, \wedge, \rightarrow, \leftrightarrow$.

A brief summary of how to write up a proof by induction on formulas is given in Appendix C.

2.4.2 The Main Result on Replacement

Now, with the help of a lemma, we are ready to prove that replacement preserves truth equivalence.

Lemma 2.4.4 Suppose $\mathsf{F}_1 \sim \mathsf{F}_2$ and $\mathsf{G}_1 \sim \mathsf{G}_2$. Then we have

a. $\neg\,\mathsf{F}_1 \;\sim\; \neg\,\mathsf{F}_2$ and

b. $\mathsf{F}_1 \,\square\, \mathsf{G}_1 \;\sim\; \mathsf{F}_2 \,\square\, \mathsf{G}_2$,

where $\square$ is any of the four binary connectives $\vee$, $\wedge$, $\rightarrow$, $\leftrightarrow$.

Proof. (a) Let e be any truth evaluation for the variables occurring in either F_1 or F_2. Then $\widehat{\neg\,\mathsf{F}_1}(\mathsf{e}) = 1$ iff $\widehat{\mathsf{F}_1}(\mathsf{e}) = 0$ iff $\widehat{\mathsf{F}_2}(\mathsf{e}) = 0$ iff $\widehat{\neg\,\mathsf{F}_2}(\mathsf{e}) = 1$. Thus $\neg\,\mathsf{F}_1 \;\sim\; \neg\,\mathsf{F}_2$.
(b) Let e be any truth evaluation for the variables occurring in any of $\mathsf{F}_1, \mathsf{F}_2, \mathsf{G}_1, \mathsf{G}_2$. Then we have $(\widehat{\mathsf{F}_1 \vee \mathsf{G}_1})(\mathsf{e}) = 1$ iff $\widehat{\mathsf{F}_1}(\mathsf{e}) = 1$ or $\widehat{\mathsf{G}_1}(\mathsf{e}) = 1$ iff $\widehat{\mathsf{F}_2}(\mathsf{e}) = 1$ or $\widehat{\mathsf{G}_2}(\mathsf{e}) = 1$ iff $(\widehat{\mathsf{F}_2 \vee \mathsf{G}_2})(\mathsf{e}) = 1$. Thus $\mathsf{F}_1 \vee \mathsf{G}_1 \;\sim\; \mathsf{F}_2 \vee \mathsf{G}_2$. The other cases are handled similarly. $\square$

Theorem 2.4.5 [REPLACEMENT THEOREM] Let F, G, G' be propositional formulas, and suppose G is a subformula of F. If $\mathsf{G} \sim \mathsf{G}'$, and F' is the result of replacing a certain occurrence of G in F with G', then $\mathsf{F} \sim \mathsf{F}'$.

Proof. For now let us say that F has the *replacement property* if the statement of the theorem holds for that particular F. We will prove by induction on formulas that every formula F has the replacement property, and hence the theorem holds.

Ground Cases:

- Let $\mathsf{F} = 0$. If G is an occurrence of a formula in F, and $\mathsf{G} \sim \mathsf{G}'$, then clearly $\mathsf{F} = \mathsf{G} = 0$, so $\mathsf{F}' = \mathsf{G}'$. But then $\mathsf{F} \sim \mathsf{F}'$. Thus 0 has the replacement property.
- Let $\mathsf{F} = 1$. If G is an occurrence of a formula in F, and $\mathsf{G} \sim \mathsf{G}'$, then clearly $\mathsf{F} = \mathsf{G} = 1$ and $\mathsf{F}' = \mathsf{G}'$, so $\mathsf{F} \sim \mathsf{F}'$. Thus 1 has the replacement property.
- Let $\mathsf{F} = P$, where P is a propositional variable. If G occurs as a subformula of F, and $\mathsf{G} \sim \mathsf{G}'$, then clearly $\mathsf{F} = \mathsf{G} = P$, so $\mathsf{F}' = \mathsf{G}'$. But then $\mathsf{F} \sim \mathsf{F}'$. So P has the replacement property.

Induction Steps:

- Suppose $\mathsf{F} = \neg\,\mathsf{F}_1$ and F_1 has the replacement property. Suppose G occurs as a subformula of F and we have selected a particular occurrence. If $\mathsf{F} = \mathsf{G}$, then clearly $\mathsf{F}' = \mathsf{G}'$, and we are finished.

  Next, suppose $\mathsf{F} \neq \mathsf{G}$. Then the occurrence of G must be an occurrence of G in F_1. Let F_1' be the result of replacing that occurrence of G in F_1 with G'. Then $\mathsf{F}' = \neg\,\mathsf{F}_1'$. By our induction hypothesis $\mathsf{F}_1 \sim \mathsf{F}_1'$, and then by Lemma 2.4.4(a) we have $\neg\,\mathsf{F}_1 \sim \neg\,\mathsf{F}_1'$, i.e., $\mathsf{F} \sim \mathsf{F}'$.
- Suppose $\mathsf{F} = \mathsf{F}_1 \,\square\, \mathsf{F}_2$ and $\mathsf{F}_1, \mathsf{F}_2$ both have the replacement property, where $\square$ is any of the four binary connectives $\vee$, $\wedge$, $\rightarrow$, $\leftrightarrow$.

  Suppose G occurs as a subformula of F and we have selected a particular occurrence. If $\mathsf{F} = \mathsf{G}$, then clearly $\mathsf{F}' = \mathsf{G}'$, and we are finished.

  Next, suppose $\mathsf{F} \neq \mathsf{G}$. Then the occurrence of G must be an occurrence in either F_1 or F_2. Let us suppose the former is the case. Let F_1' be the result of replacing that occurrence of G in F_1 with G'. Then $\mathsf{F}' = \mathsf{F}_1' \,\square\, \mathsf{F}_2$. By our induction hypothesis $\mathsf{F}_1 \sim \mathsf{F}_1'$, and then by Lemma 2.4.4(b) we have $\mathsf{F}_1 \,\square\, \mathsf{F}_2 \sim \mathsf{F}_1' \,\square\, \mathsf{F}_2$, i.e., $\mathsf{F} \sim \mathsf{F}'$. The case that the occurrence of G is in F_2 is handled similarly.

$\square$

2.4.3 Simplification of Formulas

The following example illustrates how one can use the fundamental laws, along with substitution and replacement, to *simplify* propositional formulas, i.e., to find truth equivalent formulas that look simpler. (There is no known *fast* algorithm for finding the simplest possible way to express a given formula or truth table.)

Example 2.4.6 Consider the formula $(P \vee Q) \wedge \neg(\neg P \wedge Q)$. Using substitution on a DeMorgan law we know

$$\neg(\neg P \wedge Q) \ \sim\ \neg\neg P \vee \neg Q,$$

so using replacement we have

$$(P \vee Q) \wedge \neg(\neg P \wedge Q) \ \sim\ (P \vee Q) \wedge (\neg\neg P \vee \neg Q).$$

Now, using replacement of $\neg\neg P$ by P (one of our fundamental laws) we have

$$(P \vee Q) \wedge (\neg\neg P \vee \neg Q) \ \sim\ (P \vee Q) \wedge (P \vee \neg Q).$$

Applying substitution to a distributive law we have

$$(P \vee Q) \wedge (P \vee \neg Q) \ \sim\ P \vee (Q \wedge \neg Q).$$

Now, replacing $Q \wedge \neg Q$ with 0 gives

$$P \vee (Q \wedge \neg Q) \ \sim\ P \vee 0.$$

And we have the fundamental law

$$P \vee 0 \ \sim\ P.$$

Because $\sim$ is an equivalence relation we have

$$(P \vee Q) \wedge \neg(\neg P \wedge Q) \ \sim\ P.$$

Exercises

2.4.1 Prove the following *without* using truth tables:

a. $(\neg(P \wedge Q) \vee (R \rightarrow S)) \ \sim\ (P \wedge Q) \rightarrow (R \rightarrow S)$
b. $\neg(\neg P \wedge \neg(Q \vee R)) \ \sim\ P \vee Q \vee R$
c. $\neg(P \wedge (Q \vee R)) \ \sim\ \neg P \vee (\neg Q \wedge \neg R)$
d. $(R \vee \neg R) \wedge S \ \sim\ (R \rightarrow ((R \wedge Q) \vee R)) \rightarrow S$
e. $(P \rightarrow \neg Q) \rightarrow Q \ \sim\ (S \vee (S \rightarrow (R \vee P))) \wedge Q$
f. $(P \vee \neg R) \wedge S \ \sim\ S \wedge (R \rightarrow ((P \wedge Q) \vee P))$
g. $(P \wedge Q) \vee Q \ \sim\ (S \vee (\neg P \vee P)) \wedge Q$
h. $\neg(S \rightarrow \neg S) \ \sim\ ((R \vee (P \vee R)) \vee S) \wedge S$
i. $(S \rightarrow (\neg P \rightarrow S)) \rightarrow P \ \sim\ \neg P \rightarrow (S \wedge ((P \wedge S) \wedge S))$.

2.4.2 For F a propositional formula in the connectives $\neg, \vee, \wedge$, with the constants 0,1, define its *dual* $\Delta(\mathsf{F})$ to be the formula obtained by replacing all $\wedge$'s with $\vee$'s, all $\vee$'s with $\wedge$'s, all 0's with 1's, and all 1's with 0's. We have an inductive definition over formulas as follows:

$$\begin{aligned} \Delta(P) &= P \\ \Delta(0) &= 1 \\ \Delta(1) &= 0 \\ \Delta(\neg \mathsf{F}) &= \neg \Delta(\mathsf{F}) \\ \Delta(\mathsf{F} \vee \mathsf{G}) &= \Delta(\mathsf{F}) \wedge \Delta(\mathsf{G}) \\ \Delta(\mathsf{F} \wedge \mathsf{G}) &= \Delta(\mathsf{F}) \vee \Delta(\mathsf{G}). \end{aligned}$$

Prove that

$$\Delta(\mathsf{F}(P_1, \ldots, P_n)) \sim \neg \mathsf{F}(\neg P_1, \ldots, \neg P_n).$$

2.5 ADEQUATE CONNECTIVES

2.5.1 The Adequacy of Standard Connectives

A remarkable property of the standard set of connectives noted by Post in 1921 is the fact that for every table

P	Q	$\cdots$	
1	1	$\cdots$	$\star$
			$\star$
			$\star$
			$\star$
0	0	$\cdots$	$\star$

there is a propositional formula $\mathsf{F}(P, Q, \cdots)$ with this truth table. One way of finding such a formula is to look at each row

a	b	$\cdots$	1

with a 1 in the $\star$ location and form the constituent $\widetilde{P} \wedge \widetilde{Q} \wedge \cdots$ by choosing $\widetilde{P}$ to be P if $a = 1$—otherwise it is $\neg P$—and doing likewise for $\widetilde{Q}$ and b, etc. Then by taking the disjunction of these constituents we have the desired formula F. (If there are no $\star$'s equal to 1, then we can simply use the formula 0.)

Example 2.5.1 Consider the table

P	Q	R	
1	1	1	0
1	1	0	1
1	0	1	0
1	0	0	0
0	1	1	1
0	1	0	1
0	0	1	0
0	0	0	0

Rows 2, 5, and 6 have the last entry equal to 1, so for our formula F we choose

$$(P \wedge Q \wedge \neg R) \vee (\neg P \wedge Q \wedge R) \vee (\neg P \wedge Q \wedge \neg R).$$

2.5.2 Proving Adequacy

Any set of connectives with the capability to express all truth tables is said to be *adequate*. As Post observed, the standard connectives are adequate. From this it follows that we can show a set $\mathcal{S}$ of connectives is adequate if we can express all the standard connectives in terms of $\mathcal{S}$.

Proposition 2.2.2 on p. 43 says that the standard connectives can be expressed in terms of the two connectives $\rightarrow$ and $\neg$. Thus $\{\rightarrow, \neg\}$ *is* adequate. Consequently, to show that any set $\mathcal{S}$ of connectives is adequate it suffices to express $\rightarrow$ and $\neg$ in terms of $\mathcal{S}$.

In 1880 Schröder showed that each of the standard connectives can be expressed as a propositional formula in which only the single binary connective $\curlywedge$ appears, where the truth table associated with $\curlywedge$ is

P	Q	$P \curlywedge Q$
1	1	0
1	0	0
0	1	0
0	0	1

We can express $\curlywedge$ in terms of the standard connectives by

$$P \curlywedge Q \ \sim \ \neg P \wedge \neg Q$$

and the standard connectives in terms of $\curlywedge$ by

$$\begin{array}{rcl}
1 & \sim & (P \curlywedge (P \curlywedge P)) \curlywedge (P \curlywedge (P \curlywedge P)) \\
0 & \sim & P \curlywedge (P \curlywedge P) \\
\neg P & \sim & P \curlywedge P \\
P \wedge Q & \sim & (P \curlywedge P) \curlywedge (Q \curlywedge Q) \\
P \vee Q & \sim & (P \curlywedge Q) \curlywedge (P \curlywedge Q) \\
P \rightarrow Q & \sim & ((P \curlywedge P) \curlywedge Q) \curlywedge ((P \curlywedge P) \curlywedge Q) \\
P \leftrightarrow Q & \sim & ((P \curlywedge P) \curlywedge Q) \curlywedge ((Q \curlywedge P) \curlywedge P).
\end{array}$$

Thus it follows that the single connective $\curlywedge$ *is* adequate. Consequently, to test a given $\mathcal{S}$ for being adequate it suffices to test if $\curlywedge$ can be expressed by $\mathcal{S}$.

In 1913 Sheffer showed that the *Sheffer stroke* $\mid$ with associated truth table

P	Q	$P \mid Q$
1	1	0
1	0	1
0	1	1
0	0	1

is also a single binary connective in terms of which the standard connectives can be expressed. Whitehead and Russell regarded this as a fundamental advance in symbolic logic, and the Sheffer stroke was used in a later edition of their *Principia Mathematica.*

Definition 2.5.2 If a single connective is adequate, then it is called a *Sheffer* connective.

The only binary Sheffer connectives on $\{0, 1\}$ are the Sheffer stroke and Schröder's $\curlywedge$ (see 2.5.7, p. 57). There are 56 ternary Sheffer connectives on $\{0, 1\}$ (see 2.5.13, p. 58).

Example 2.5.3 Let $\mathcal{S} = \{\tau\}$, where τ is the ternary connective with truth table given by

P	Q	R	τ
1	1	1	0
1	1	0	0
1	0	1	0
1	0	0	1
0	1	1	0
0	1	0	1
0	0	1	0
0	0	0	1

Thus we have

$$\begin{aligned} \neg P &\sim \tau(P, P, P) \\ P \to Q &\sim \tau(P, \tau(P, P, P), \tau(P, Q, \tau(P, P, P))). \end{aligned}$$

Thus τ *is* adequate, so it is a ternary Sheffer connective.

2.5.3 Proving Inadequacy

How do we show that a set $\mathcal{S}$ of connectives is not adequate? Basically we want to show that some standard connective cannot be expressed by $\mathcal{S}$. First, we check that the constants can be expressed. To express 0 we need a formula $\mathsf{F}(P)$ made up from the connectives in $\mathcal{S}$ such that $\mathsf{F}(P) \sim 0$. (If there are constants in $\mathcal{S}$, then P need not appear in F.) We need to do likewise, for 1. To express $\neg$ we need a formula $\mathsf{F}(P)$ such that $\mathsf{F}(P) \sim \neg P$, and to express a binary connective $\Box$ we need a formula $\mathsf{F}(P, Q)$ such that $\mathsf{F}(P, Q) \sim P \Box Q$.

Example 2.5.4 The set $\mathcal{S} = \{\neg\}$ is *not* adequate. To see this, note that a formula $\mathsf{F}(P, Q)$ that uses only the connective $\neg$ is of the form

$$\neg(\cdots(\neg P)\cdots)$$

or

$$\neg(\cdots(\neg Q)\cdots).$$

In the first case this is equivalent to P if the number of $\neg$'s is even, and it is equivalent to $\neg P$ if the number of $\neg$'s is odd. A similar comment holds in the second case. Thus we cannot express any of the standard binary connectives using $\mathcal{S}$.

Example 2.5.5 The set $\mathcal{S} = \{\wedge\}$ is *not* adequate. To see this, note that a formula $\mathsf{F}(P)$ that uses only the connective $\wedge$ has the property that

$$\widehat{\mathsf{F}}(0) = 0.$$

But then we cannot have $\mathsf{F}(P) \sim \neg P$, so it is not possible to express $\neg$ using $\mathcal{S}$.

Example 2.5.6 Let us use the symbol τ for the ternary connective whose truth table is given by

P	Q	R	τ
1	1	1	1
1	1	0	1
1	0	1	0
1	0	0	0
0	1	1	1
0	1	0	0
0	0	1	1
0	0	0	0

It is easy to see that for any truth evaluation $\mathsf{e} = (e_1, e_2, e_3)$ we have

$$\tau(e_1, e_2, e_3) = \begin{cases} e_2 & \text{if } e_1 = 1 \\ e_3 & \text{if } e_1 = 0. \end{cases}$$

Thus we see that $\tau(e_1, e_2, e_3)$ is e_2 if $e_1 = 1$, else it is e_3. This is the familiar *if–then–else* connective from computer science, namely, *if* e_1 *then* e_2 *else* e_3. It is *not* adequate because we cannot express $\neg$ with it.

Exercises

2.5.1 Find a formula $\mathsf{F}(P, Q)$ with the fewest symbols possible, not counting parentheses, using the standard connectives, such that F has the following truth table, i.e., F expresses the Sheffer stroke.

P	Q	F
1	1	0
1	0	1
0	1	1
0	0	1

2.5.2 Show that $\sim$ is an equivalence relation on the set $\mathcal{F}_n$ of propositional formulas in the n variables $P_1, \ldots, P_n$, i.e., $\mathcal{F}_n = \mathcal{F}(\{P_1, \ldots, P_n\})$, with 2^{2^n} equivalence classes.

2.5.3 Express each of the standard connectives in terms of the Sheffer stroke.

2.5.4 Prove that each of the following sets of connectives *is* adequate.

a. $\neg$ and $\vee$
b. $\neg$ and $\wedge$
c. $\rightarrow$ and 0

2.5.5 Show that each of the following sets of connectives is *not* adequate.

a. $\vee$ and 0,1
b. $\rightarrow$
c. $\leftrightarrow$
d. $\vee$ and $\wedge$
e. $\neg$ and $\leftrightarrow$
f. $\rightarrow$ and 1

2.5.6 Show that each binary connective $\beta(P,Q)$ is equivalent to exactly one of the following 16 formulas:

$$\begin{array}{llllll} 0 & 1 & P & Q & \neg P & \neg Q \\ P \vee Q & \neg P \vee Q & P \vee \neg Q & \neg P \vee \neg Q & & \\ P \wedge Q & \neg P \wedge Q & P \wedge \neg Q & \neg P \wedge \neg Q & & \\ P \leftrightarrow Q & \neg(P \leftrightarrow Q). & & & & \end{array}$$

[*Hint*: Look at the 16 truth tables for two variables P, Q.]

2.5.7 Fill in the details of the following argument:

> $\curlywedge$ and $\mid$ are the only *binary* connectives that are adequate. For let $f(P,Q)$ be such a binary connective. Then $f(P,P)$ must be equivalent to $\neg P$. This eliminates $P \rightarrow Q$, $Q \rightarrow P$, and their negations; $P \leftrightarrow Q$ and its negation; $P \wedge Q$ and $P \vee Q$. The constant functions are obviously not adequate, as well as the two projections (namely, $f(P,Q) = P$ and $f(P,Q) = Q$) and their negations. This completes the proof.

2.5.8 Show that $\{\vee, \wedge, \rightarrow, \leftrightarrow, 1\}$ is not adequate. Conclude that any adequate set of standard connectives must have either $\neg$ or 0 in it.

2.5.9 Among the subsets of $\{\vee, \wedge, \neg, \rightarrow, \leftrightarrow, 0, 1\}$ that are adequate, find all those that are *minimal*, i.e., an adequate set of connectives with the property that if any connective is removed from the set, it is no longer adequate.

2.5.10 If a ternary connective τ is a Sheffer connective, show that τ evaluates to 1 at $(0,0,0)$, and to 0 at $(1,1,1)$.

2.5.11 Show that a ternary connective τ is adequate iff the following two conditions hold:

a. $\tau(P,P,P) \sim \neg P$, and

b. at least one of $\tau(P,Q,Q)$, $\tau(P,P,Q)$, and $\tau(P,Q,P)$ is not equivalent to either of $\neg P$ and $\neg Q$.

2.5.12 Determine which of the following ternary connectives are Sheffer connectives.

a.

P	Q	R	τ
1	1	1	0
1	1	0	0
1	0	1	0
1	0	0	1
0	1	1	0
0	1	0	0
0	0	1	0
0	0	0	1

b.

P	Q	R	τ
1	1	1	0
1	1	0	0
1	0	1	0
1	0	0	1
0	1	1	1
0	1	0	1
0	0	1	0
0	0	0	1

c.

P	Q	R	τ
1	1	1	0
1	1	0	1
1	0	1	1
1	0	0	1
0	1	1	0
0	1	0	0
0	0	1	0
0	0	0	1

2.5.13 Show that of the 256 possible ternary connectives exactly 56 are adequate, i.e., there are 56 ternary Sheffer connectives.

2.5.14 Show that an n–ary connective γ is a Sheffer connective (i.e., is adequate) iff the following hold:

a. $\gamma(P,\ldots,P) \sim \neg P$, and

b. at least one of the formulas $\gamma(\widetilde{P}_1,\ldots,\widetilde{P}_n)$, where each $\widetilde{P}_i$ is either P or Q, is not equivalent to either of $\neg P$ and $\neg Q$.

2.5.15 Show that of the 2^{2^n} possible n–ary connectives (we assume that distinct connectives have distinct truth tables), $2^k(2^k-1)$ are Sheffer connectives, where $k = 2^{n-1}-1$. Consequently, we have

$$\lim_{n\to\infty} \frac{\#\text{ of } n\text{–ary Sheffer connectives}}{\#\text{ of } n\text{–ary connectives}} = \frac{1}{4},$$

i.e., the proportion of n–ary connectives that are Sheffer connectives tends to $1/4$.

2.6 DISJUNCTIVE AND CONJUNCTIVE FORMS

Notational Convention 2.6.1 Since the associative law holds for $\vee$ and $\wedge$ it is common practice to write a sequence of conjunctions, or disjunctions, without using parentheses. For example, instead of

$$((P_1 \wedge (P_2 \wedge P_3)) \wedge P_4),$$

we prefer to write

$$P_1 \wedge P_2 \wedge P_3 \wedge P_4.$$

Any propositional formula F can be transformed into *disjunctive form*, i.e., F is equivalent to a disjunction of formulas, each of which is a conjunction of propositional variables and their negations. For example, we have

$$P_1 \leftrightarrow P_3 \;\sim\; (P_1 \wedge P_3) \vee (\neg P_1 \wedge \neg P_3).$$

Likewise, we have *conjunctive forms* such as

$$P_1 \leftrightarrow P_3 \;\sim\; (\neg P_1 \vee P_3) \wedge (P_1 \vee \neg P_3).$$

If we also specify a finite set of variables containing the variables of F, say we want to work with $\mathsf{F}(P_1, \ldots, P_n)$, then we can put this in *disjunctive normal form (DNF)* by finding an equivalent formula that is either 0 or a disjunction of certain conjunctions of the form

$$\widetilde{P_1} \wedge \cdots \wedge \widetilde{P_n},$$

called *DNF–constituents*, where $\widetilde{P_i}$ is either P_i or $\neg P_i$.

Likewise, there is a *conjunctive normal form (CNF)* that is either 1 or a conjunction of *CNF–constituents* of the form

$$\widetilde{P_1} \vee \cdots \vee \widetilde{P_n}.$$

When we are asked to put a formula $\mathsf{F}(P_1, \ldots, P_n)$ into normal form, it is assumed that we will use the full set of indicated variables $P_1, \ldots, P_n$ for the constituents.

Example 2.6.2 Let $\mathsf{F}(P_1, P_3)$ be $P_1 \leftrightarrow P_3$. Then we have the normal forms:

$$(P_1 \wedge P_3) \vee (\neg P_1 \wedge \neg P_3)$$

and

$$(P_1 \vee \neg P_3) \wedge (\neg P_1 \vee P_3).$$

We emphasize that when we want to put a formula F into one of these normal forms it is essential to know the set of variables that we want to use for the normal form. The set of variables used must include all the variables that occur in F, but it could be a larger set, as in the next example.

Example 2.6.3 Let $\mathsf{F}(P_1, P_2, P_3)$ be $P_1 \leftrightarrow P_3$. Then we have the normal forms:

$$(P_1 \wedge P_2 \wedge P_3) \vee (P_1 \wedge \neg P_2 \wedge P_3) \vee (\neg P_1 \wedge P_2 \wedge \neg P_3) \\ \vee (\neg P_1 \wedge \neg P_2 \wedge \neg P_3),$$

and

$$(P_1 \vee P_2 \vee \neg P_3) \wedge (P_1 \vee \neg P_2 \vee \neg P_3) \wedge (\neg P_1 \vee P_2 \vee P_3) \\ \wedge (\neg P_1 \vee \neg P_2 \vee P_3).$$

We will look at two ways to find a disjunctive [normal] form for a propositional formula F. *Similar results hold for conjunctive [normal] forms.*

2.6.1 Rewrite Rules to Obtain Normal Forms

In the following discussion we use the symbol $\rightsquigarrow$ to mean that one can *replace the left–hand side with the right–hand side.* These rules can be applied to any subformula of a given formula to create a new formula.

Method 1: Use the truth equivalence rules in Sec. 2.2 to transform the formula into the desired form. For this purpose the following *rewrite rules* can be applied in any order:

$$\begin{aligned}
\mathsf{F} \rightarrow \mathsf{G} &\rightsquigarrow \neg\mathsf{F} \vee \mathsf{G} \\
\mathsf{F} \leftrightarrow \mathsf{G} &\rightsquigarrow (\mathsf{F} \rightarrow \mathsf{G}) \wedge (\mathsf{G} \rightarrow \mathsf{F}) \\
\neg(\mathsf{F} \vee \mathsf{G}) &\rightsquigarrow \neg\mathsf{F} \wedge \neg\mathsf{G} \\
\neg(\mathsf{F} \wedge \mathsf{G}) &\rightsquigarrow \neg\mathsf{F} \vee \neg\mathsf{G} \\
\neg\neg\mathsf{F} &\rightsquigarrow \mathsf{F} \\
\mathsf{F} \wedge (\mathsf{G} \vee \mathsf{H}) &\rightsquigarrow (\mathsf{F} \wedge \mathsf{G}) \vee (\mathsf{F} \wedge \mathsf{H}) \\
(\mathsf{F} \vee \mathsf{G}) \wedge \mathsf{H} &\rightsquigarrow (\mathsf{F} \wedge \mathsf{H}) \vee (\mathsf{G} \wedge \mathsf{H}).
\end{aligned}$$

Applying these rules to any subformula of a formula F will, by Theorem 2.4.5, give an equivalent formula F'. The first two rules eliminate all occurrences of $\rightarrow$ and $\leftrightarrow$. The next three rules push the negations next to the propositional variables and eliminate double negations. The next two rules make sure no $\vee$ occurs inside the scope of a $\wedge$. These rules are applied until no further applications are possible.

Example 2.6.4

$$\begin{aligned} P \leftrightarrow Q &\leadsto (P \rightarrow Q) \wedge (Q \rightarrow P) \\ &\leadsto (\neg P \vee Q) \wedge (\neg Q \vee P) \\ &\leadsto (\neg P \wedge (\neg Q \vee P)) \vee (Q \wedge (\neg Q \vee P)) \\ &\leadsto ((\neg P \wedge \neg Q) \vee (\neg P \wedge P)) \vee ((Q \wedge \neg Q) \vee (Q \wedge P)). \end{aligned}$$

Now this formula clearly gives a disjunctive form, but not a normal form. We can simplify it considerably, but to do this we need to invoke additional rewrite rules.

Here is a further list of useful rules for simplification.

$$\begin{aligned} 0 \wedge \mathsf{F} &\leadsto 0 \\ \mathsf{F} \wedge 0 &\leadsto 0 \\ 0 \vee \mathsf{F} &\leadsto \mathsf{F} \\ \mathsf{F} \vee 0 &\leadsto \mathsf{F} \\ \neg 0 &\leadsto 1 \\ \neg 1 &\leadsto 0 \\ 1 \wedge \mathsf{F} &\leadsto \mathsf{F} \\ \mathsf{F} \wedge 1 &\leadsto \mathsf{F} \\ 1 \vee \mathsf{F} &\leadsto 1 \\ \mathsf{F} \vee 1 &\leadsto 1 \\ \cdots \wedge \mathsf{F} \wedge \cdots \wedge \neg \mathsf{F} \wedge \cdots &\leadsto 0 \\ \cdots \wedge \neg \mathsf{F} \wedge \cdots \wedge \mathsf{F} \wedge \cdots &\leadsto 0 \\ \cdots \wedge \mathsf{F} \wedge \cdots \wedge \mathsf{F} \wedge \cdots &\leadsto \cdots \wedge \mathsf{F} \wedge \cdots \\ \cdots \vee \mathsf{F} \vee \cdots \vee \neg \mathsf{F} \vee \cdots &\leadsto 1 \\ \cdots \vee \neg \mathsf{F} \vee \cdots \vee \mathsf{F} \vee \cdots &\leadsto 1 \\ \cdots \vee \mathsf{F} \vee \cdots \vee \mathsf{F} \vee \cdots &\leadsto \cdots \vee \mathsf{F} \vee \cdots \end{aligned}$$

The second and third last rules say that if we have a disjunction with both a formula F and its negation $\neg \mathsf{F}$ as disjuncts, then the disjunction is equivalent to 1. The last rule says that if a formula F

occurs more than once as a disjunct, then we can eliminate repeat occurrences. The previous three rules are similar, but for conjunctions. Applying these rules to the conclusion of our previous example we have:

Example 2.6.5

$$
\begin{aligned}
&((\neg P \wedge \neg Q) \vee (\neg P \wedge P)) \vee ((Q \wedge \neg Q) \vee (Q \wedge P)) \\
&\qquad\qquad\qquad \rightsquigarrow ((\neg P \wedge \neg Q) \vee 0) \vee (0 \vee (Q \wedge P)) \\
&\qquad\qquad\qquad \rightsquigarrow (\neg P \wedge \neg Q) \vee (Q \wedge P).
\end{aligned}
$$

We now have a fairly simple disjunctive form for $P \leftrightarrow Q$, which happens to be normal. In general, this procedure will not give a normal form. Consider the formula $\mathsf{F}(P, Q)$ given by $(P \wedge Q) \vee \neg P$. This is in disjunctive form, and none of the transformations already mentioned apply. The problem is clearly the second disjunct, namely, $\neg P$, which has no occurrence of the variable Q. We can remedy this by replacing $\neg P$ with the conjunction $\neg P \wedge (Q \vee \neg Q)$ (this conjunction is truth equivalent to $\neg P$):

Example 2.6.6

$$
\begin{aligned}
(P \wedge Q) \vee \neg P \quad &\sim \quad (P \wedge Q) \vee (\neg P \wedge (Q \vee \neg Q)) \\
&\rightsquigarrow \quad (P \wedge Q) \vee ((\neg P \wedge Q) \vee (\neg P \wedge \neg Q)).
\end{aligned}
$$

We prefer to write the last formula as

$$(P \wedge Q) \vee (\neg P \wedge Q) \vee (\neg P \wedge \neg Q).$$

Now we have a disjunctive normal form.

One more rule is needed, to handle the exceptional case that the above rules reduce the formula to simply the constant 1. In this case we rewrite 1 as a join of *all* the DNF–constituents.

2.6.2 Using Truth Tables to Find Normal Forms

Method 2: This method takes us to the disjunctive normal form of a formula F by first forming the truth table for F and then directly reading off the disjunctive normal form. Suppose $\mathsf{F}(P_1, \ldots, P_n)$ has the disjunctive normal form $\mathsf{F}_1 \vee \cdots \vee \mathsf{F}_k$, where each F_i is a DNF–constituent. Note that a DNF–constituent has the key property that it is true for exactly one truth evaluation of the propositional variables $P_1, \ldots, P_n$, and different DNF–constituents are not both true for the same evaluation. Thus the DNF–constituents F_i are in one–to–one correspondence with the lines of the truth table of F for which F is true. (The CNF–constituents are in one–to–one correspondence with the lines of the truth table of F for which F is false.)

Example 2.6.7 The next table has the DNF– and CNF–constituents for the various rows of a truth table in the variables P, Q, R:

P	Q	R	DNF–constituent	CNF–constituent
1	1	1	$P \wedge Q \wedge R$	$\neg P \vee \neg Q \vee \neg R$
1	1	0	$P \wedge Q \wedge \neg R$	$\neg P \vee \neg Q \vee R$
1	0	1	$P \wedge \neg Q \wedge R$	$\neg P \vee Q \vee \neg R$
1	0	0	$P \wedge \neg Q \wedge \neg R$	$\neg P \vee Q \vee R$
0	1	1	$\neg P \wedge Q \wedge R$	$P \vee \neg Q \vee \neg R$
0	1	0	$\neg P \wedge Q \wedge \neg R$	$P \vee \neg Q \vee R$
0	0	1	$\neg P \wedge \neg Q \wedge R$	$P \vee Q \vee \neg R$
0	0	0	$\neg P \wedge \neg Q \wedge \neg R$	$P \vee Q \vee R$

Example 2.6.8 Let $\mathsf{F}(P, Q)$ be the formula $(P \wedge Q) \vee \neg P$. Then the truth table for F is given by

P	Q	$(P \wedge Q) \vee \neg P$
1	1	1
1	0	0
0	1	1
0	0	1

We see that the DNF–constituents of F correspond to the first, third, and fourth rows and thus must be $P \wedge Q$, $\neg P \wedge Q$, and $\neg P \wedge \neg Q$. Thus the disjunctive normal form for $(P \wedge Q) \vee \neg P$ is $(P \wedge Q) \vee (\neg P \wedge Q) \vee (\neg P \wedge \neg Q)$.

Example 2.6.9 This example shows how to use a truth table to find a conjunctive normal form of a formula. Let $\mathsf{F}(P, Q)$ be the formula $(P \to Q) \wedge (Q \to P)$. Then the truth table for F is given by

P	Q	$(P \to Q) \wedge (Q \to P)$
1	1	1
1	0	0
0	1	0
0	0	1

We see that the CNF–constituents of F correspond to the second and third rows and thus must be $\neg P \vee Q$ (for the second row) and

$P \vee \neg Q$ (for the third row). Thus the conjunctive normal form for $(P \to Q) \wedge (Q \to P)$ is $(\neg P \vee Q) \wedge (P \vee \neg Q)$.

Since Method 2 looks so straightforward, you might wonder why we would use Method 1. The answer is simply that constructing truth tables can require an enormous, if not impossible, amount of work for relatively simple formulas. After all, if F has n propositional variables, then the truth table for F will have 2^n lines.

2.6.3 Uniqueness of Normal Forms

Notational Convention 2.6.10 We assume that the variables in a disjunctive normal form of a formula $\mathsf{F}(P_1, \ldots, P_n)$ have been ordered according to their appearance in the expression used to name the formula, namely, in $\mathsf{F}(P_1, \ldots, P_n)$, and not necessarily in the order in which they actually appear in the formula (indeed, some of the variables may not appear in the formula).

This convention dictates that the DNF–constituents are of the form $\widetilde{P_1} \wedge \cdots \wedge \widetilde{P_n}$. However, we have not set up any convention to order the DNF–constituents. This leads to the following result, that we have the disjunctive normal form determined up to the ordering of the DNF–constituents.

Theorem 2.6.11 A propositional formula $\mathsf{F}(P_1, \ldots, P_n)$ is truth equivalent to a unique disjunctive normal form (in the specified variables $P_1, \ldots, P_n$) up to the ordering of the DNF–constituents.

Proof. This follows from the fact that we can find a disjunctive normal form that is equivalent to F using the truth table for F as in Method 2. Furthermore, any disjunctive normal form for F is so determined by the truth table for F. □

Because of this theorem it is common to speak of *the* disjunctive normal form. And when we say that two disjunctive normal forms are the same we mean *up to the ordering* of the DNF–constituents. (Later, when we are working with proof systems, it will be necessary to specify the ordering that we want for the disjunctive normal form.)

Theorem 2.6.12 Two propositional formulas, say $\mathsf{F}(P_1, \ldots, P_n)$ and $\mathsf{G}(P_1, \ldots, P_n)$, are equivalent iff they have the same disjunctive normal form (in the variables $P_1, \ldots, P_n$).

Proof. If they are equivalent, then clearly they have the same truth tables and hence the same disjunctive normal form. Conversely, if they have the same disjunctive normal form, then, since a formula is equivalent to its disjunctive normal form, and since $\sim$ is an equivalence relation, the two formulas are equivalent. □

Exercises

2.6.1 Find the smallest disjunctive form possible for the following. (Do not count parentheses.) Also find the disjunctive normal form with respect to the variables occurring in the formula.

a. $P \vee Q$
b. $P \rightarrow Q$
c. $P \leftrightarrow Q$
d. $P \rightarrow (Q \rightarrow R)$
e. $(P \rightarrow Q) \rightarrow (Q \rightarrow R)$
f. $(P \wedge Q) \vee (R \wedge S)$
g. $P \rightarrow (Q \rightarrow (R \rightarrow S))$.

2.6.2 Find the smallest conjunctive form possible for the formulas in the previous exercise. Also find the conjunctive normal form with respect to the variables occurring in the formula.

2.6.3

a. If $\mathsf{L}_1 \wedge \cdots \wedge \mathsf{L}_n$ is a constituent of the disjunctive normal form of $\mathsf{F}(P_1, \ldots, P_n)$, where L_i is either P_i or $\neg P_i$, prove that $\overline{\mathsf{L}}_1 \vee \cdots \vee \overline{\mathsf{L}}_n$ is a constituent of the conjunctive normal form of $\neg \mathsf{F}$, where $\overline{P}_i = \neg P_i$, and $\overline{\neg P_i} = P_i$.
b. Prove that the number of constituents $\widetilde{P}_1 \wedge \cdots \wedge \widetilde{P}_n$ of the disjunctive normal form of F plus the number of constituents $\widetilde{P}_1 \vee \cdots \vee \widetilde{P}_n$ of the conjunctive normal form of F is 2^n.
c. Conclude that if F has 5 variables, then either the conjunctive normal form or the disjunctive normal form of F involves at least 16 constituents and hence at least 144 symbols.

2.7 VALID ARGUMENTS, TAUTOLOGIES, AND SATISFIABILITY

A *(logical) argument*, in mathematics as in law, is a claim that from certain *premisses* one can draw a certain *conclusion*. An essential duty of logic is to determine what constitutes a valid argument. As in Chapter 1, we use the symbol $\therefore$ for the word *therefore*.

Definition 2.7.1 An argument $\mathsf{F}_1, \cdots, \mathsf{F}_n \;\therefore\; \mathsf{F}$, also written

$$\begin{array}{l} \mathsf{F}_1 \\ \vdots \\ \mathsf{F}_n \\ \therefore \mathsf{F} \end{array}$$

is *valid* (or *correct*) in propositional logic provided every truth evaluation e (for the variables occurring in $\mathsf{F}_1, \ldots, \mathsf{F}_n, \mathsf{F}$) that makes $\mathsf{F}_1, \ldots, \mathsf{F}_n$ true also makes F true, i.e.,

$$\widehat{\mathsf{F}}_1(\mathsf{e}) = \cdots = \widehat{\mathsf{F}}_n(\mathsf{e}) = 1 \qquad \text{implies} \qquad \widehat{\mathsf{F}}(\mathsf{e}) = 1.$$

Proposition 2.7.2 An argument

$$\mathsf{F}_1, \cdots, \mathsf{F}_n \qquad \therefore \mathsf{F}$$

in propositional logic is valid iff

$$\mathsf{F}_1 \wedge \cdots \wedge \mathsf{F}_n \to \mathsf{F}$$

is a tautology.

Proof. Let e be a truth evaluation for the variables occurring in $\mathsf{F}_1, \ldots, \mathsf{F}_n, \mathsf{F}$. Then, with G being the formula $\mathsf{F}_1 \wedge \cdots \wedge \mathsf{F}_n \to \mathsf{F}$,

$$\widehat{\mathsf{G}}(\mathsf{e}) \;=\; \widehat{\mathsf{F}}_1(\mathsf{e}) \wedge \cdots \wedge \widehat{\mathsf{F}}_n(\mathsf{e}) \to \widehat{\mathsf{F}}(\mathsf{e}).$$

Thus $\mathsf{F}_1 \wedge \cdots \wedge \mathsf{F}_n \to \mathsf{F}$ is a tautology iff for every such truth evaluation e, some $\widehat{\mathsf{F}}_i(\mathsf{e}) = 0$ or $\widehat{\mathsf{F}}(\mathsf{e}) = 1$. But this is precisely the condition for the argument $\mathsf{F}_1, \cdots, \mathsf{F}_n \;\therefore \mathsf{F}$ to be valid. □

Example 2.7.3 According to Chrysippus a good hunting dog has basic skills in reasoning.

> When running after a rabbit, the dog found that the path suddenly split in three directions. The dog sniffed the first path and found no scent; then it sniffed the second path and found no scent; then, without bothering to sniff the third path, it ran down that path.

We can summarize the canine's fine reasoning as follows:

- The rabbit went this way or that way or the other way.
- Not this way.
- Not that way.
- Therefore the other way.

Using propositional formulas we can express the argument as

$$\begin{gathered} P \vee Q \vee R \\ \neg P \\ \neg Q \\ \therefore R. \end{gathered}$$

By Proposition 2.7.2 this argument is valid iff the formula

$$((P \vee Q \vee R) \wedge \neg P \wedge \neg Q) \to R$$

is a tautology. This can easily be checked using a truth table:

P	Q	R	$P \vee Q \vee R$	$\neg P$	$\neg Q$	R	$((P \vee Q \vee R) \wedge \neg P \wedge \neg Q) \to R$
1	1	1	1	0	0	1	1
1	1	0	1	0	0	0	1
1	0	1	1	0	1	1	1
1	0	0	1	0	1	0	1
0	1	1	1	1	0	1	1
0	1	0	1	1	0	0	1
0	0	1	1	1	1	1	1
0	0	0	0	1	1	0	1

Now we come to our second characterization of valid arguments. From now on we will admit truth evaluations of an infinite as well as a finite number of variables, i.e., if we have an infinite set of variables $\{P_1, P_2, \ldots\}$, then a corresponding truth evaluation e will be given by an infinite sequence $(e_1, e_2, \ldots)$ of 0's and 1's.

Definition 2.7.4 A set $\mathcal{S}$ of propositional formulas is *satisfiable* if there is a truth evaluation e for the variables in $\mathcal{S}$ that makes every formula in $\mathcal{S}$ true. We say that e *satisfies* $\mathcal{S}$. The expression $\mathsf{Sat}(\mathcal{S})$ means that $\mathcal{S}$ is satisfiable; the expression $\neg\,\mathsf{Sat}(\mathcal{S})$ means that $\mathcal{S}$ is not satisfiable.

Thus a finite set $\{\mathsf{F}_1, \ldots, \mathsf{F}_n\}$ of formulas is satisfiable iff when we look at the combined truth tables for the F_i we can find a line that looks as follows:

P_1	$\cdots$	P_n	F_1	$\cdots$	F_n
e_1	$\cdots$	e_n	1	$\cdots$	1

Example 2.7.5 Let $\mathcal{S}$ be the following set of propositional formulas:

$$\begin{array}{l} \neg(T \wedge S) \\ (T \wedge P) \vee S \\ Q \vee \neg P \\ \neg T \vee (Q \wedge (S \vee R)) \\ \neg R \vee T \\ \neg(R \wedge (P \wedge T)). \end{array}$$

Then $\mathcal{S}$ is satisfied by precisely the following:

P	Q	R	S	T
1	1	0	1	0
0	1	0	1	0
0	0	0	1	0

Example 2.7.6 Let $\mathcal{S}$ be the following set of propositional formulas:

$$\begin{array}{l} \neg(S \wedge T) \\ (R \wedge Q) \vee (T \wedge Q) \\ P \vee (S \wedge \neg T) \\ \neg T \vee (Q \wedge (S \vee R)) \\ \neg R \vee T \\ \neg(R \wedge (P \wedge T)). \end{array}$$

Then $\mathcal{S}$ is not satisfiable. We can see this by making a combined truth table and checking that on every line some $\widehat{\mathsf{F}}_i(\mathsf{e})$ has the value 0. This example is minimal in the sense that if we remove any of the six formulas, then the remainder will be satisfiable.

Proposition 2.7.7 An argument

$$\mathsf{F}_1, \cdots, \mathsf{F}_n \qquad \therefore \mathsf{F}$$

in the propositional logic is valid iff the set of propositional formulas

$$\{\mathsf{F}_1, \cdots, \mathsf{F}_n, \neg\,\mathsf{F}\}$$

is not satisfiable.

Proof. The set $\{\mathsf{F}_1, \cdots, \mathsf{F}_n, \neg\,\mathsf{F}\}$ is not satisfiable iff every evaluation e for the variables in $\mathsf{F}_1, \ldots, \mathsf{F}_n, \mathsf{F}$ that makes $\{\mathsf{F}_1, \cdots, \mathsf{F}_n\}$ true also makes F true. But this is precisely what it means for the argument $\mathsf{F}_1, \cdots, \mathsf{F}_n \ \therefore \mathsf{F}$ to be valid. □

This is the basic connection between validity and nonsatisfiability. We will return to it in Sec. 2.10. In summary, we now have the following characterizations of valid arguments.

Theorem 2.7.8 The following are equivalent:

- The argument $\mathsf{F}_1, \cdots, \mathsf{F}_n \quad \therefore \mathsf{F}$ is valid.
- The formula $\mathsf{F}_1 \wedge \cdots \wedge \mathsf{F}_n \rightarrow \mathsf{F}$ is a tautology.
- $\neg \mathsf{Sat}(\{\mathsf{F}_1, \cdots, \mathsf{F}_n, \neg \mathsf{F}\})$.
- $\neg \mathsf{Sat}(\mathsf{F}_1 \wedge \cdots \wedge \mathsf{F}_n \wedge \neg \mathsf{F})$.

Example 2.7.9 Consider the following 10 propositional formulas:

F1: $\neg(R \rightarrow \neg(R \vee P))$
F2: $R \wedge (R \leftrightarrow (P \vee \neg Q))$
F3: $(Q \leftrightarrow (R \rightarrow \neg Q)) \wedge Q$
F4: $(Q \leftrightarrow (Q \leftrightarrow P)) \leftrightarrow (R \wedge Q)$
F5: $(Q \vee P) \vee ((R \rightarrow Q) \wedge \neg Q)$
F6: $(((P \vee (P \vee R)) \wedge P) \leftrightarrow Q) \rightarrow P$
F7: $(P \wedge (R \rightarrow Q)) \wedge ((Q \leftrightarrow R) \vee P)$
F8: $(R \wedge Q) \rightarrow (R \vee ((R \wedge (Q \wedge R)) \leftrightarrow P))$
F9: $((P \leftrightarrow Q) \rightarrow ((Q \leftrightarrow P) \leftrightarrow P)) \rightarrow \neg(P \vee R)$
F10: $(((P \vee Q) \vee R) \vee (R \rightarrow (\neg Q \rightarrow R))) \rightarrow R$

and their (combined) truth table:

	P	Q	R	F1	F2	F3	F4	F5	F6	F7	F8	F9	F10
1.	1	1	1	1	1	0	1	1	1	1	1	0	1
2.	1	1	0	0	0	1	0	1	1	1	1	0	0
3.	1	0	1	1	1	0	0	1	1	0	1	0	1
4.	1	0	0	0	0	0	0	1	1	1	1	0	0
5.	0	1	1	1	0	0	0	1	1	0	1	0	1
6.	0	1	0	0	0	1	1	1	1	0	1	1	0
7.	0	0	1	1	1	0	1	0	0	0	1	1	1
8.	0	0	0	0	0	0	1	1	0	0	1	1	0

From this table we can read off the following information:

a. The disjunctive and conjunctive normal forms:

$(P \wedge Q \wedge R) \vee (P \wedge Q \wedge \neg R) \vee (P \wedge \neg Q \wedge \neg R)$ is the DNF of formula F7.

The CNF for F6 is $(P \vee Q \vee \neg R) \wedge (P \vee Q \vee R)$.

b. Formulas F1 and F10 are equivalent.

Note that the columns below them are identical.

c. Formulas F5 and F6 are not equivalent.

The columns below them disagree in row 8.

d. Formula F8 is a tautology,

as all entries in its column are 1.

e. None of these formulas is a contradiction,

as no formula has a column of zeros below it.

f. The set of formulas F3, F4, F5, F6, F8, F9 is satisfiable,

as all these formulas are true in row 6.

g. The set of formulas F6, F9, F10 is not satisfiable,

because in every row one of the three formulas has the value 0.

h. The argument "F1, F2, F4, F5 ∴ F10" is valid,

since the only row with F1, F2, F4, F5 true is row 1, and for this row F10 is also true.

i. The argument "F4, F5, F6, F7 ∴ F3" is not valid,

since in row 1 all the formulas F4, F5, F6, F7 are true, but formula F3 is false.

Let us apply the propositional logic to two examples that we considered in Chapter 1, followed by an example concerning two tribes, one whose members always lie, and the other whose members always tell the truth.

Example 2.7.10 (See Example 1.3.3 on p. 14)
The study of a class of substances has led to the following results:

- If a and b appear, so does precisely one of c or d.
- If b and c appear, then both or neither of a, d appears.
- If neither of a, b appears, then neither of c, d appears.
- If neither of c, d appears, then neither of a, b appears.

Show that not all three of a, b, c can occur, and if a, b are both missing, then so is c.

Let A denote the proposition asserting that *a appears*, etc. The translation into propositional logic is

$$
\begin{array}{l}
(A \wedge B) \rightarrow (C \leftrightarrow \neg D) \\
(B \wedge C) \rightarrow (A \leftrightarrow D) \\
(\neg A \wedge \neg B) \rightarrow (\neg C \wedge \neg D) \\
(\neg C \wedge \neg D) \rightarrow (\neg A \wedge \neg B) \\
\therefore (\neg (A \wedge B \wedge C)) \wedge ((\neg A \wedge \neg B) \rightarrow \neg C).
\end{array}
$$

Here is the combined truth table, where F1 is the first of the five formulas, and so forth.

	A	B	C	D	F1	F2	F3	F4	F5
1.	1	1	1	1	0	1	1	1	0
2.	1	1	1	0	1	0	1	1	0
3.	1	1	0	1	1	1	1	1	1
4.	1	1	0	0	0	1	1	0	1
5.	1	0	1	1	1	1	1	1	1
6.	1	0	1	0	1	1	1	1	1
7.	1	0	0	1	1	1	1	1	1
8.	1	0	0	0	1	1	1	0	1
9.	0	1	1	1	1	0	1	1	1
10.	0	1	1	0	1	1	1	1	1
11.	0	1	0	1	1	1	1	1	1
12.	0	1	0	0	1	1	1	0	1
13.	0	0	1	1	1	1	0	1	0
14.	0	0	1	0	1	1	0	1	0
15.	0	0	0	1	1	1	0	1	1
16.	0	0	0	0	1	1	1	1	1

The premisses $\mathsf{F}_1, \ldots, \mathsf{F}_4$ are true in lines 3,5,6,7,9,11,16 (and no others), and for these lines the conclusion F_5 is also true. Thus the argument is valid.

Example 2.7.11 (See Example 1.3.4 on p. 16)

1. Good–natured tenured mathematics professors are dynamic.
2. Grumpy student advisors play slot machines.
3. Smokers wearing Hawaiian shirts are phlegmatic.
4. Comical student advisors are mathematics professors.
5. Untenured faculty who smoke are nervous.
6. Phlegmatic tenured faculty members who wear Hawaiian shirts are comical.
7. Student advisors who are not stock market players are scholars.
8. Relaxed student advisors are creative.
9. Creative scholars who do not play slot machines wear Hawaiian shirts.
10. Nervous smokers play slot machines.
11. Student advisors who play slot machines are nonsmokers.
12. Creative stock market players who are good–natured wear Hawaiian shirts.
13. Therefore no student advisors are smokers.

If we let x be an unspecified faculty member and let A represent the proposition *x is good–natured*, etc., then we have the following proposition names:

	Defined by	Negation
A	good–natured	grumpy
B	tenured	untenured
C	mathematics professor	
D	dynamic	phlegmatic
E	wears Hawaiian shirts	
F	smoker	nonsmoker
G	comical	
H	relaxed	nervous
I	stock market player	
J	scholar	
K	creative	
L	plays slot machines	
M	student advisor	

Using these names we can translate the listed 13 statements into propositional logic.

1. $A \wedge B \wedge C \rightarrow D$
2. $\neg A \wedge M \rightarrow L$
3. $F \wedge E \rightarrow \neg D$
4. $G \wedge M \rightarrow C$
5. $\neg B \wedge F \rightarrow \neg H$
6. $\neg D \wedge B \wedge E \rightarrow G$
7. $M \wedge \neg I \rightarrow J$
8. $H \wedge M \rightarrow K$
9. $K \wedge J \wedge \neg L \rightarrow E$
10. $\neg H \wedge F \rightarrow L$
11. $M \wedge L \rightarrow \neg F$
12. $K \wedge I \wedge A \rightarrow E$
13. $\therefore M \rightarrow \neg F$

We will not construct a combined truth table for this example as it would require $2^{13} = 8192$ rows. The table would start off as follows:

	A	B	C	D	E	F	G	H	I	J	K	L	M	F_1	F_2	F_3	F_4	F_5	F_6	F_7	F_8	F_9	F_{10}	F_{11}	F_{12}	F_{13}
1.	1	1	1	1	1	1	1	1	1	1	1	1	1	1	1	0	1	1	1	1	1	1	1	0	1	0
2.	1	1	1	1	1	1	1	1	1	1	1	1	0	1	1	0	1	1	1	1	1	1	1	1	1	1

and it would fill about 125 pages using this font size.

Example 2.7.12 There are two tribes on the island of Tufa—the Tu's, who always tell the truth, and the Fa's, who always lie. A traveler encountered three residents of the island, A,B, and C, and each made a statement to the traveler:

- A said, " B and C tell the truth iff C tells the truth."

- B said, "If A and C tell the truth, then it is not the case that if B and C tell the truth, then A tells the truth."
- C said, " B is lying iff A or B is telling the truth."

From this information can you determine which tribes A,B, and C belong to? Let A be the statement that A is telling the truth (and thus A is a Tu), etc. Then we have

$$A \leftrightarrow ((B \wedge C) \leftrightarrow C)$$
$$B \leftrightarrow ((A \wedge C) \rightarrow \neg((B \wedge C) \rightarrow A))$$
$$C \leftrightarrow (\neg B \leftrightarrow (A \vee B)).$$

Letting these three propositional formulas be F, G, and H, we have the combined truth table:

	A	B	C	F	G	H
1.	1	1	1	1	0	0
2.	1	1	0	1	1	1
3.	1	0	1	0	1	1
4.	1	0	0	1	0	0
5.	0	1	1	0	1	0
6.	0	1	0	0	1	1
7.	0	0	1	1	0	0
8.	0	0	0	0	0	1

Thus A and B must be Tu's, and C a Fa, as line 2 is the only one that *makes all three statements true*.

Exercises

2.7.1 Express each of the following in symbolic form; and determine if they are valid arguments:

a. (*L. Carroll*[2]) No professors are ignorant.
All ignorant people are vain.
Therefore no professors are vain.

b. (*L. Carroll*[3]) No doctors are enthusiastic.
You are enthusiastic.
Therefore you are not a doctor.

2.7.2 (*L. Carroll*[4]) Express the following as an argument in the propositional calculus, and determine if it is valid (the universe of discourse is the class of writers):

---

[2]This is item 18 on p. 108 of [**7**].
[3]This is item 1 on p. 107 of [**7**].
[4]This is item 44 on p. 120 of [**7**].

a. All writers, who understand human nature, are clever.
b. No writer is a true poet unless he can stir the hearts of men.
c. Shakespeare wrote *Hamlet.*
d. No writer who does not understand human nature can stir the hearts of men.
e. None but a true poet could have written *Hamlet.*
f. Therefore Shakespeare is clever.

Use the following when analyzing the preceding argument:

A is able to stir the hearts of men
C is clever
S is Shakespeare
P is a true poet
N understands human nature
H is the writer of *Hamlet*

2.7.3 Determine if the following argument is valid:

$$\neg C \wedge D$$
$$\neg(\neg B \wedge C \wedge D)$$
$$\neg(\neg B \vee (\neg A \wedge B)) \wedge \neg C \wedge \neg D$$
$$\therefore\ A \wedge \neg B.$$

2.7.4 Determine if the following argument is valid (the universe of discourse is people):

a. Those who do not need 12 hours sleep are the life of a party.
b. Those who need 12 hours sleep are not skiers nor do they like junk food.
c. Anyone who is not a skier likes TV soap operas.
d. A person who is the life of a party does not like TV soap operas nor ride a motorcycle.
e. Skiers do not like junk food.
f. People who do not ride motorcycles like junk food.
g. Thus everyone likes TV soap operas and rides a motorcycle.

Use the following when analyzing the preceding argument:

N needs to sleep 12 hours
L is the life of a party
S is a skier
J likes junk food
T likes TV soap operas
M rides a motorcycle

2.7.5 If $F_1 \sim G_1, \ldots, F_n \sim G_n$, and $F \sim G$, then show

a. $F_1, \ldots, F_n \ \therefore\ F$ is a valid argument iff $G_1, \ldots, G_n \ \therefore\ G$ is a valid argument.

b. $F_1, \ldots, F_n$ is satisfiable iff $G_1, \ldots, G_n$ is satisfiable.

2.7.6 Determine if the following arguments are valid.

a.
$\neg(P \wedge R)$
$\therefore \neg(Q \wedge (P \wedge R))$

b.
$(Q \vee S) \leftrightarrow P$
$\therefore (S \rightarrow Q) \vee P$

c.
$(S \rightarrow Q) \vee P$
$\therefore S \rightarrow (S \leftrightarrow R)$

d.
$(Q \vee S) \leftrightarrow P$
$((P \leftrightarrow S) \wedge R) \leftrightarrow S$
$\therefore (P \wedge (R \leftrightarrow Q)) \vee S$

e.
$(R \vee S) \rightarrow (Q \wedge P)$
$((S \rightarrow P) \leftrightarrow R) \leftrightarrow \neg S$
$(R \vee P) \leftrightarrow (R \vee (Q \leftrightarrow R))$
$\therefore (S \rightarrow Q) \wedge R$

f.
$(\neg P \rightarrow S) \wedge R$
$Q \leftrightarrow (S \vee (R \wedge Q))$
$\therefore (P \rightarrow (P \rightarrow Q)) \wedge Q$

g.
$(S \rightarrow Q) \vee P$
$(P \wedge (R \leftrightarrow Q)) \vee S$
$((P \wedge R) \vee (Q \wedge S)) \leftrightarrow P$
$\therefore (Q \vee S) \leftrightarrow P$

2.8 COMPACTNESS

A number of results in mathematics have the flavor *since such and such is true for all finite parts, it must be true of the whole thing.* Such results are referred to as *applications of compactness.* One of the most basic compactness theorems occurs in propositional logic.

2.8.1 The Compactness Theorem for Propositional Logic

Theorem 2.8.1 [COMPACTNESS FOR PROPOSITIONAL LOGIC]
Let $\mathcal{S}$ be a set of propositional formulas. Then $\mathsf{Sat}(\mathcal{S})$ iff $\mathsf{Sat}(\mathcal{S}_0)$ for every finite $\mathcal{S}_0 \subseteq \mathcal{S}$.

Proof. We assume that every formula in $\mathcal{S}$ is built from the propositional variables $\{P_n : n = 1, 2, 3, \cdots\}$, and we enumerate the set $\mathcal{S}$ as $\{F_1, F_2, F_3, \ldots\}$. For each $n \geq 1$, let $\mathcal{S}_n$ be the set of those formulas from $\{F_1, \ldots, F_n\}$ whose propositional variables come from

$\{P_1, \ldots, P_n\}$. (Thus $\mathcal{S}_n \subseteq \{\mathsf{F}_1, \ldots, \mathsf{F}_n\}$, and it is possible that $\mathcal{S}_n = \emptyset$.)

If $e_1 e_2 \cdots e_n$ is a string of 0's and 1's of length n, then we say that $e_1 e_2 \cdots e_n$ is *good for* $\mathcal{S}$ if the corresponding truth evaluation $\mathsf{e} = (e_1, e_2, \cdots, e_n)$ makes each formula in $\mathcal{S}_n$ true.

Before continuing with the proof let us illustrate this definition with an example.

Example 2.8.2 Let $\mathcal{S}$ be the set of formulas

$$\{P_1, P_1 \vee \neg P_2, P_1 \vee \neg P_2 \vee P_3, P_1 \vee \neg P_2 \vee P_3 \vee \neg P_4, \ldots\}.$$

Then $e_1 e_2 e_3 e_4 = 1010$ is good for $\mathcal{S}$, whereas $e_1 e_2 e_3 e_4 = 0110$ is not good for $\mathcal{S}$.

If $e_1 e_2 \cdots e_k$ is a good string of 0's and 1's of length k, then we say that $e_1 e_2 \cdots e_k$ is *very good for* $\mathcal{S}$ if for every $n > k$ there is a good extension $e_1 e_2 \cdots e_n$ of length n of $e_1 e_2 \cdots e_k$.

Note that by assumption, for every n there is a string of length n that is good for $\mathcal{S}$, since $\mathcal{S}_n$ is a finite subset of $\mathcal{S}$ and is therefore satisfiable. Also note that if $e_1 e_2 \cdots e_n$ is good for $\mathcal{S}$, then each inital segment $e_1 e_2 \cdots e_k$ $(1 \leq k \leq n)$ is also good for $\mathcal{S}$.

Now we define a truth evaluation $\mathsf{e} = (e_1, e_2, \ldots, e_n, \ldots)$ that will simultaneously satisfy all the formulas in $\mathcal{S}$. The idea is to define its values consecutively so that for every n, $e_1 e_2 \cdots e_n$ will be good for $\mathcal{S}$.

First, we observe that either

a. for every n there is a good string of length n starting with 1; or

b. for every n there is a good string of length n starting with 0.

We define

$$e_1 = \begin{cases} 1 & \text{in case a} \\ 0 & \text{if not case a.} \end{cases}$$

This choice of e_1 is clearly very good for $\mathcal{S}$.

Now suppose we define $e_1, \ldots, e_k$ so that $e_1 \cdots e_k$ is very good for $\mathcal{S}$. We observe that either

a. for every $n > k$ there exists a good string $e_1 \cdots e_n$ extending $e_1 \cdots e_k$ with $e_{k+1} = 1$; or

b. for every $n > k$ there exists a good string $e_1 \cdots e_n$ extending $e_1 \cdots e_k$ with $e_{k+1} = 0$.

Inductively we define

$$e_{k+1} = \begin{cases} 1 & \text{in case a} \\ 0 & \text{if not case a.} \end{cases}$$

This choice of e_{k+1} makes $e_1 e_2 \cdots e_{k+1}$ very good for $\mathcal{S}$.

Then the sequence $\mathsf{e} = (e_1, e_2, \ldots, e_n, \ldots)$ satisfies all the formulas in $\mathcal{S}$. $\square$

2.8.2 Applications of Compactness

Example 2.8.3 [MATCHING PROBLEM] Let A and B be two given sets, and suppose that R is a relation from A to B, i.e., $R \subseteq A \times B$, such that every element of A is related to at least one element of B, and every element of A is related to only finitely many elements of B. Suppose that for every finite subset A_0 of A it is possible to find a *matching* $f : A_0 \to B$, i.e., a one–to–one function f such that for each $a \in A_0$ we have $aRf(a)$. Then we claim that it is possible to find a matching for all of A.

To prove this we introduce propositional variables P_{ab} for each pair (a, b) in $A \times B$. We think of P_{ab} as asserting that "a is matched to b." Then we form a set $\mathcal{S}$ of propositional formulas consisting of

1. $P_{ab_1} \vee \cdots \vee P_{ab_n}$ where b_i ranges over all (of the finitely many) b's in B for which aRb holds;
2. $\neg P_{ab_1} \vee \neg P_{ab_2}$ for $b_1 \neq b_2$ in B, and for any $a \in A$;
3. $\neg P_{a_1 b} \vee \neg P_{a_2 b}$ for $a_1 \neq a_2$ in A, and for any $b \in B$.

Item 1 says that each a is matched to at least one of the b's such that aRb holds. Item 2 says that each a is matched to at most one b. And item 3 says that different a's are not matched to the same b. These three conditions are exactly what is needed for a matching.

The assumptions guarantee that every finite subset of $\mathcal{S}$ is satisfiable. Thus, by compactness, so is $\mathcal{S}$. Any assignment of truth values that satisfies $\mathcal{S}$ gives a matching.

One interpretation of the matching problem is to suppose that the set A of young ladies and the set B of young men are such that given any finite subset of the set of ladies, the matchmaker can (simultaneously) find each of them a match, each with a man she likes. Then the matchmaker can actually (simultaneously) find a match for all the ladies, such that each is matched with a man she likes. (This interpretation is interesting only if one has an infinite set A of young ladies.)

Example 2.8.4 [GRAPH COLORING] Let $\mathbf{G} = (G, r)$ be a *graph*, that is, r is an irreflexive and symmetric binary relation on the nonempty set G, i.e., we have $(a, a) \notin r$ for every $a \in G$, and $(a, b) \in r$ implies $(b, a) \in r$. Members of G are called *vertices*, and r is called the *edge* relation. If $(a, b) \in r$, then we say that (a, b) is an *edge* and that a and b belong to this edge. We also say that a and b are *adjacent*. A *subgraph* $\mathbf{G}' = (G', r')$ of $\mathbf{G} = (G, r)$ consists of a nonempty subset G' of G, and r' must be the relation r restricted to G', i.e., for $a, b \in G'$, $(a, b) \in r'$ iff $(a, b) \in r$.

Suppose k is a positive integer. A *k–coloring* of $\mathbf{G} = (G, r)$ is an assignment of colors to the vertices G, from a specified collection of k colors, such that adjacent vertices do not have the same color.

($\star$) [ERDÖS–DEBRUIJN] Let k be fixed. Suppose that every finite subgraph of the graph $\mathbf{G} = (G, r)$ has a k–coloring. Then $\mathbf{G}$ has a k–coloring.

To prove ($\star$) we take propositional variables P_{ai} for $a \in G$ and $1 \le i \le k$. We think of P_{ai} as asserting "the vertex a gets the ith color." We form the set $\mathcal{S}$ of propositional formulas as follows:

1. $P_{a1} \vee \cdots \vee P_{ak}$ for each $a \in G$;
2. $\neg P_{ai} \vee \neg P_{aj}$ for all $a \in G$ and $1 \le i < j \le k$;
3. $\neg P_{ai} \vee \neg P_{bi}$ for all $a, b \in G$ with $(a, b) \in r$, and $1 \le i \le k$.

Item 1 says that each a is given at least one of the k colors. Item 2 says that no a is given two different colors. And item 3 says that adjacent a's are not given the same color. These are the three conditions we need for a k–coloring to exist.

The assumptions guarantee that every finite subset of $\mathcal{S}$ is satisfiable. Thus, by compactness, so is $\mathcal{S}$. Any assignment of truth values that satisfies $\mathcal{S}$ gives a k–coloring of the graph.

Exercises

2.8.1 Let $\mathcal{S}$ be an arbitrary infinite set of natural numbers, presented in binary notation (e.g., 6 is presented as 110). Prove that there is an infinite sequence of different binary numbers $b_1, b_2, \ldots$ such that each b_i is a prefix of b_{i+1} and also a prefix of some element of $\mathcal{S}$.

A *homomorphism* f from a graph $\mathbf{G}$ to a graph $\mathbf{H}$ is a map $f : G \to H$ such that if a and b are adjacent in $\mathbf{G}$ then $f(a)$ and $f(b)$ are adjacent in $\mathbf{H}$. (We say f *preserves* the adjacency relation, or it *preserves* edges.)

2.8.2 For **H** a *finite* graph prove that there is a homomorphism from **G** to **H** iff for every finite subgraph $\mathbf{G}_0$ of **G** there is a homomorphism from $\mathbf{G}_0$ to **H**.

2.8.3 Show that the Erdös–deBruijn theorem is a special case of Exercise 2.8.2.

2.9 THE PROPOSITIONAL PROOF SYSTEM PC

By 1877 mathematicians were beginning to consider purely syntactic approaches to logic, taking certain propositional formulas as axioms and adding to these certain rules for producing new formulas called *rules of inference.* We will refer to such a combination of axioms and rules as a *propositional proof system*. Before describing the system PC we need a few definitions.

2.9.1 Simple Equivalences

Definition 2.9.1 The *depth* of a formula is defined inductively by

$$\begin{aligned}
\mathsf{depth}(0) &= 0\\
\mathsf{depth}(1) &= 0\\
\mathsf{depth}(P) &= 0\\
\mathsf{depth}(\neg\,\mathsf{F}) &= \mathsf{depth}(\mathsf{F}) + 1\\
\mathsf{depth}(\mathsf{F} \,\square\, \mathsf{G}) &= \max(\mathsf{depth}(\mathsf{F}), \mathsf{depth}(\mathsf{G})) + 1,
\end{aligned}$$

where $\square$ is any of the four binary connectives.

Example 2.9.2 Thus we have

$$\begin{aligned}
\mathsf{depth}(P) = \mathsf{depth}(Q) = \mathsf{depth}(R) &= 0\\
\mathsf{depth}(Q \vee R) = \mathsf{depth}(P \wedge Q) = \mathsf{depth}(P \wedge R) &= 1\\
\mathsf{depth}(P \wedge (Q \vee R)) = \mathsf{depth}((P \wedge Q) \vee (P \wedge R)) &= 2.
\end{aligned}$$

The depth of a formula is just the largest number of edges in a path from the top node to a bottom node of the tree of the formula. The *tree* of a formula is obtained by parsing the formula and putting the "outermost" connective at the top, and so forth. The inductive definition of the tree $\mathsf{Tree}(\mathsf{F})$ of a formula F is as follows:

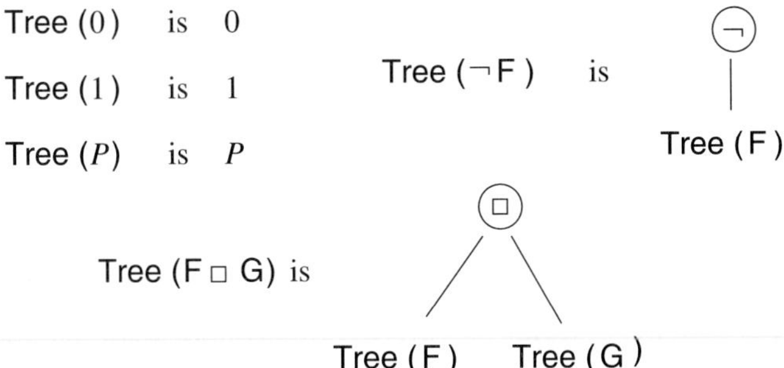

Figure 2.9.3 Inductive definition of Tree(F)

where □ above is any of the binary connectives.

Thus Tree($(P \to (Q \vee R)) \wedge \neg(Q \leftrightarrow \neg P)$) is

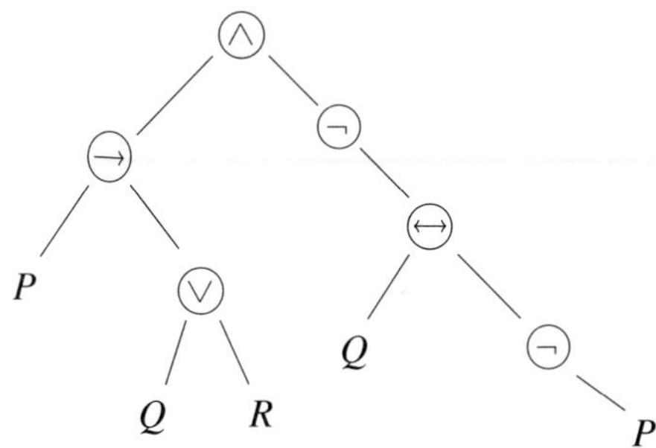

Figure 2.9.4 Example of a formula tree

We see immediately that the formula $(P \to (Q \vee R)) \wedge \neg(Q \leftrightarrow \neg P)$ has depth 4 since the path along the right–hand side has length 4, and there are no longer paths in the tree.

Definition 2.9.5

a. A formula F is *small* if the depth of F is at most 2, i.e., depth(F) ≤ 2.
b. A truth equivalence F $\sim$ G is *small* if both F and G are small.
c. A truth equivalence F $\sim$ G is *simple* if it is a substitution instance of a small truth equivalence F$'$ $\sim$ G$'$.

Observe that all the fundamental equivalences on p. 44 are *small* equivalences, with the exception of items 30 and 31; the substitution instances of these small equivalences, given below, are our standard *simple* equivalences. There are other simple equivalences, for example, $F \lor (1 \land H) \sim (F \lor 0) \lor H$, but this list suffices for our treatment of the propositional logic.

1.	$F \lor F$	$\sim$	F	idempotent
2.	$F \land F$	$\sim$	F	idempotent
3.	$F \lor G$	$\sim$	$G \lor F$	commutative
4.	$F \land G$	$\sim$	$G \land F$	commutative
5.	$F \lor (G \lor H)$	$\sim$	$(F \lor G) \lor H$	associative
6.	$F \land (G \land H)$	$\sim$	$(F \land G) \land H$	associative
7.	$F \land (F \lor G)$	$\sim$	F	absorption
8.	$F \lor (F \land G)$	$\sim$	F	absorption
9.	$F \land (G \lor H)$	$\sim$	$(F \land G) \lor (F \land H)$	distributive
10.	$F \lor (G \land H)$	$\sim$	$(F \lor G) \land (F \lor H)$	distributive
11.	$F \lor \neg F$	$\sim$	1	
12.	$F \land \neg F$	$\sim$	0	
13.	$\neg \neg F$	$\sim$	F	
14.	$F \lor 1$	$\sim$	1	
15.	$F \land 1$	$\sim$	F	
16.	$F \lor 0$	$\sim$	F	
17.	$F \land 0$	$\sim$	0	
18.	$\neg (F \lor G)$	$\sim$	$\neg F \land \neg G$	DeMorgan's law
19.	$\neg (F \land G)$	$\sim$	$\neg F \lor \neg G$	DeMorgan's law
20.	$F \to G$	$\sim$	$\neg F \lor G$	
21.	$F \to G$	$\sim$	$\neg G \to \neg F$	contrapositive
22.	$F \to (G \to H)$	$\sim$	$(F \land G) \to H$	
23.	$F \to (G \to H)$	$\sim$	$(F \to G) \to (F \to H)$	
24.	$F \leftrightarrow F$	$\sim$	1	
25.	$F \leftrightarrow G$	$\sim$	$G \leftrightarrow F$	
26.	$(F \leftrightarrow G) \leftrightarrow H$	$\sim$	$F \leftrightarrow (G \leftrightarrow H)$	
27.	$F \leftrightarrow \neg G$	$\sim$	$\neg (F \leftrightarrow G)$	
28.	$F \leftrightarrow (G \leftrightarrow F)$	$\sim$	G	
29.	$F \leftrightarrow G$	$\sim$	$(F \to G) \land (G \to F)$	

2.9.2 The Proof System

Now we are ready to describe the propositional proof system PC.

PC

Connectives $\vee, \wedge, \rightarrow, \leftrightarrow, \neg, 0, 1$

Axioms All formulas of the form $\mathsf{F} \vee \neg\, \mathsf{F}$

Rules of inference

S (simplification) From $\mathsf{F} \wedge \mathsf{G}$ one can infer F; and from $\mathsf{F} \wedge \mathsf{G}$ one can infer G.

C (conjunction) From F and G one can infer $\mathsf{F} \wedge \mathsf{G}$.

A (addition) From F one can infer $\mathsf{F} \vee \mathsf{G}$; and one can infer $\mathsf{G} \vee \mathsf{F}$.

MP (modus ponens) From F and $\mathsf{F} \rightarrow \mathsf{G}$ one can infer G.

MT (modus tollens) From $\mathsf{F} \rightarrow \mathsf{G}$ and $\neg\, \mathsf{G}$ one can infer $\neg\, \mathsf{F}$.

HS (hypothetical syllogism) From $\mathsf{F} \rightarrow \mathsf{G}$ and $\mathsf{G} \rightarrow \mathsf{H}$ one can infer $\mathsf{F} \rightarrow \mathsf{H}$.

DS (disjunctive syllogism) From $\mathsf{F} \vee \mathsf{G}$ and $\neg\, \mathsf{F}$ one can infer G; and from $\mathsf{F} \vee \mathsf{G}$ and $\neg\, \mathsf{G}$ one can infer F.

CD (constructive dilemma) From $\mathsf{F}_1 \rightarrow \mathsf{G}_1$, $\mathsf{F}_2 \rightarrow \mathsf{G}_2$, and $\mathsf{F}_1 \vee \mathsf{F}_2$ one can infer $\mathsf{G}_1 \vee \mathsf{G}_2$.

SR (simple replacement) If F' is a subformula of F and $\mathsf{F}' \sim \mathsf{G}'$ is a *simple* equivalence, then from F one can infer any G that results from replacing an occurrence of F' in F with G'.

As we mentioned in the preface, this system is related to, but quite distinct from, the proof system RS used by Copi.

The next box has a brief symbolic description of PC.

PC

Connectives $\vee, \wedge, \rightarrow, \leftrightarrow, \neg, 0, 1$

Axioms $\mathsf{F} \vee \neg \mathsf{F}$

Rules of inference

(S) $\dfrac{\mathsf{F} \wedge \mathsf{G}}{\mathsf{F}}$; $\dfrac{\mathsf{F} \wedge \mathsf{G}}{\mathsf{G}}$ (C) $\dfrac{\mathsf{F},\ \mathsf{G}}{\mathsf{F} \wedge \mathsf{G}}$

(A) $\dfrac{\mathsf{F}}{\mathsf{F} \vee \mathsf{G}}$; $\dfrac{\mathsf{F}}{\mathsf{G} \vee \mathsf{F}}$ (MP) $\dfrac{\mathsf{F},\ \mathsf{F} \rightarrow \mathsf{G}}{\mathsf{G}}$

(MT) $\dfrac{\mathsf{F} \rightarrow \mathsf{G},\ \neg \mathsf{G}}{\neg \mathsf{F}}$ (HS) $\dfrac{\mathsf{F} \rightarrow \mathsf{G},\ \mathsf{G} \rightarrow \mathsf{H}}{\mathsf{F} \rightarrow \mathsf{H}}$

(DS) $\dfrac{\mathsf{F} \vee \mathsf{G},\ \neg \mathsf{F}}{\mathsf{G}}$ $\dfrac{\mathsf{F} \vee \mathsf{G},\ \neg \mathsf{G}}{\mathsf{F}}$

(CD) $\dfrac{\mathsf{F}_1 \rightarrow \mathsf{G}_1,\ \mathsf{F}_2 \rightarrow \mathsf{G}_2,\ \mathsf{F}_1 \vee \mathsf{F}_2}{\mathsf{G}_1 \vee \mathsf{G}_2}$

(SR) $\dfrac{\mathsf{F}(\cdots \mathsf{F}' \cdots)}{\mathsf{F}(\cdots \mathsf{G}' \cdots)}$ *for* $\mathsf{F}' \sim \mathsf{G}'$ *a simple equivalence*

In the last item, (SR), note that $\mathsf{F}(\cdots \mathsf{F}' \cdots)$ means that F' is referring to *a particular occurrence* of F' in the formula F. This is not to be confused with the earlier notation $\mathsf{F}(\cdots, \mathsf{F}', \cdots)$ that means F' is being substituted for some variable in F.

Definition 2.9.6 $\mathcal{S} \vdash \mathsf{F}$ (read: F *can be derived from* $\mathcal{S}$) means there is a sequence of propositional formulas $\mathsf{F}_1, \ldots, \mathsf{F}_n$, with $\mathsf{F} = \mathsf{F}_n$ such that for each i

- either F_i is an axiom,
- or F_i is in $\mathcal{S}$,
- or F_i is obtained from previous F_j's by an application of one of the rules of inference.

Such a sequence is an $\mathcal{S}$–*derivation* (or $\mathcal{S}$–*proof*) of F. A $\emptyset$–derivation is simply called a *derivation*.

Let us get a little practice with this proof system before moving on to the main theorems. On the left side of the examples of derivations given next we list the steps of the derivation, and on the right the reason justifying each step.

Following the reason there are line numbers, if needed, that refer to the line(s) used as input to the rule of inference, and in the case of using simple replacement, a parenthesized number to indicate the simple equivalence used from the list on p. 81. Of course, given a simple equivalence $\mathsf{F} \sim \mathsf{G}$ we also have the simple equivalence $\mathsf{G} \sim \mathsf{F}$, so each simple equivalence on p. 81 does double duty.

In line 4 of the derivation in Example 2.9.7, the reason given is **A: 3**, meaning that addition rule A is applied to line 3 (we have added $\neg P$ to the formula in line 3). And in line 5 we have **SR: 4 (20)**, meaning that this is a simple replacement applied to line 4, using the simple equivalence (20).

Example 2.9.7 The following is a derivation of the tautology $P \to (P \to P)$.

1.	$P \vee \neg P$	axiom
2.	$\neg P \vee P$	SR: 1 (3)
3.	$P \to P$	SR: 2 (20)
4.	$\neg P \vee (P \to P)$	A: 3
5.	$P \to (P \to P)$	SR: 4 (20)

To be precise the derivation is just the following sequence of five formulas:

$$P \vee \neg P,\ \neg P \vee P,\ P \to P,\ \neg P \vee (P \to P),\ P \to (P \to P).$$

However, it is much easier to check the correctness of a derivation if it is written as shown, with reasons for each step.

Example 2.9.8 The following is a derivation of the tautology $P \to (Q \to P)$. This requires a little more work.

1.	$P \vee \neg P$	axiom
2.	$\neg P \vee P$	SR: 1 (3)
3.	$(\neg P \vee P) \vee \neg Q$	A: 2
4.	$\neg P \vee (P \vee \neg Q)$	SR: 3 (5)
5.	$\neg P \vee (\neg Q \vee P)$	SR: 4 (3)
6.	$\neg P \vee (Q \to P)$	SR: 5 (20)
7.	$P \to (Q \to P)$	SR: 6 (20)

You may wonder how to go about finding such a derivation. Actually, PC is so rich that there are often many good ways to carry out a derivation. In this example we really started with the conclusion and put it in disjunctive form. Then we reversed the steps to obtain the derivation.

2.9.3 Soundness and Completeness

One of the things we expect from a proof system is that it yield only true conclusions, and we would like it to yield all true conclusions.

Definition 2.9.9 A propositional proof system is

- *sound* if one can derive only tautologies, i.e., $\vdash \mathsf{F}$ implies F is a tautology;
- *complete* if one can derive all tautologies.

Theorem 2.9.10 [SOUNDNESS] If $\vdash \mathsf{F}$ in the PC propositional proof system, then F is a tautology.

Proof. Let $\mathsf{F}_1, \ldots, \mathsf{F}_n$ be a derivation of F. First, note that F_1 must be an axiom (as there are no premisses). Then, for each F_i that is an axiom, we see that F_i is a tautology. If F_i is a tautology and $\mathsf{F}_i \to \mathsf{F}_j$ is a tautology, then F_j is a tautology. Thus modus ponens preserves tautologies. Likewise, we can show that the other rules preserve tautologies. Thus all F_i are tautologies. □

Theorem 2.9.11 [COMPLETENESS] Every tautology can be derived in the PC propositional proof system, i.e., if F is a tautology then $\vdash \mathsf{F}$.

Proof. Let F' be the disjunctive normal form of $\mathsf{F}(P_1, \ldots, P_n)$. Using the simple replacement rules we can derive F' from F as we did in Sec. 2.6.1. The steps are reversible because simple replacement rules are reversible. Thus we have $\mathsf{F}' \vdash \mathsf{F}$.

Now, the tautology $\mathsf{G} = (P_1 \vee \neg P_1) \wedge \cdots \wedge (P_n \vee \neg P_n)$ can be derived using just the axioms $P_i \vee \neg P_i$ with the conjunction rule applied $n-1$ times. So we have $\vdash \mathsf{G}$. Applying the simple replacement rules we can then derive G', the disjunctive normal form of G, so $\vdash \mathsf{G}'$.

But now we observe that F' is the same as G', up to order of constituents, as they are both tautologies in disjunctive normal form. Thus we have $\vdash \mathsf{G}'$, $\mathsf{G}' \vdash \mathsf{F}'$, and $\mathsf{F}' \vdash \mathsf{F}$. Thus $\vdash \mathsf{F}$, i.e., F is derivable. □

If we look at the proof of the completeness theorem, then it is clear that we used only the (SR) rule and the (C) rule. The system PC on p. 83 is highly redundant, i.e., there are far more rules than actually needed. So why do we have the excessive collection of rules of inference? This is simply to increase the likelihood of finding short derivations.

2.9.4 Derivations with Premisses

So far we have concentrated on $\vdash \mathsf{F}$ for tautologies F. Now we want to look at some examples of $\mathcal{S} \vdash \mathsf{F}$. The set $\mathcal{S}$ is referred to as the *premisses*, or the *given* formulas.

Example 2.9.12 The following is a derivation to witness

$$\mathsf{F} \rightarrow (\mathsf{G} \wedge \mathsf{H}) \vdash \mathsf{F} \rightarrow \mathsf{G}.$$

1.	$\mathsf{F} \rightarrow (\mathsf{G} \wedge \mathsf{H})$	given
2.	$\neg \mathsf{F} \vee (\mathsf{G} \wedge \mathsf{H})$	SR: 1 (20)
3.	$(\neg \mathsf{F} \vee \mathsf{G}) \wedge (\neg \mathsf{F} \vee \mathsf{H})$	SR: 2 (10)
4.	$\neg \mathsf{F} \vee \mathsf{G}$	S: 3
5.	$\mathsf{F} \rightarrow \mathsf{G}$	SR: 4 (20)

Example 2.9.13 The following is a derivation to witness

$$(\mathsf{F} \vee \mathsf{G}) \rightarrow \mathsf{H} \vdash \mathsf{F} \rightarrow \mathsf{H}.$$

1.	$(F \vee G) \rightarrow H$	given
2.	$\neg(F \vee G) \vee H$	SR: 1 (20)
3.	$(\neg F \wedge \neg G) \vee H$	SR: 2 (18)
4.	$H \vee (\neg F \wedge \neg G)$	SR: 3 (3)
5.	$(H \vee \neg F) \wedge (H \vee \neg G)$	SR: 4 (10)
6.	$H \vee \neg F$	S: 5
7.	$\neg F \vee H$	SR: 6 (3)
8.	$F \rightarrow H$	SR: 7 (20)

Example 2.9.14 The following is a derivation to witness

$$F \wedge \neg F \vdash G.$$

In effect it says that from $F \wedge \neg F$ you can derive anything!

1.	$F \wedge \neg F$	given
2.	$F \vee \neg F$	axiom
3.	$\neg F \vee F$	SR: 2 (3)
4.	$(\neg F \vee F) \vee G$	A: 3
5.	$\neg\neg(\neg F \vee F) \vee G$	SR: 4 (13)
6.	$\neg(\neg F \vee F) \rightarrow G$	SR: 5 (20)
7.	$(\neg\neg F \wedge \neg F) \rightarrow G$	SR: 6 (18)
8.	$(F \wedge \neg F) \rightarrow G$	SR: 7 (13)
9.	G	MP: 1,8

Example 2.9.15 The following is a derivation to witness

$$F \rightarrow G,\ F \rightarrow \neg G \vdash \neg F.$$

This is a version of proof by contradiction. If F leads to a contradiction, then $\neg\,\mathsf{F}$ must hold.

1.	$\mathsf{F} \to \mathsf{G}$	given
2.	$\mathsf{F} \to \neg\,\mathsf{G}$	given
3.	$\neg\,\mathsf{G} \to \neg\,\mathsf{F}$	SR: 1 (21)
4.	$\neg\neg\,\mathsf{G} \to \neg\,\mathsf{F}$	SR: 2 (21)
5.	$\neg\,\mathsf{G} \vee \neg\neg\,\mathsf{G}$	axiom
6.	$\neg\,\mathsf{F} \vee \neg\,\mathsf{F}$	CD: 3,4,5
7.	$\neg\,\mathsf{F}$	SR: 6 (1)

Example 2.9.16 The following is a derivation to witness

$$\mathsf{F} \to \mathsf{G},\ \mathsf{F} \to \mathsf{H} \vdash \mathsf{F} \to (\mathsf{G} \wedge \mathsf{H}).$$

1.	$\mathsf{F} \to \mathsf{G}$	given
2.	$\mathsf{F} \to \mathsf{H}$	given
3.	$\neg\,\mathsf{F} \vee \mathsf{G}$	SR: 1 (20)
4.	$\neg\,\mathsf{F} \vee \mathsf{H}$	SR: 2 (20)
5.	$(\neg\,\mathsf{F} \vee \mathsf{G}) \wedge (\neg\,\mathsf{F} \vee \mathsf{H})$	C: 3,4
6.	$\neg\,\mathsf{F} \vee (\mathsf{G} \wedge \mathsf{H})$	SR: 5 (10)
7.	$\mathsf{F} \to (\mathsf{G} \wedge \mathsf{H})$	SR: 6 (20)

Example 2.9.17 The following is a derivation to witness

$$\mathsf{F} \to \mathsf{H},\ \mathsf{G} \to \mathsf{L} \vdash (\mathsf{F} \wedge \mathsf{G}) \to (\mathsf{H} \wedge \mathsf{L}).$$

(This is essentially a generalization of the derivation in the previous example. Can you see where we have borrowed steps from it?)

1.	$\mathsf{F} \rightarrow \mathsf{H}$	given
2.	$\mathsf{G} \rightarrow \mathsf{L}$	given
3.	$\neg \mathsf{F} \vee \mathsf{H}$	SR: 1 (20)
4.	$(\neg \mathsf{F} \vee \mathsf{H}) \vee \neg \mathsf{G}$	A: 3
5.	$\neg \mathsf{F} \vee (\mathsf{H} \vee \neg \mathsf{G})$	SR: 4 (5)
6.	$\neg \mathsf{F} \vee (\neg \mathsf{G} \vee \mathsf{H})$	SR: 5 (3)
7.	$(\neg \mathsf{F} \vee \neg \mathsf{G}) \vee \mathsf{H}$	SR: 6 (5)
8.	$\neg \mathsf{G} \vee \mathsf{L}$	SR: 2 (20)
9.	$(\neg \mathsf{G} \vee \mathsf{L}) \vee \neg \mathsf{F}$	A: 8
10.	$\neg \mathsf{F} \vee (\neg \mathsf{G} \vee \mathsf{L})$	SR: 9 (3)
11.	$(\neg \mathsf{F} \vee \neg \mathsf{G}) \vee \mathsf{L}$	SR: 10 (5)
12.	$((\neg \mathsf{F} \vee \neg \mathsf{G}) \vee \mathsf{H}) \wedge ((\neg \mathsf{F} \vee \neg \mathsf{G}) \vee \mathsf{L})$	C: 7,11
13.	$(\neg \mathsf{F} \vee \neg \mathsf{G}) \vee (\mathsf{H} \wedge \mathsf{L})$	SR: 12 (10)
14.	$\neg\neg(\neg \mathsf{F} \vee \neg \mathsf{G}) \vee (\mathsf{H} \wedge \mathsf{L})$	SR: 13 (13)
15.	$\neg(\neg \mathsf{F} \vee \neg \mathsf{G}) \rightarrow (\mathsf{H} \wedge \mathsf{L})$	SR: 14 (20)
16.	$\neg\neg(\mathsf{F} \wedge \mathsf{G}) \rightarrow (\mathsf{H} \wedge \mathsf{L})$	SR: 15 (19)
17.	$(\mathsf{F} \wedge \mathsf{G}) \rightarrow (\mathsf{H} \wedge \mathsf{L})$	SR: 16 (13)

Example 2.9.18 As a final example in this section let us return to Boole's argument from Example 2.7.10 on p. 70, replacing the $\therefore$ with a $\vdash$:

$$\begin{array}{l} (A \wedge B) \rightarrow (C \leftrightarrow \neg D) \\ (B \wedge C) \rightarrow (A \leftrightarrow D) \\ (\neg A \wedge \neg B) \rightarrow (\neg C \wedge \neg D) \\ (\neg C \wedge \neg D) \rightarrow (\neg A \wedge \neg B) \\ \vdash (\neg((A \wedge B) \wedge C)) \wedge ((\neg A \wedge \neg B) \rightarrow \neg C). \end{array}$$

Note that when carrying out derivations it is important to have precisely defined formulas. Hence, we have added the parentheses in the preceding conclusion. The following is a derivation to witness the claim (that we have a derivation). Notice that our strategy for dealing with a conclusion consisting of a conjunction is to derive each conjunct separately (lines 8 and 34) and then to use rule (C).

Some of the pieces of the derivation are borrowed from the preceding examples. Can you spot them?

1.	$(A \wedge B) \rightarrow (C \leftrightarrow \neg D)$	given
2.	$(B \wedge C) \rightarrow (A \leftrightarrow D)$	given
3.	$(\neg A \wedge \neg B) \rightarrow (\neg C \wedge \neg D)$	given
4.	$(\neg C \wedge \neg D) \rightarrow (\neg A \wedge \neg B)$	given
5.	$\neg(\neg A \wedge \neg B) \vee (\neg C \wedge \neg D)$	SR: 3 (20)
6.	$(\neg(\neg A \wedge \neg B) \vee \neg C) \wedge (\neg(\neg A \wedge \neg B) \vee \neg D)$	SR: 5 (10)
7.	$\neg(\neg A \wedge \neg B) \vee \neg C$	S: 6
8.	$(\neg A \wedge \neg B) \rightarrow \neg C$	SR: 7 (20)
9.	$\neg(A \wedge B) \vee (C \leftrightarrow \neg D)$	SR: 1 (20)
10.	$\neg(A \wedge B) \vee ((C \rightarrow \neg D) \wedge (\neg D \rightarrow C))$	SR: 9 (29)
11.	$(\neg(A \wedge B) \vee (C \rightarrow \neg D)) \wedge (\neg(A \wedge B) \vee (\neg D \rightarrow C))$	SR: 10 (10)
12.	$\neg(A \wedge B) \vee (C \rightarrow \neg D)$	S: 11
13.	$\neg(A \wedge B) \vee (\neg C \vee \neg D)$	SR: 12 (20)
14.	$(\neg(A \wedge B) \vee \neg C) \vee \neg D$	SR: 13 (5)
15.	$((\neg A \vee \neg B) \vee \neg C) \vee \neg D$	SR: 14 (19)
16.	$\neg(B \wedge C) \vee (A \leftrightarrow D)$	SR: 2 (20)
17.	$\neg(B \wedge C) \vee ((A \rightarrow D) \wedge (D \rightarrow A))$	SR: 16 (29)
18.	$(\neg(B \wedge C) \vee (A \rightarrow D)) \wedge (\neg(B \wedge C) \vee (D \rightarrow A))$	SR: 17 (10)
19.	$\neg(B \wedge C) \vee (A \rightarrow D)$	S: 18
20.	$\neg(B \wedge C) \vee (\neg A \vee D)$	SR: 19 (20)
21.	$(\neg(B \wedge C) \vee \neg A) \vee D$	SR: 20 (5)
22.	$(\neg A \vee \neg(B \wedge C)) \vee D$	SR: 21 (3)
23.	$(\neg A \vee (\neg B \vee \neg C)) \vee D$	SR: 22 (19)
24.	$((\neg A \vee \neg B) \vee \neg C) \vee D$	SR: 23 (5)
25.	$\neg D \vee ((\neg A \vee \neg B) \vee \neg C)$	SR: 15 (3)
26.	$D \rightarrow ((\neg A \vee \neg B) \vee \neg C)$	SR: 25 (20)
27.	$D \vee ((\neg A \vee \neg B) \vee \neg C)$	SR: 24 (3)

28.	$\neg\neg D \vee ((\neg A \vee \neg B) \vee \neg C)$	SR: 27 (13)
29.	$\neg D \rightarrow ((\neg A \vee \neg B) \vee \neg C)$	SR: 28 (20)
30.	$D \vee \neg D$	axiom
31.	$((\neg A \vee \neg B) \vee \neg C) \vee ((\neg A \vee \neg B) \vee \neg C)$	CD: 26,29,30
32.	$(\neg A \vee \neg B) \vee \neg C$	SR: 31 (1)
33.	$\neg (A \wedge B) \vee \neg C$	SR: 32 (19)
34.	$\neg ((A \wedge B) \wedge C)$	SR: 33 (19)
35.	$\neg ((A \wedge B) \wedge C) \wedge ((\neg A \wedge \neg B) \rightarrow \neg C)$	C: 8,34

2.9.5 Proving Theorems about $\vdash$

In this section we concentrate on proving general theorems about working with $\mathcal{S} \vdash \mathsf{F}$, rather than finding particular derivations. The first theorem is rather obvious: that if you increase the hypotheses, then you can derive at least as much as before.

Theorem 2.9.19 Suppose $\mathcal{S} \vdash \mathsf{F}$ and $\mathcal{S}' \supseteq \mathcal{S}$. Then $\mathcal{S}' \vdash \mathsf{F}$.

Proof. Let $\mathsf{F}_1, \ldots, \mathsf{F}_n$ be a derivation of F from $\mathcal{S}$. Then this is also a derivation of F from $\mathcal{S}'$. □

Next we claim that there are analogs of all the rules of inference for $\vdash$.

Theorem 2.9.20 Suppose

$$\frac{\mathsf{H}_1, \ldots, \mathsf{H}_n}{\mathsf{H}}$$

is a rule of inference of PC. Then from the hypotheses

$$\mathcal{S} \vdash \mathsf{H}_1, \cdots, \mathcal{S} \vdash \mathsf{H}_n$$

we can conclude that

$$\mathcal{S} \vdash \mathsf{H}.$$

Proof. We give a proof for one of the rules, modus ponens, and leave the others as exercises, since the pattern of proof will be very similar.

So let us suppose we have

$$\mathcal{S} \vdash \mathsf{F} \text{ and } \mathcal{S} \vdash \mathsf{F} \to \mathsf{G}.$$

This says that we have *two separate derivations*, say, $\mathsf{F}_1, \ldots, \mathsf{F}_m$ is a derivation of F from $\mathcal{S}$, and $\mathsf{G}_1, \ldots, \mathsf{G}_n$ is a derivation of $\mathsf{F} \to \mathsf{G}$ from $\mathcal{S}$. We need to find a *single* derivation of G from $\mathcal{S}$. The idea is simple. We just append one derivation to the other, say $\mathsf{F}_1, \ldots, \mathsf{F}_m, \mathsf{G}_1, \ldots, \mathsf{G}_n$. This is still a derivation from $\mathcal{S}$, and in this derivation we see both F (as F_m) and $\mathsf{F} \to \mathsf{G}$ (as G_n). Now that we have both formulas in a *single* derivation we can apply the modus ponens rule from PC to detach G, i.e.,

$$\mathsf{F}_1, \ldots, \mathsf{F}_m, \mathsf{G}_1, \ldots, \mathsf{G}_n, \mathsf{G}$$

is the desired derivation of G from $\mathcal{S}$. □

The next theorem about derivations, Herbrand's famous Deduction Lemma, is a standard workhorse in the study of proof theory. It says that one can derive G from $\mathcal{S}$ with the help of F iff one can derive $\mathsf{F} \to \mathsf{G}$ from just $\mathcal{S}$.

Theorem 2.9.21 [DEDUCTION LEMMA] Let $\mathcal{S}$ be a set of propositional formulas, and let F and G be propositional formulas. Then

$$\mathcal{S} \cup \{\mathsf{F}\} \vdash \mathsf{G} \quad \text{iff} \quad \mathcal{S} \vdash \mathsf{F} \to \mathsf{G}. \tag{21}$$

Proof. One direction, namely, ($\Leftarrow$), is rather obvious. Suppose we have a derivation $\mathsf{F}_1, \ldots, \mathsf{F}_n$ of $\mathsf{F} \to \mathsf{G}$ from $\mathcal{S}$. Then the following is a derivation of G from $\mathcal{S} \cup \{\mathsf{F}\}$:

$$\mathsf{F}_1, \ldots, \mathsf{F}_n, \mathsf{F}, \mathsf{G}.$$

The justifications for the last two steps are (1) the formula F is in $\mathcal{S} \cup \{\mathsf{F}\}$, and we are always allowed to use a premiss in a derivation; and (2) the formula F_n is just $\mathsf{F} \to \mathsf{G}$, so G, the last formula, follows by applying modus ponens to the previous two steps.

The other direction ($\Rightarrow$) is more subtle. We give a proof based on the completeness theorem.

Now we suppose that $\mathcal{S} \cup \{\mathsf{F}\} \vdash \mathsf{G}$ holds. Since proofs are finite, let us choose $\mathsf{H}_1, \ldots, \mathsf{H}_k$ from $\mathcal{S}$ such that $\mathsf{H}_1, \ldots, \mathsf{H}_k, \mathsf{F} \vdash \mathsf{G}$. Let e be a truth evaluation for the variables that occur in $\mathsf{H}_1, \ldots, \mathsf{H}_k, \mathsf{F}, \mathsf{G}$. If the formulas $\mathsf{H}_1, \ldots, \mathsf{H}_k, \mathsf{F}$ are true at e, then so is G. This follows from the fact that all our axioms are true at e, and if any rule from PC is applied to formulas that are true at e, then the resulting formula is also true at e. But what this really says is that $\mathsf{H}_1, \ldots, \mathsf{H}_k, \mathsf{F} \therefore \mathsf{G}$ is a valid argument.

But then $\mathsf{H}_1 \to (\mathsf{H}_2 \to \cdots \to (\mathsf{H}_k \to (\mathsf{F} \to \mathsf{G}))\cdots)$ is a tautology (see Theorem 2.7.8 on p. 69 and equivalence (22) on p. 44). So from the completeness theorem for PC we know that there is a derivation of

$$\mathsf{H}_1 \to (\mathsf{H}_2 \to \cdots \to (\mathsf{H}_k \to (\mathsf{F} \to \mathsf{G}))\cdots)$$

with no assumptions, i.e.,

$$\vdash \mathsf{H}_1 \to (\mathsf{H}_2 \to \cdots \to (\mathsf{H}_k \to (\mathsf{F} \to \mathsf{G}))\cdots).$$

Now, applying the first half of the Herbrand theorem k times, to pull the H_i's over to the left side of the $\vdash$, but without pulling F over, we have

$$\mathsf{H}_1, \ldots, \mathsf{H}_k \vdash \mathsf{F} \to \mathsf{G}.$$

As $\{\mathsf{H}_1, \ldots, \mathsf{H}_k\} \subseteq \mathcal{S}$, it follows that $\mathcal{S} \vdash \mathsf{F} \to \mathsf{G}$. □

Corollary 2.9.22 For formulas $\mathsf{F}_1, \ldots, \mathsf{F}_n$ and F we have

$$\mathsf{F}_1, \ldots, \mathsf{F}_n \vdash \mathsf{F} \quad \text{iff} \quad \vdash \mathsf{F}_1 \to (\mathsf{F}_2 \to \cdots \to (\mathsf{F}_n \to \mathsf{F})\cdots).$$

2.9.6 Generalized Soundness and Completeness

Now we turn to an important extension of the soundness and completeness of PC.

Definition 2.9.23 We say that a propositional formula F is a *consequence* of a (possibly infinite) set $\mathcal{S}$ of propositional formulas, written $\mathcal{S} \models \mathsf{F}$, if every truth evaluation that satisfies $\mathcal{S}$ makes F true.

The assertion $\models \mathsf{F}$ means that $\mathcal{S}$ is empty, and thus every truth evaluation makes F true, i.e., the assertion $\models \mathsf{F}$ is equivalent to saying that F is a tautology. And the assertion $\mathsf{F}_1, \ldots, \mathsf{F}_n \therefore \mathsf{F}$ is equivalent to the assertion $\mathsf{F}_1, \ldots, \mathsf{F}_n \models \mathsf{F}$.

Now we can formulate a strengthening of the completeness theorem.

Theorem 2.9.24 In the PC propositional proof system we have

$$\mathcal{S} \vdash \mathsf{F} \quad \text{iff} \quad \mathcal{S} \models \mathsf{F}.$$

Proof. ($\Longrightarrow$: *Generalized soundness*) Suppose $\mathcal{S} \vdash \mathsf{F}$. Then, since derivations are finite, there are $\mathsf{F}_1, \ldots, \mathsf{F}_n \in \mathcal{S}$ such that

$$\mathsf{F}_1, \ldots, \mathsf{F}_n \vdash \mathsf{F}.$$

Then by Corollary 2.9.22

$$\vdash \mathsf{F}_1 \to (\mathsf{F}_2 \to \cdots (\mathsf{F}_n \to \mathsf{F}) \cdots),$$

and thus by the completeness of PC, Theorem 2.9.11 (p. 85),

$$\mathsf{F}_1 \to (\mathsf{F}_2 \to \cdots (\mathsf{F}_n \to \mathsf{F}) \cdots)$$

is a tautology. Consequently, any assignment that satisfies $\mathcal{S}$ will make each of $\mathsf{F}_1, \ldots, \mathsf{F}_n$ true and thus will also make F true. Thus $\mathcal{S} \models \mathsf{F}$.

($\Longleftarrow$: *Generalized completeness*) Suppose $\mathcal{S} \nvdash \mathsf{F}$. Then for any $\mathsf{F}_1, \ldots, \mathsf{F}_n \in \mathcal{S}$ certainly $\mathsf{F}_1, \ldots, \mathsf{F}_n \nvdash \mathsf{F}$. Using Corollary 2.9.22 we have

$$\nvdash \mathsf{F}_1 \to (\mathsf{F}_2 \to \cdots (\mathsf{F}_n \to \mathsf{F}) \cdots),$$

so, by the completeness of PC,

$$\mathsf{F}_1 \to (\mathsf{F}_2 \to \cdots (\mathsf{F}_n \to \mathsf{F}) \cdots)$$

is not a tautology. Thus it is possible to find an assignment of truth values that satisfies $\mathsf{F}_1, \ldots, \mathsf{F}_n$, but makes F false. Thus

$$\mathsf{F}_1, \ldots, \mathsf{F}_n \not\models \mathsf{F}.$$

In other words, every finite subset of $\mathcal{S} \cup \{\neg \mathsf{F}\}$ is satisfiable. So by the compactness theorem (Theorem 2.8.1, p. 75), $\mathcal{S} \cup \{\neg \mathsf{F}\}$ itself is satisfiable. But then $\mathcal{S} \not\models \mathsf{F}$. □

Exercises

2.9.1 Find derivations for the following tautologies:

a. $\neg (P \land R) \lor P$ **b.** $(\neg Q \leftrightarrow Q) \to S$

c. $(\neg Q \land P) \to P$ **d.** $R \lor (R \to \neg R)$

e. $((S \lor R) \land R) \leftrightarrow R$ **f.** $(S \to (S \lor P)) \lor S$

g. $((S \leftrightarrow R) \land S) \to S$ **h.** $R \to ((S \leftrightarrow Q) \lor R)$

i. $S \to ((P \leftrightarrow Q) \to S)$ **j.** $(\neg Q \lor R) \lor (R \lor Q)$

k. $(S \to \neg P) \lor P$ **l.** $R \lor (R \to (P \leftrightarrow \neg Q))$

2.9.2 Finish the proof of soundness, Theorem 2.9.10 (p. 85), by showing that all the rules of PC preserve tautologies.

2.9.3 For the following choices of binary connective □ determine if the claim

$$\mathsf{F} \to \mathsf{G} \vdash (\mathsf{F} \mathbin{\Box} \mathsf{H}) \to (\mathsf{G} \mathbin{\Box} \mathsf{H})$$

is valid, and if so find a derivation.

a. $\square = \vee$
b. $\square = \wedge$
c. $\square = \rightarrow$
d. $\square = \leftrightarrow$

2.9.4 For the following choices of binary connective $\square$ determine if the claim

$$\mathsf{F} \rightarrow \mathsf{G} \vdash (\mathsf{H} \square \mathsf{F}) \rightarrow (\mathsf{H} \square \mathsf{G})$$

is valid, and if so find a derivation.

a. $\square = \vee$
b. $\square = \wedge$
c. $\square = \rightarrow$
d. $\square = \leftrightarrow$

2.9.5 For the following choices of binary connective $\square$ determine if the claim

$$\mathsf{F} \leftrightarrow \mathsf{G} \vdash (\mathsf{F} \square \mathsf{H}) \leftrightarrow (\mathsf{G} \square \mathsf{H})$$

is valid, and if so find a derivation.

a. $\square = \vee$
b. $\square = \wedge$
c. $\square = \rightarrow$
d. $\square = \leftrightarrow$

2.9.6 For the following choices of binary connective $\square$ determine if the claim

$$\mathsf{F} \leftrightarrow \mathsf{G} \vdash (\mathsf{H} \square \mathsf{F}) \leftrightarrow (\mathsf{H} \square \mathsf{G})$$

is valid, and if so find a derivation.

a. $\square = \vee$
b. $\square = \wedge$
c. $\square = \rightarrow$
d. $\square = \leftrightarrow$

2.9.7 Find derivations to justify the following claims:

a. $P \wedge (R \vee Q)$
$\vdash P \leftrightarrow (P \vee \neg Q)$

b. $P \leftrightarrow Q$
$\vdash \neg P \leftrightarrow \neg Q$

c. $\neg R \leftrightarrow Q$
$\vdash P \rightarrow (Q \vee R)$

d. $\neg R \rightarrow S$
$\vdash (P \rightarrow R) \vee S$

e. $\neg (Q \rightarrow S)$
$\vdash \neg R \vee Q$

f. $\neg (Q \leftrightarrow R)$
$\vdash (Q \vee S) \vee ((Q \wedge R) \rightarrow S)$

g. $\neg(P \vee R)$
$\vdash (R \wedge Q) \to (Q \leftrightarrow P)$

h. $Q \to \neg R$
$\vdash Q \to (R \to P)$

i. $S \wedge (Q \vee P)$
$\vdash \neg Q \vee S$

j. $\neg(P \wedge R)$
$\vdash \neg(Q \wedge (P \wedge R))$

k. $\neg(P \leftrightarrow \neg R)$
$R \wedge (P \leftrightarrow Q)$
$\vdash P \leftrightarrow (P \vee Q)$

2.9.8 (*Consistency and Satisfiability*) A set $\mathcal{S}$ of propositional formulas is *consistent* if one cannot derive the contradiction $P \wedge \neg P$ from $\mathcal{S}$. $\mathsf{Consis}(\mathcal{S})$ means that $\mathcal{S}$ is consistent. Otherwise $\mathcal{S}$ is *inconsistent*, and we write $\neg\,\mathsf{Consis}(\mathcal{S})$.

a. Prove that $\mathcal{S}$ is consistent iff there is *some* formula F that one cannot derive from $\mathcal{S}$.

b. Prove that $\mathcal{S} \vdash \mathsf{F}$ iff $\mathcal{S} \cup \{\neg\mathsf{F}\}$ is inconsistent.

c. Prove that $\mathsf{Consis}(\mathcal{S})$ holds iff $\mathsf{Sat}(\mathcal{S})$ holds.

d. Prove that $\mathcal{S} \models \mathsf{F}$ iff $\neg\,\mathsf{Sat}(\mathcal{S} \cup \{\neg\mathsf{F}\})$ iff $\mathcal{S} \vdash \mathsf{F}$ iff $\neg\,\mathsf{Consis}(\mathcal{S} \cup \{\neg\mathsf{F}\})$.

2.9.9

a. Show that the following propositional proof system, PCa, is sound but not complete.

PCa

Connectives (and constants) $\vee, \wedge, \to, \leftrightarrow, \neg, 0, 1$
Axioms All tautologies of the form $\mathsf{F} \to \mathsf{F}$
Rules of inference Hypothetical Syllogism

b. Show that the following propositional proof system, PCb, is not sound, but complete.

PCb

Connectives (and Constants) $\vee, \wedge, \to, \leftrightarrow, \neg, 0, 1$
Axioms All formulas of the form $\mathsf{F} \to \mathsf{G}$
Rules of inference modus ponens

c. Show that the following propositional proof system, PCc, is neither sound nor complete.

PCc

Connectives (and Constants) $\vee, \wedge, \to, \leftrightarrow, \neg, 0, 1$
Axioms All formulas of the form $\mathsf{F} \to \mathsf{G}$
Rules of inference Hypothetical syllogism

d. Show that the following propositional proof system, PCd, is both sound and complete.

PCd

Connectives (and Constants) $\vee, \wedge, \rightarrow, \leftrightarrow, \neg, 0, 1$
Axioms All tautologies F
Rules of inference None!

e. Show that the following propositional proof system, PCe, is both sound and complete.

PCe

Connectives (and Constants) $\vee, \wedge, \rightarrow, \leftrightarrow, \neg, 0, 1$
Axioms None!
Rules of inference You can infer any tautology.

2.9.10

a. If $F_1, \ldots, F_k$ is a derivation to witness $\vdash L \rightarrow (F \wedge G)$, then show how to find a derivation to witness $\vdash L \rightarrow F$; to witness $\vdash L \rightarrow G$.

b. If $F_1, \ldots, F_k$ is a derivation to witness $\vdash L \rightarrow F$, and $G_1, \ldots, G_m$ is a derivation to witness $\vdash L \rightarrow G$, then show how to find a derivation to witness $\vdash L \rightarrow (F \wedge G)$.

c. If $F_1, \ldots, F_k$ is a derivation to witness $\vdash L \rightarrow F$, then show how to find a derivation to witness $\vdash L \rightarrow (F \vee G)$; to witness $\vdash L \rightarrow (G \vee F)$.

d. If $F_1, \ldots, F_k$ is a derivation to witness $\vdash L \rightarrow F$, and $G_1, \ldots, G_m$ is a derivation to witness $\vdash L \rightarrow (F \rightarrow G)$, then show how to find a derivation to witness $\vdash L \rightarrow G$.

e. If $F_1, \ldots, F_k$ is a derivation to witness $\vdash L \rightarrow (F \rightarrow G)$, and $G_1, \ldots, G_m$ is a derivation to witness $\vdash L \rightarrow \neg G$, then show how to find a derivation to witness $\vdash L \rightarrow \neg F$.

f. If $F_1, \ldots, F_k$ is a derivation to witness $\vdash L \rightarrow (F \rightarrow G)$, and $G_1, \ldots, G_m$ is a derivation to witness $\vdash L \rightarrow (G \rightarrow H)$, then show how to find a derivation to witness $\vdash L \rightarrow (F \rightarrow H)$.

g. If $F_1, \ldots, F_k$ is a derivation to witness $\vdash L \rightarrow (F \vee G)$, and $G_1, \ldots, G_m$ is a derivation to witness $\vdash L \rightarrow \neg F$, then show how to find a derivation to witness $\vdash L \rightarrow G$. Also, if $F_1, \ldots, F_k$ is a derivation to witness $\vdash L \rightarrow (F \vee G)$, and $G_1, \ldots, G_m$ is a derivation to witness $\vdash L \rightarrow \neg G$, then show how to find a derivation to witness $\vdash L \rightarrow F$.

h. If $F_1, \ldots, F_k$ is a derivation to witness $\vdash L \rightarrow (F_1 \rightarrow G_1)$, and $G_1, \ldots, G_m$ is a derivation to witness $\vdash L \rightarrow (F_2 \rightarrow G_2)$, and $H_1, \ldots, H_n$ is a derivation to witness $\vdash L \rightarrow (F_1 \vee F_2)$, then show how to find a derivation to witness $\vdash L \rightarrow (G_1 \vee G_2)$.

i. If $\mathsf{F}_1, \ldots, \mathsf{F}_k$ is a derivation to witness $\vdash \mathsf{L} \to \mathsf{F}(\cdots \mathsf{F}' \cdots)$, and $\mathsf{F}' \sim \mathsf{G}'$ is a simple equivalence, then show how to find a derivation to witness $\vdash \mathsf{L} \to \mathsf{F}(\cdots \mathsf{G}' \cdots)$.

j. Show how to use parts a–i to convert a derivation $\mathsf{H}_1, \ldots, \mathsf{H}_k$ witnessing $\mathcal{S} \cup \{\mathsf{F}\} \vdash \mathsf{G}$ into a derivation witnessing $\mathcal{S} \vdash \mathsf{F} \to \mathsf{G}$.

2.10 RESOLUTION

Resolution is a rule of inference used to show a set of propositional formulas of the form $\widetilde{P}_1 \vee \cdots \vee \widetilde{P}_m$ is *not* satisfiable. Here $\widetilde{P}_i$ means P_i or $\neg P_i$, where P_i is a propositional variable.

2.10.1 A Motivation

Let us give some justification for the study of such formulas.

In Theorem 2.7.8 (p. 69) we have

$$\mathsf{F}_1, \cdots, \mathsf{F}_n \ \therefore \mathsf{F} \qquad \text{holds iff} \qquad \mathsf{F}_1 \wedge \cdots \wedge \mathsf{F}_n \wedge \neg \mathsf{F} \ \text{ is not satisfiable.}$$

The next step is to put each of the F_i and $\neg \mathsf{F}$ into *conjunctive* form, say,

$$\begin{aligned} \mathsf{F}_i &\sim \mathsf{G}_{i1} \wedge \cdots \wedge \mathsf{G}_{ik_i}, \qquad 1 \le i \le n \\ \neg \mathsf{F} &\sim \mathsf{G}_1 \wedge \cdots \wedge \mathsf{G}_k, \end{aligned}$$

where each G_{ij} and G_ℓ is of the form $\widetilde{P}_1 \vee \cdots \vee \widetilde{P}_m$. Then

$$(\mathsf{F}_1 \wedge \cdots \wedge \mathsf{F}_n \wedge \neg \mathsf{F}) \ \sim \ (\mathsf{G}_{11} \wedge \cdots \wedge \mathsf{G}_{nk_n} \wedge \mathsf{G}_1 \wedge \cdots \wedge \mathsf{G}_k),$$

and thus

$$\mathsf{F}_1, \cdots, \mathsf{F}_n \ \therefore \mathsf{F} \qquad \text{holds iff} \qquad \neg \mathsf{Sat}(\{\mathsf{G}_{11}, \cdots, \mathsf{G}_{nk_n}, \mathsf{G}_1, \cdots, \mathsf{G}_k\}).$$

Thus saying that an argument is valid is equivalent to saying that a certain set of disjunctions of variables and negated variables is not satisfiable.

Example 2.10.1 To determine the validity of the argument

$$P \to Q, \ \neg P \to R, \ Q \vee R \to S \qquad \therefore S$$

we consider the satisfiability of

$$\{P \to Q, \ \neg P \to R, \ Q \vee R \to S, \ \neg S\}.$$

Converting this into the desired disjunctions gives

$$\{\neg P \vee Q, \ P \vee R, \ \neg Q \vee S, \ \neg R \vee S, \ \neg S\}.$$

It is easily checked that this is not satisfiable, so the propositional argument is valid.

Example 2.10.2 Let us apply this to the problem from Boole as formulated in Example 2.7.10 on p. 70:

$$\begin{array}{l}
(A \land B) \to (C \leftrightarrow \neg D) \\
(B \land C) \to (A \leftrightarrow D) \\
(\neg A \land \neg B) \to (\neg C \land \neg D) \\
(\neg C \land \neg D) \to (\neg A \land \neg B) \\
\therefore (\neg (A \land B \land C)) \land ((\neg A \land \neg B) \to \neg C).
\end{array}$$

First, we convert each of these premisses, and the negation of the conclusion, into a conjunctive form. For economy of space we use juxtaposition for $\land$, with $\land$ given binding precedence over $\lor$, and we use $\overline{A}$ for $\neg A$:

$$\begin{aligned}
AB \to (C \leftrightarrow \overline{D}) \quad &\sim \quad \overline{AB} \lor (C \leftrightarrow \overline{D}) \\
&\sim \quad \overline{A} \lor \overline{B} \lor (C \lor D)(\overline{C} \lor \overline{D}) \\
&\sim \quad (\overline{A} \lor \overline{B} \lor C \lor D)(\overline{A} \lor \overline{B} \lor \overline{C} \lor \overline{D}),
\end{aligned}$$

$$\begin{aligned}
BC \to (A \leftrightarrow D) \quad &\sim \quad \overline{BC} \lor (A \lor \overline{D})(\overline{A} \lor D) \\
&\sim \quad (\overline{B} \lor \overline{C}) \lor (A \lor \overline{D})(\overline{A} \lor D) \\
&\sim \quad (\overline{B} \lor \overline{C} \lor A \lor \overline{D})(\overline{B} \lor \overline{C} \lor \overline{A} \lor D) \\
&\sim \quad (A \lor \overline{B} \lor \overline{C} \lor \overline{D})(\overline{A} \lor \overline{B} \lor \overline{C} \lor D),
\end{aligned}$$

$$\begin{aligned}
\overline{A}\,\overline{B} \to \overline{C}\,\overline{D} \quad &\sim \quad \overline{\overline{A}\,\overline{B}} \lor (\overline{C}\,\overline{D}) \\
&\sim \quad (A \lor B) \lor (\overline{C}\,\overline{D}) \\
&\sim \quad (A \lor B \lor \overline{C})(A \lor B \lor \overline{D}),
\end{aligned}$$

$$\begin{aligned}
\overline{C}\,\overline{D} \to \overline{A}\,\overline{B} \quad &\sim \quad \overline{\overline{C}\,\overline{D}} \lor \overline{A}\,\overline{B} \\
&\sim \quad (C \lor D) \lor \overline{A}\,\overline{B} \\
&\sim \quad (C \lor D \lor \overline{A})(C \lor D \lor \overline{B}) \\
&\sim \quad (\overline{A} \lor C \lor D)(\overline{B} \lor C \lor D),
\end{aligned}$$

and for the conclusion,

$$\begin{aligned}
\overline{ABC(\overline{A}\,\overline{B} \to \overline{C})} &\sim ABC \vee \overline{(\overline{A}\,\overline{B} \to \overline{C})} \\
&\sim ABC \vee \overline{(\overline{\overline{A}\,\overline{B}} \vee \overline{C})} \\
&\sim ABC \vee \overline{A}\,\overline{B}\,C \\
&\sim (AB \vee \overline{A}\,\overline{B})C \\
&\sim (A \vee \overline{B})(\overline{A} \vee B)C.
\end{aligned}$$

Thus we end up with the following collection of disjunctions that is not satisfiable iff the original argument is valid:

$$\begin{array}{l}
\neg A \vee \neg B \vee C \vee D \\
\neg A \vee \neg B \vee \neg C \vee \neg D \\
A \vee \neg B \vee \neg C \vee \neg D \\
\neg A \vee \neg B \vee \neg C \vee D \\
A \vee B \vee \neg C \\
A \vee B \vee \neg D \\
\neg A \vee C \vee D \\
\neg B \vee C \vee D \\
A \vee \neg B \\
\neg A \vee B \\
C.
\end{array}$$

Example 2.10.3 Let us apply this conversion, of an argument to a set of disjunctions, to the problem formulated in Example 2.7.11 on p. 71. These formulas are rather easy to convert to the desired collection of disjunctions (the right–hand sides of the following equivalences), yielding a collection of disjunctions that is not satisfiable iff the original argument is valid:

$$\begin{aligned}
A \wedge B \wedge C \to D &\sim \neg A \vee \neg B \vee \neg C \vee D \\
\neg A \wedge M \to L &\sim A \vee \neg M \vee L \\
F \wedge E \to \neg D &\sim \neg F \vee \neg E \vee \neg D \\
G \wedge M \to C &\sim \neg G \vee \neg M \vee C \\
\neg B \wedge F \to \neg H &\sim B \vee \neg F \vee \neg H \\
\neg D \wedge B \wedge E \to G &\sim D \vee \neg B \vee \neg E \vee G \\
M \wedge \neg I \to J &\sim \neg M \vee I \vee J \\
H \wedge M \to K &\sim \neg H \vee \neg M \vee K \\
K \wedge J \wedge \neg L \to E &\sim \neg K \vee \neg J \vee L \vee E \\
\neg H \wedge F \to L &\sim H \vee \neg F \vee L
\end{aligned}$$

$$\begin{array}{rcl} M \wedge L \rightarrow \neg F & \sim & \neg M \vee \neg L \vee \neg F \\ K \wedge I \wedge A \rightarrow E & \sim & \neg K \vee \neg I \vee \neg A \vee E \\ \neg(M \rightarrow \neg F) & \sim & M \wedge F. \end{array}$$

2.10.2 Clauses

When working with formulas of the form $\widetilde{P}_1 \vee \cdots \vee \widetilde{P}_m$ we are not concerned about the ordering of the $\widetilde{P}_i$, nor do we care about the number of repeat occurrences of a given $\widetilde{P}_i$. All the information that we need is present in the set $\{\widetilde{P}_1, \cdots, \widetilde{P}_m\}$.

Definition 2.10.4 A propositional variable P, as well as a negated propositional variable $\neg P$, is called a *literal*. Finite sets $\{\mathsf{L}_1, \ldots, \mathsf{L}_m\}$ of literals are called *clauses* . The *literals of a clause* are the members of the clause.

Definition 2.10.5 [SATISFIABLE]

- A clause $\{\mathsf{L}_1, \cdots, \mathsf{L}_n\}$ is *satisfiable* (by a truth evaluation e) iff the propositional formula $\mathsf{L}_1 \vee \cdots \vee \mathsf{L}_n$ is satisfiable (by e).
- By definition the *empty clause* $\{\ \}$ is not satisfiable.
- A set $\mathcal{S}$ of clauses is *satisfiable* iff there is a truth evaluation e that satisfies each clause in $\mathcal{S}$.

Remark 2.10.6 We use sets of literals, i.e., clauses, as a trick to get around having to fuss about the ordering of the literals, and their repeat occurrences. In a set there is no ordering, and every element occurs just once.

In the previous section we saw that the validity of an argument in the propositional logic could be converted to the question of the nonsatisfiability of a collection of formulas, each formula being expressed as a disjunction of variables and negated variables. Now, we can convert each of the formulas in this collection to a clause, and then the argument is valid iff the collection of clauses (that we obtain in this manner) is not satisfiable.

Example 2.10.7 The argument from Example 2.10.1 (p. 98),

$$P \rightarrow Q,\ \neg P \rightarrow R,\ Q \vee R \rightarrow S \qquad \therefore S,$$

is valid iff the set of propositional formulas

$$\{\neg P \vee Q,\ P \vee R,\ \neg Q \vee S,\ \neg R \vee S,\ \neg S\}$$

is not satisfiable, and this holds iff the set of clauses obtained from these formulas,

$$\{\neg P,\ Q\}\quad \{P,\ R\}\quad \{\neg Q,\ S\}$$
$$\{\neg R,\ S\}\quad \{\neg S\},$$

is not satisfiable.

Definition 2.10.8 If L is a literal, then the *complement* $\overline{\mathsf{L}}$ of L is defined by $\overline{P} = \neg P$, $\overline{\neg P} = P$, for P a propositional variable. A literal is *positive* if it is a propositional variable; otherwise it is *negative* .

2.10.3 Resolution

Now that we have seen why we are interested in showing that a set of clauses is not satisfiable, namely, because we can use this property to determine if an argument is valid, let us look at the most popular method for showing that a set of clauses is not satisfiable: the use of resolution.

Definition 2.10.9 *Resolution* is the rule

$$\frac{\mathsf{C} \cup \{\mathsf{L}\},\ \mathsf{D} \cup \{\overline{\mathsf{L}}\}}{\mathsf{C} \cup \mathsf{D}},$$

where C and D are clauses and L is a literal. We say that we are *resolving the two clauses over* L; and $\mathsf{C} \cup \mathsf{D}$ is the *resolvent*.

A *(resolution) derivation* from a set of clauses $\mathcal{S}$ is a finite sequence of clauses such that each is in $\mathcal{S}$ or comes from previous clauses in the sequence by resolution.

Example 2.10.10 A resolution derivation of the empty clause from the set of clauses

$$\{\neg P, Q\}\quad \{\neg Q, \neg R, S\}\quad \{P\}\quad \{R\}\quad \{\neg S\}$$

is

1.	$\{\neg S\}$	given
2.	$\{\neg Q, \neg R, S\}$	given
3.	$\{\neg Q, \neg R\}$	resolution 1,2
4.	$\{R\}$	given
5.	$\{\neg Q\}$	resolution 3,4
6.	$\{\neg P, Q\}$	given
7.	$\{\neg P\}$	resolution 5,6
8.	$\{P\}$	given
9.	$\{\,\}$	resolution 7,8.

Example 2.10.11 For the derivation in Example 2.10.10 we could use the following picture:

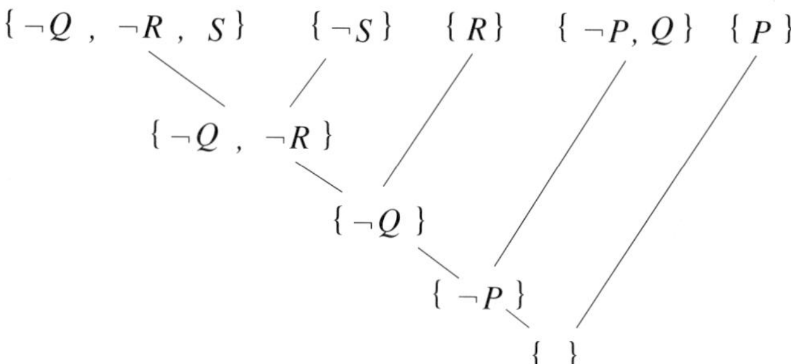

Figure 2.10.12 An example of resolution

Lemma 2.10.13 Resolution preserves satisfiability, i.e., if a set of clauses is satisfiable by a truth evaluation e, then any resolvent of a pair of clauses from $\mathcal{S}$ is also satisfiable by e.

Proof. Suppose $\mathcal{S}$ is a set of clauses that is satisfiable by e. Let $\mathsf{C}_1 \cup \{\mathsf{L}\}$ and $\mathsf{C}_2 \cup \{\overline{\mathsf{L}}\}$ be two members of $\mathcal{S}$. Now, either e does not satisfy L or e does not satisfy $\overline{\mathsf{L}}$. Let us say $\widehat{\mathsf{L}}(\mathsf{e}) = 0$. Then e satisfies C_1, as it satisfies $\mathsf{C}_1 \cup \{\mathsf{L}\}$. But then it satisfies $\mathsf{C}_1 \cup \mathsf{C}_2$. □

Warning

A common mistake in using resolution is to apply it to more than one variable. This is not correct. For example the following is an invalid use of resolution and will likely lead to a false conclusion:

$$\frac{\mathsf{C} \cup \{\mathsf{L}_1, \overline{\mathsf{L}_2}\},\ \mathsf{D} \cup \{\overline{\mathsf{L}_1}, \mathsf{L}_2\}}{\mathsf{C} \cup \mathsf{D}}.$$

In particular, the use of

$$\frac{\{P, \neg Q\},\ \{\neg P, Q\}}{\{\,\}}$$

disagrees with our previous lemma, since we can satisfy the premisses by setting P and Q equal to 1, but not the conclusion. This is not resolution.

2.10.4 The Davis–Putnam Procedure

Resolution was employed in the Davis–Putnam procedure (DPP) theorem prover in the 1950s. We will use the notation $\uplus$ for *disjoint union*, i.e., when we write $X \uplus Y$ we are referring to the union $X \cup Y$ and also asserting that $X \cap Y = \emptyset$, i.e., X and Y are disjoint.

The Davis–Putnam Procedure

Given a nonempty set $\mathcal{S}$ of nonempty clauses in the propositional variables $P_1, \cdots, P_n$ the Davis–Putnam procedure repeats the following steps until there are no variables left:

- Throw out all clauses that have both a literal L and its complement $\overline{\mathsf{L}}$ in them.
- Choose a variable P appearing in one of the clauses.
- Add all possible resolvents using resolution over P to the set of clauses.
- Throw out all clauses with P or $\neg P$ in them.

We refer to the preceding sequence as *eliminating* the variable P.

If in some step one resolves $\{P\}$ and $\{\neg P\}$ then one has the empty clause, and it will be the only clause at the end of the procedure. If one never has a pair $\{P\}$ and $\{\neg P\}$ to resolve, then all the clauses will be thrown out, and the output will be no clauses. So the output is either the empty clause or no clauses.[5]

Now we give a more precise version of the steps involved.

- Let $\mathcal{S}_1 = \mathcal{S}$.
- Let $i = 1$.
- LOOP until $i = n + 1$.
- Discard members of $\mathcal{S}_i$ in which a literal and its complement appear, to obtain $\mathcal{S}'_i$.
- Let $\mathcal{T}_i$ be the set of clauses in $\mathcal{S}'_i$ in which P_i or $\neg P_i$ appears.
- Let $\mathcal{U}_i$ be the set of resolvent clauses obtained by resolving (over P_i) every pair of clauses $\mathsf{C} \uplus \{P_i\}$ and $\mathsf{D} \uplus \{\neg P_i\}$ in $\mathcal{T}_i$.
- Set $\mathcal{S}_{i+1}$ equal to $(\mathcal{S}'_i \setminus \mathcal{T}_i) \cup \mathcal{U}_i$.
- Let i be increased by 1.
- ENDLOOP.
- Output $\mathcal{S}_{n+1}$.

Example 2.10.14 Let us apply the DPP to the clauses

$$\{\neg P, Q\} \qquad \{P\}.$$

- Eliminating P gives $\{Q\}$.
- Eliminating Q gives no clauses.

[5] This may seem rather subtle, but just think of the difference between arriving in the library with (1) an empty backpack and (2) no backpack.

Thus the output is "no clauses."

Example 2.10.15 Let us apply the DPP to Example 2.10.10 (p. 102), namely, to the clauses

$$\{\neg P, Q\} \qquad \{\neg Q, \neg R, S\} \qquad \{P\} \qquad \{R\} \qquad \{\neg S\}.$$

- Eliminating P gives $\{Q\}$, $\{\neg Q, \neg R, S\}$, $\{R\}$, $\{\neg S\}$. (This is $\mathcal{S}_2$ and $\mathcal{S}'_2$.)
- Eliminating Q gives $\{\neg R, S\}$, $\{R\}$, $\{\neg S\}$. (This is $\mathcal{S}_3$ and $\mathcal{S}'_3$.)
- Eliminating R gives $\{S\}$, $\{\neg S\}$. (This is $\mathcal{S}_4$ and $\mathcal{S}'_4$.)
- Eliminating S gives $\{\ \}$. (This is $\mathcal{S}_5$.)

So the output is the empty clause.

Example 2.10.16 A more challenging example for the DPP is the following set of clauses. The $\mathcal{S}_i$'s are left out for $i > 1$. Before each phase of applying resolution we number the clauses (the $\mathcal{T}_i$ steps), and in the next phase (the $\mathcal{U}_i$ steps) we provide two numbers with each clause, to describe the two clauses used to provide that resolvent:

$\mathcal{S}'_1 = \mathcal{S}_1$:

$$\begin{array}{llll} \{P, Q\} & \{\neg P, \neg Q\} & \{\neg Q, R, T\} & \{Q, \neg R, T\} \\ \{Q, R, \neg T\} & \{\neg Q, \neg R, \neg T\} & \{\neg R, S\} & \{R, \neg S\} \\ \{\neg P, S, T\} & \{P, \neg S, T\} & \{P, S, \neg T\} & \{\neg P, \neg S, \neg T\}. \end{array}$$

(Resolving over P)

$\mathcal{T}_1$:

$$\begin{array}{lll} (1)\,\{P, Q\} & (2)\,\{\neg P, \neg Q\} & (3)\,\{\neg P, S, T\} \\ (4)\,\{P, \neg S, T\} & (5)\,\{P, S, \neg T\} & (6)\,\{\neg P, \neg S, \neg T\}. \end{array}$$

$\mathcal{U}_1$:

$$\begin{array}{lll} (1,2)\,\{Q, \neg Q\} & (1,3)\,\{Q, S, T\} & (1,6)\,\{Q, \neg S, \neg T\} \\ (2,4)\,\{\neg Q, \neg S, T\} & (2,5)\,\{\neg Q, S, \neg T\} & (3,4)\,\{S, \neg S, T\} \\ (3,5)\,\{S, T, \neg T\} & (4,6)\,\{\neg S, T, \neg T\} & (5,6)\,\{S, \neg S, \neg T\}. \end{array}$$

$\mathcal{S}'_2$:

$$\begin{array}{llll} \{\neg Q, R, T\} & \{Q, \neg R, T\} & \{Q, R, \neg T\} & \{\neg Q, \neg R, \neg T\} \\ \{\neg R, S\} & \{R, \neg S\} & \{Q, S, T\} & \{Q, \neg S, \neg T\} \\ \{\neg Q, \neg S, T\} & \{\neg Q, S, \neg T\}. & & \end{array}$$

(Resolving over Q)
$\mathcal{T}_2$:

$(1)\,\{\neg Q, R, T\}$ $(2)\,\{Q, \neg R, T\}$ $(3)\,\{Q, R, \neg T\}$
$(4)\,\{\neg Q, \neg R, \neg T\}$ $(5)\,\{Q, S, T\}$ $(6)\,\{Q, \neg S, \neg T\}$
$(7)\,\{\neg Q, \neg S, T\}$ $(8)\,\{\neg Q, S, \neg T\}$.

$\mathcal{U}_2$:

$(1,2)\,\{R, \neg R, T\}$ $(1,3)\,\{R, T, \neg T\}$ $(1,5)\,\{R, S, T\}$
$(1,6)\,\{R, \neg S, T, \neg T\}$ $(2,4)\,\{\neg R, T, \neg T\}$ $(2,7)\,\{\neg R, \neg S, T\}$
$(2,8)\,\{\neg R, S, T, \neg T\}$ $(3,4)\,\{R, \neg R, \neg T\}$ $(3,7)\,\{R, \neg S, T, \neg T\}$
$(3,8)\,\{R, S, \neg T\}$ $(4,5)\,\{\neg R, S, T, \neg T\}$ $(4,6)\,\{\neg R, \neg S, \neg T\}$
$(5,7)\,\{S, \neg S, T\}$ $(5,8)\,\{S, T, \neg T\}$ $(6,7)\,\{\neg S, \neg T, T\}$
$(6,8)\,\{S, \neg S, \neg T\}$.

$\mathcal{S}_3'$:

$\{\neg R, S\}$ $\{R, \neg S\}$ $\{R, S, T\}$ $\{\neg R, \neg S, T\}$
$\{R, S, \neg T\}$ $\{\neg R, \neg S, \neg T\}$.

(Resolving over R)
$\mathcal{T}_3$:

$(1)\,\{\neg R, S\}$ $(2)\,\{R, \neg S\}$ $(3)\,\{R, S, T\}$
$(4)\,\{\neg R, \neg S, T\}$ $(5)\,\{R, S, \neg T\}$ $(6)\,\{\neg R, \neg S, \neg T\}$.

$\mathcal{U}_3$:

$(1,2)\,\{S, \neg S\}$ $(1,3)\,\{S, T\}$ $(1,5)\,\{S, \neg T\}$
$(2,4)\,\{\neg S, T\}$ $(2,6)\,\{\neg S, \neg T\}$ $(3,4)\,\{S, \neg S, T\}$
$(3,6)\,\{S, \neg S, T, \neg T\}$ $(4,5)\,\{S, \neg S, T, \neg T\}$ $(5,6)\,\{S, \neg S, \neg T\}$.

$\mathcal{S}_4'$:

$\{S, T\}$ $\{S, \neg T\}$ $\{\neg S, T\}$ $\{\neg S, \neg T\}$.

(Resolving over S)
$\mathcal{T}_4$: $(1)\,\{S, T\}$ $(2)\,\{S, \neg T\}$ $(3)\,\{\neg S, T\}$ $(4)\,\{\neg S, \neg T\}$.

$\mathcal{U}_4$: $(1,3)\,\{T\}$ $(1,4)\,\{T, \neg T\}$ $(2,3)\,\{T, \neg T\}$ $(2,4)\,\{\neg T\}$.
$\mathcal{S}_5'$: $\{T\}$ $\{\neg T\}$.

(Resolving over T)
$\mathcal{T}_5 :\ (1)\,\{T\}\quad (2)\,\{\neg T\}.$

$\mathcal{U}_5 : \{\ \}.$

$\mathcal{S}_6 : \{\ \}.$

Again the output is the empty clause.

2.10.5 Soundness and Completeness for the DPP

The following theorem shows that the Davis–Putnam procedure is both sound and complete.

Theorem 2.10.17 [THE DPP IS SOUND AND COMPLETE.] Let $\mathcal{S}$ be a finite set of clauses. Then $\mathcal{S}$ is not satisfiable iff the output of the Davis–Putnam procedure is the empty clause.

Proof. Observe that there are two possible outputs $\mathcal{S}_{n+1}$ from this program: (a) $\mathcal{S}_{n+1}$ contains the empty clause only, or (b) $\mathcal{S}_{n+1}$ contains no clauses at all, i.e., $\mathcal{S}_{n+1} = \emptyset$. In the first case the output is not satisfiable; and in the second case it is satisfiable.

It is easy to show, by induction on i, that if C is a clause in $\mathcal{S}_i$, then there is a resolution derivation of C from $\mathcal{S}$. Consequently, if $\mathcal{S}_{n+1}$ contains the empty clause, then $\mathcal{S}$ is unsatisfiable by Lemma 2.10.13 (p. 103). It only remains to show that the original set of clauses is satisfiable if $\mathcal{S}_{n+1}$ contains no clauses.

So suppose that $\mathcal{S}_{n+1}$ has no clauses. Then $\mathcal{S}_{n+1}$ is satisfiable by the empty evaluation e_{n+1}. To prove the set of clauses $\mathcal{S}$ is satisfiable it suffices to show that if $\mathcal{S}_{i+1}$ is satisfiable by $\mathsf{e}_{i+1} = (e_{i+1}, \cdots, e_n)$, then $\mathcal{S}_i$ is satisfiable by $\mathsf{e}_i = (e_i, e_{i+1}, \cdots, e_n)$ for some choice of e_i. It will be convenient to abbreviate the latter as (e_i, e_{i+1}). (Note that the propositional variables occurring in $\mathcal{S}_{i+1}$ are from $P_{i+1}, \cdots, P_n$, and those in $\mathcal{S}_i$ are from $P_i, \cdots, P_n$.) Clearly, $\mathcal{S}_i$ is satisfiable iff $\mathcal{S}_i'$ is satisfiable. Now $\mathcal{S}_i' \subseteq \mathcal{S}_{i+1} \cup \mathcal{T}_i$. Consequently it suffices to show that if $\mathcal{S}_{i+1}$ is satisfiable by e_{i+1}, then $\mathcal{T}_i$ is satisfiable by some (e_i, e_{i+1}).

So suppose $\mathcal{S}_{i+1}$ is satisfiable by a truth evaluation e_{i+1}. If neither of the evaluations $(0, \mathsf{e}_{i+1})$ and $(1, \mathsf{e}_{i+1})$ satisfies $\mathcal{T}_i$, then we must have clauses $\mathsf{D} \uplus \{P_i\}$ and $\mathsf{E} \uplus \{\neg P_i\}$ in $\mathcal{T}_i$ such that $(0, \mathsf{e}_{i+1})$ does not satisfy $\mathsf{D} \uplus \{P_i\}$ and $(1, \mathsf{e}_{i+1})$ does not satisfy $\mathsf{E} \uplus \{\neg P_i\}$. Then e_{i+1} does not satisfy either D or E, and thus e_{i+1} does not satisfy

$\mathsf{D} \cup \mathsf{E}$. Now $\mathsf{D} \cup \mathsf{E}$ is in $\mathcal{S}_{i+1}$. This gives a contradiction. Consequently one of the evaluations $(0, \mathsf{e}_{i+1})$ and $(1, \mathsf{e}_{i+1})$ must satisfy $\mathcal{T}_i$.

Thus we have demonstrated that the satisfiability of $\mathcal{S}_{i+1}$ implies the satisfiability of $\mathcal{S}_i$, and the theorem is proved. □

2.10.6 Applications of the DPP

Example 2.10.18 Let us apply the DPP to Boole's problem as presented in Example 2.10.2 (p. 99), namely, to the clauses

$$\begin{array}{lll} \{\neg A, \neg B, C, D\} & \{\neg A, \neg B, \neg C, \neg D\} & \{A, \neg B, \neg C, \neg D\} \\ \{\neg A, \neg B, \neg C, D\} & \{A, B, \neg C\} & \{A, B, \neg D\} \\ \{\neg A, C, D\} & \{\neg B, C, D\} & \{A, \neg B\} \\ \{\neg A, B\} & \{C\}. & \end{array}$$

However, this time we will suppress some of the details and just show the results, the $\mathcal{S}'_i$ steps, for resolving over each variable.

After resolving over C we have

$$\begin{array}{lll} \{\neg A, \neg B, \neg D\} & \{A, \neg B, \neg D\} & \{\neg A, \neg B, D\} \\ \{A, B\} & \{A, B, \neg D\} & \{A, \neg B\} \\ \{\neg A, B\}. & & \end{array}$$

Then we resolve over A to obtain

$$\{\neg B, \neg D\} \quad \{\neg B, D\} \quad \{B\} \quad \{B, \neg D\}.$$

Next, we resolve over B:

$$\{\neg D\} \quad \{D\}.$$

Finally, resolving over D gives the empty clause:

$$\{\ \}.$$

Thus Boole's argument is valid.

Example 2.10.19 As a final example in this section let us apply the DPP to Example 1.3.4 as formulated in Example 2.10.3 on p. 100, namely, we want to determine if the following set of clauses is satisfiable:

$$\begin{array}{lll} \{\neg A, \neg B, \neg C, D\} & \{A, \neg M, L\} & \{\neg F, \neg E, \neg D\} \\ \{\neg G, \neg M, C\} & \{B, \neg F, \neg H\} & \{D, \neg B, \neg E, G\} \\ \{\neg M, I, J\} & \{\neg H, \neg M, K\} & \{\neg K, \neg J, L, E\} \\ \{H, \neg F, L\} & \{\neg M, \neg L, \neg F\} & \{\neg K, \neg I, \neg A, E\} \\ \{M\} & \{F\}. & \end{array}$$

We have two unit clauses to start with, $\{M\}$ and $\{F\}$, so let us resolve over M and F first, starting with M:

$$\begin{array}{lll}
\{\neg A, \neg B, \neg C, D\} & \{A, L\} & \{\neg F, \neg E, \neg D\} \\
\{\neg G, C\} & \{B, \neg F, \neg H\} & \{D, \neg B, \neg E, G\} \\
\{I, J\} & \{\neg H, K\} & \{\neg K, \neg J, L, E\} \\
\{H, \neg F, L\} & \{\neg L, \neg F\} & \{\neg K, \neg I, \neg A, E\} \\
\{F\}. & &
\end{array}$$

Then resolving over F gives

$$\begin{array}{lll}
\{\neg A, \neg B, \neg C, D\} & \{A, L\} & \{\neg E, \neg D\} \\
\{\neg G, C\} & \{B, \neg H\} & \{D, \neg B, \neg E, G\} \\
\{I, J\} & \{\neg H, K\} & \{\neg K, \neg J, L, E\} \\
\{H, L\} & \{\neg L\} & \{\neg K, \neg I, \neg A, E\}.
\end{array}$$

Now we have picked up a new unit clause $\{\neg L\}$, so let us resolve over $\neg L$:

$$\begin{array}{lll}
\{\neg A, \neg B, \neg C, D\} & \{A\} & \{\neg E, \neg D\} \\
\{\neg G, C\} & \{B, \neg H\} & \{D, \neg B, \neg E, G\} \\
\{I, J\} & \{\neg H, K\} & \{\neg K, \neg J, E\} \\
\{H\} & \{\neg K, \neg I, \neg A, E\}. &
\end{array}$$

Now let us resolve over A:

$$\begin{array}{lll}
\{\neg B, \neg C, D\} & \{\neg E, \neg D\} & \{\neg G, C\} \\
\{B, \neg H\} & \{D, \neg B, \neg E, G\} & \{I, J\} \\
\{\neg H, K\} & \{\neg K, \neg J, E\} & \{H\} \\
\{\neg K, \neg I, E\}. & &
\end{array}$$

And then over H:

$$\begin{array}{lll}
\{\neg B, \neg C, D\} & \{\neg E, \neg D\} & \{\neg G, C\} \\
\{B\} & \{D, \neg B, \neg E, G\} & \{I, J\} \\
\{K\} & \{\neg K, \neg J, E\} & \{\neg K, \neg I, E\}.
\end{array}$$

And over B:

$$\begin{array}{lll}
\{\neg C, D\} & \{\neg E, \neg D\} & \{\neg G, C\} \\
\{D, \neg E, G\} & \{I, J\} & \{K\} \\
\{\neg K, \neg J, E\} & \{\neg K, \neg I, E\}. &
\end{array}$$

And over K:

$$\begin{array}{lll}
\{\neg C, D\} & \{\neg E, \neg D\} & \{\neg G, C\} \\
\{D, \neg E, G\} & \{I, J\} & \{\neg J, E\} \\
\{\neg I, E\}. & &
\end{array}$$

We have run out of unit clauses, so let us resolve over C:

$$\begin{array}{lll} \{\neg G, D\} & \{\neg E, \neg D\} & \{D, \neg E, G\} \\ \{I, J\} & \{\neg J, E\} & \{\neg I, E\}. \end{array}$$

And over J:

$$\begin{array}{lll} \{\neg G, D\} & \{\neg E, \neg D\} & \{D, \neg E, G\} \\ \{I, E\} & \{\neg I, E\}. & \end{array}$$

And over I: $\{\neg G, D\}$ $\{\neg E, \neg D\}$ $\{D, \neg E, G\}$ $\{E\}$.

And over E: $\{\neg G, D\}$ $\{\neg D\}$ $\{D, G\}$.

And over D: $\{\neg G\}$ $\{G\}$.

And over G: $\{\ \}$.

Thus the clauses are unsatisfiable, and we have verified the correctness of the argument in Example 1.3.4.

2.10.7 Soundness and Completeness for Resolution

The following theorem shows that the resolution proof system is both sound and complete.

Theorem 2.10.20 A finite nonempty set $\mathcal{S}$ of clauses is unsatisfiable iff there is a derivation of the empty clause from $\mathcal{S}$ using resolution.

Proof. Use Lemma 2.10.13 (p. 103) and the result for the DPP proved above. □

Remark 2.10.21 From Theorem 2.10.20, a given formula F is a tautology iff when one puts $\neg\,\mathsf{F}$ in conjunctive form, the set of clauses obtained from the conjuncts has a refutation by resolution.

Example 2.10.22 Consider the propositional formula

$$((P \to Q) \wedge (Q \to R) \wedge (R \to S)) \to (P \to S).$$

Converting its negation into clauses gives

$$\{\neg P, Q\},\ \{\neg Q, R\},\ \{\neg R, S\},\ \{P\},\ \{\neg S\}.$$

As this set of clauses is not satisfiable, it follows that the formula is indeed a tautology.

2.10.8 Generalized Soundness and Completeness for Resolution

We also have the generalized version of Theorem 2.10.20, paralleling the generalized results for PC in Theorem 2.9.24 (p. 93).

Theorem 2.10.23 A nonempty set $\mathcal{S}$ of clauses is unsatisfiable iff there is a derivation of the empty clause using resolution.

Proof. From compactness, Theorem 2.8.1 (p. 75), we know that $\mathcal{S}$ is not satisfiable iff some finite subset $\mathcal{S}_0$ is not satisfiable. But by the above a finite $\mathcal{S}_0$ is unsatisfiable iff one can derive the empty clause from $\mathcal{S}_0$. Thus $\mathcal{S}$ is not satisfiable iff one can derive the empty clause from it. □

Exercises

2.10.1 Convert each of the following arguments into a set of clauses such that the argument is valid iff the set of clauses is not satisfiable.

a. $(R \leftrightarrow S) \rightarrow \neg P$
$(S \wedge Q) \leftrightarrow ((S \leftrightarrow P) \rightarrow S)$
$(P \wedge (R \leftrightarrow P)) \vee \neg (P \vee Q)$
$\therefore\ \neg ((S \leftrightarrow Q) \leftrightarrow S)$

b. $(Q \vee P) \rightarrow (S \wedge R)$
$((P \rightarrow R) \leftrightarrow Q) \leftrightarrow \neg P$
$(Q \vee R) \leftrightarrow (Q \vee (S \leftrightarrow Q))$
$\therefore\ (P \rightarrow S) \wedge Q$

c. $R \leftrightarrow (P \vee Q)$
$S \leftrightarrow \neg (S \rightarrow R)$
$(S \vee R) \leftrightarrow (P \leftrightarrow S)$
$\therefore (P \wedge (P \leftrightarrow Q)) \rightarrow \neg P$

d. $\neg Q \vee S$
$\neg (S \wedge P)$
$(P \rightarrow Q) \wedge R$
$(P \rightarrow R) \rightarrow P$
$\therefore \neg (S \rightarrow P)$

e. $(S \rightarrow \neg Q) \leftrightarrow R$
$((R \leftrightarrow P) \wedge Q) \leftrightarrow P$
$(R \wedge (Q \leftrightarrow S)) \vee P$
$((R \wedge Q) \vee (S \wedge P)) \leftrightarrow R$
$\therefore (S \wedge (R \leftrightarrow P)) \wedge P$

2.10.2 Apply the DPP to determine if each of the following sets of clauses is satisfiable.

a. $\{P, \neg Q, \neg R, S, \neg T\}$ $\{\neg Q\}$ $\{P, \neg R, \neg S\}$
$\{Q, R, \neg S\}$ $\{P, \neg Q, R, S, \neg T\}$ $\{\neg P, \neg Q, \neg S, \neg T\}$
$\{P, \neg Q, R, \neg T\}$ $\{Q, \neg R, S, T\}$ $\{P, \neg Q, R, S\}$
$\{P, \neg R, \neg S\}$ $\{\neg P, S\}$ $\{P, \neg Q, T\}$

b. $\{P, \neg R, \neg S, \neg T\}$ $\{P, \neg Q, R, \neg S, T\}$ $\{\neg Q, \neg R, \neg T\}$
$\{Q, S\}$ $\{P, \neg Q, R, \neg S, \neg T\}$ $\{Q, \neg R, \neg S, T\}$
$\{\neg R, \neg S, \neg T\}$ $\{Q, \neg R, \neg T\}$ $\{\neg R, S, T\}$
$\{P, \neg Q, \neg T\}$ $\{\neg R, T\}$ $\{R, \neg T\}$
$\{\neg R\}$ $\{Q, \neg R, \neg S, T\}$ $\{\neg Q, \neg R, \neg S, T\}$
$\{P, Q, R, \neg S, T\}$ $\{P, \neg Q, S\}$ $\{\neg P, \neg Q, \neg S, T\}$
$\{Q, \neg R, \neg S, \neg T\}$ $\{\neg P, S\}$ $\{\neg S, T\}$

c. $\{\neg P, \neg Q, R, S, T\}$ $\{S\}$ $\{\neg P, Q, \neg R, \neg S, \neg T\}$
$\{P, \neg Q, R, T\}$ $\{P, \neg Q, R, S, \neg T\}$ $\{\neg Q, \neg S, \neg T\}$
$\{Q, \neg S\}$ $\{Q\}$ $\{\neg Q, \neg S\}$
$\{P, \neg Q, R, \neg S, T\}$ $\{\neg P, \neg Q, \neg S\}$ $\{P, \neg Q, R\}$
$\{P, Q, \neg R, T\}$ $\{\neg P, \neg Q, S, \neg T\}$ $\{\neg P, Q\}$
$\{P, \neg Q, R\}$ $\{\neg P, R\}$ $\{\neg P, \neg R, \neg T\}$

d. $\{S\}$ $\{\neg R, \neg S, \neg T\}$ $\{P, \neg Q, \neg R, S, \neg T\}$
$\{P, \neg Q, R, \neg S, \neg T\}$ $\{P, \neg R, \neg S\}$ $\{\neg P, \neg Q, R, S, T\}$
$\{P, \neg Q, T\}$ $\{\neg Q, \neg S\}$ $\{Q, \neg S, \neg T\}$
$\{P, Q, \neg S\}$ $\{\neg Q, S, \neg T\}$ $\{\neg R, \neg S, \neg T\}$
$\{\neg Q, \neg R, S, T\}$ $\{P, Q, \neg R, \neg T\}$ $\{\neg Q, S, T\}$
$\{P, \neg Q, \neg R, \neg S\}$ $\{\neg P, \neg S, T\}$ $\{P, Q, R, \neg S\}$

e. $\{P, Q, \neg S\}$ $\{P, \neg Q, \neg R, S, \neg T\}$ $\{\neg P, Q, \neg R, \neg S, T\}$
$\{P, \neg Q, R, \neg S, \neg T\}$ $\{\neg P, Q, S\}$ $\{P, \neg R, \neg S, T\}$
$\{S\}$ $\{\neg P, \neg Q, \neg R, \neg S, T\}$ $\{Q, \neg T\}$
$\{P, Q, R, S\}$ $\{\neg P, \neg Q, R, \neg S\}$ $\{\neg Q, \neg R, \neg S, T\}$
$\{\neg Q, \neg R\}$ $\{P, \neg Q, \neg R, \neg S, T\}$ $\{\neg Q, \neg R, S\}$
$\{\neg P, Q, R, S\}$ $\{P, Q, \neg R, S, \neg T\}$ $\{P, \neg Q, \neg R, S\}$

2.10.3 Apply resolution to show that each of the following sets of clauses is not satisfiable.

a. $\{P, \neg R\}$ $\{Q, \neg R\}$ $\{Q, \neg S\}$ $\{\neg P, T\}$
$\{\neg Q, \neg T\}$ $\{\neg Q, R, T\}$ $\{P, S, \neg T\}$ $\{\neg P, Q, R\}$
$\{Q, R, S, T\}$

b. $\{P, \neg R\}$ $\{Q, \neg R\}$ $\{\neg P, S\}$ $\{\neg P, T\}$
$\{\neg Q, \neg T\}$ $\{\neg Q, R, T\}$ $\{P, R, \neg S\}$ $\{P, S, \neg T\}$
$\{Q, R, S, T\}$ $\{R, \neg S, \neg T\}$

c. $\{P, \neg R\}$ $\{Q, \neg R\}$ $\{\neg P, T\}$ $\{\neg Q, \neg T\}$
$\{\neg Q, R, T\}$ $\{P, R, \neg S\}$ $\{P, S, \neg T\}$ $\{\neg P, Q, R\}$
$\{Q, R, S, T\}$

d. $\{P, \neg R\}$ $\{\neg P, S\}$ $\{\neg P, T\}$ $\{P, Q, S\}$
$\{\neg P, \neg R\}$ $\{\neg Q, \neg T\}$ $\{\neg Q, R, T\}$ $\{P, R, \neg S\}$
$\{R, \neg S, \neg T\}$

e. $\{P, \neg R\}$ $\{\neg P, S\}$ $\{\neg P, \neg R\}$ $\{\neg Q, R, T\}$
$\{P, R, \neg S\}$ $\{P, S, \neg T\}$ $\{\neg P, Q, R\}$ $\{Q, R, S, T\}$
$\{R, \neg S, \neg T\}$

f. $\{P, \neg S\}$ $\{P, \neg T\}$ $\{Q, \neg U\}$ $\{\neg Q, R\}$
$\{R, S, U\}$ $\{\neg P, \neg R\}$ $\{R, \neg S, U\}$ $\{P, \neg R, S, T\}$

2.10.4 Determine if each of the following sets of clauses is satisfiable. If so, give a truth evaluation that satisfies the set. Otherwise, use resolution to prove that the set is not satisfiable.

a. $\{P, \neg Q, \neg R, S, \neg T\}$ $\{\neg Q\}$ $\{P, \neg R, \neg S\}$
$\{Q, R, \neg S\}$ $\{P, \neg Q, R, S, \neg T\}$ $\{\neg P, \neg Q, \neg S, \neg T\}$
$\{P, \neg Q, R, \neg T\}$ $\{Q, \neg R, S, T\}$ $\{P, \neg Q, R, S\}$
$\{P, \neg R, \neg S\}$ $\{\neg P, S\}$ $\{P, \neg Q, T\}$

b. $\{P, \neg R, \neg S, \neg T\}$ $\{P, \neg Q, R, \neg S, T\}$ $\{\neg Q, \neg R, \neg T\}$
$\{Q, S\}$ $\{P, \neg Q, R, \neg S, \neg T\}$ $\{Q, \neg R, \neg S, T\}$
$\{\neg R, \neg S, \neg T\}$ $\{Q, \neg R, \neg T\}$ $\{\neg R, S, T\}$
$\{P, \neg Q, \neg T\}$ $\{\neg R, T\}$ $\{Q, R, \neg S\}$
$\{\neg R\}$ $\{Q, \neg R, \neg S, T\}$ $\{\neg Q, \neg R, \neg S, T\}$
$\{P, Q, R, \neg S, T\}$ $\{P, \neg Q, S\}$ $\{\neg P, \neg Q, \neg S, T\}$
$\{Q, \neg R, \neg S, \neg T\}$ $\{\neg P, S\}$ $\{\neg S, \neg T\}$

c. $\{\neg P, \neg Q, R, S, T\}$ $\{Q, S\}$ $\{\neg P, Q, \neg R, \neg S, \neg T\}$
$\{P, \neg Q, R, T\}$ $\{P, \neg Q, R, S, \neg T\}$ $\{\neg Q, \neg S, \neg T\}$
$\{Q, \neg S\}$ $\{Q\}$ $\{\neg Q, \neg S\}$
$\{P, \neg Q, R, \neg S, T\}$ $\{\neg P, \neg Q, \neg S\}$ $\{P, \neg Q, R\}$
$\{P, Q, \neg R, T\}$ $\{\neg P, \neg Q, S, \neg T\}$ $\{\neg P, Q\}$
$\{P, \neg Q, R\}$ $\{\neg P, R\}$ $\{\neg P, \neg R, \neg T\}$

d. $\{S\}$ $\{\neg R, \neg S, \neg T\}$ $\{P, \neg Q, \neg R, S, \neg T\}$
$\{P, \neg Q, R, \neg S, \neg T\}$ $\{P, \neg R, \neg S\}$ $\{\neg P, \neg Q, R, S, T\}$
$\{P, \neg Q, T\}$ $\{\neg Q, \neg S\}$ $\{Q, \neg S, \neg T\}$
$\{P, Q, \neg S\}$ $\{\neg Q, S, \neg T\}$ $\{\neg R, \neg S, \neg T\}$
$\{\neg Q, \neg R, S, T\}$ $\{P, Q, \neg R, \neg T\}$ $\{\neg Q, S, T\}$
$\{P, \neg Q, \neg R, \neg S\}$ $\{\neg P, \neg S, T\}$ $\{P, Q, R, \neg S\}$

e. $\{P, Q, \neg S\}$ $\{P, \neg Q, \neg R, S, \neg T\}$ $\{\neg P, Q, \neg R, \neg S, T\}$
$\{P, \neg Q, R, \neg S, \neg T\}$ $\{\neg P, Q, S\}$ $\{P, \neg R, \neg S, T\}$
$\{S\}$ $\{\neg P, \neg Q, \neg R, \neg S, T\}$ $\{Q, \neg T\}$
$\{P, Q, R, S\}$ $\{\neg P, \neg Q, R, \neg S\}$ $\{\neg Q, \neg R, \neg S, T\}$
$\{\neg Q, \neg R\}$ $\{P, \neg Q, \neg R, \neg S, T\}$ $\{\neg R, S\}$
$\{\neg P, Q, R, S\}$ $\{P, Q, \neg R, S, \neg T\}$ $\{P, \neg Q, \neg R, S\}$

2.11 HORN CLAUSES

In the next two sections (2.12 and 2.13) we will look at some clauses for which resolution is rather slow. However, for certain types of clauses it is known to be reasonably fast.

Definition 2.11.1 A *Horn clause* is a clause with at most one positive literal.

Example 2.11.2 Examples of Horn clauses are

$$\{\neg P, \neg Q, \neg R, \neg S\}$$
$$\{P, \neg Q, \neg R, \neg S\}$$
$$\{\neg P, Q, \neg R, \neg S\}$$
$$\{\neg P, \neg Q, R, \neg S\}$$
$$\{\neg P, \neg Q, \neg R, S\}.$$

Lemma 2.11.3 A resolvent of two Horn clauses is always a Horn clause.

Proof. The number of positive literals in the resolvent of two clauses is always less than the sum of the numbers of positive literals in the two clauses. □

Horn clauses have been popular in *logic programming*, e.g., in Prolog. Many special kinds of resolution have been developed for Horn clauses—one of the simplest uses unit clauses.

Definition 2.11.4 A *unit* clause is a clause $\{\mathsf{L}\}$ with a single literal. *Unit resolution* refers to resolution derivations in which at least one of the clauses used in each resolution step is a unit clause.

In Example 2.10.15 (p. 105) we see that the clauses are all Horn clauses and that indeed the resolution used is unit resolution. However, in Example 2.10.16 (p. 105), 7 of the 12 clauses are not Horn clauses, and the resolution used is not unit resolution.

Theorem 2.11.5 Unit resolution is sound and complete for Horn clauses.

Proof. We know that resolution is sound, so we need only to show that unit resolution is complete for Horn clauses. This follows from showing that if we cannot derive the empty clause by unit resolution from a set of Horn clauses, then the set of Horn clauses is satisfiable.

Let $\mathcal{S}$ be a set of Horn clauses, and let $\mathcal{S}'$ be the set of all clauses that one can derive from $\mathcal{S}$ using unit resolution. If the empty clause is not in $\mathcal{S}'$, then we observe, for each variable P, that

a. if $\{P\} \in \mathcal{S}'$ then $\{\neg P\} \notin \mathcal{S}'$;
b. if $\{\neg P\} \in \mathcal{S}'$ then $\{P\} \notin \mathcal{S}'$.

So let us suppose that the unit clauses in $\mathcal{S}'$ are $\{P_1\}, \ldots, \{P_r\}$, $\{\neg P_{r+1}\}, \ldots, \{\neg P_s\}$ and that the other variables appearing in the clauses of $\mathcal{S}'$ are $P_{s+1}, \ldots$. Let e be the evaluation given by $1 = e_1 = \cdots = e_r$, $0 = e_{r+1} = \cdots$. Then we claim e satisfies $\mathcal{S}'$. Otherwise, choose a clause $\mathsf{C} \in \mathcal{S}'$, with the fewest literals, for which $\widehat{\mathsf{C}}(\mathsf{e}) = 0$. Such a C must have at least two literals in it. We cannot resolve C with any of the unit clauses of $\mathcal{S}'$, otherwise we would have a resolvent $\mathsf{C}' \in \mathcal{S}'$ with fewer literals, and $\widehat{\mathsf{C}'}(\mathsf{e}) = 0$, contradicting the minimality of C. Also, no literal from the unit clauses can appear in C, since e makes them true. Thus all the variables appearing in C must come from $P_{s+1}, \ldots$, and because C is not a unit clause, it must have a negative literal in it. But then $\widehat{\mathsf{C}}(\mathsf{e}) = 1$, as $e_i = 0$ for $i > s$. Thus we have our contradiction. □

One advantage of resolution with unit clauses is that the resolvents do not grow in size, as can happen with ordinary resolution. Resolution with a unit clause will delete one of the literals from the other clause.

Exercises

2.11.1 Apply unit resolution to the following sets of Horn clauses to determine if they are satisfiable.

a. $\{R, \neg S\}$ $\{\neg P, Q, \neg S\}$ $\{P, \neg R, \neg T\}$
$\{\neg Q, \neg S\}$ $\{\neg P, \neg Q, \neg S, T\}$ $\{P\}$
$\{S\}$ $\{\neg P, Q, \neg R, \neg T\}$ $\{P, \neg R, \neg T\}$

b. $\{\neg P, R, \neg T\}$ $\{\neg P, \neg R, \neg T\}$ $\{Q\}$
$\{\neg Q, \neg R\}$ $\{P\}$ $\{R, \neg T\}$
$\{P, \neg S\}$ $\{\neg P, \neg R\}$ $\{\neg P, Q\}$
$\{\neg P, Q, \neg R, \neg S\}$ $\{\neg R, \neg T\}$ $\{P, \neg R, \neg S\}$

c. $\{P, \neg T\}$ $\{\neg P, R\}$ $\{P, \neg R, \neg S, \neg T\}$
$\{R\}$ $\{\neg T\}$ $\{Q, \neg R, \neg S, \neg T\}$
$\{\neg P\}$ $\{S\}$ $\{\neg Q, \neg R, T\}$

d. $\{P\}$ $\{P, \neg Q, \neg R, \neg S\}$ $\{R\}$
$\{T\}$ $\{\neg P, \neg R, \neg S, \neg T\}$ $\{P, \neg T\}$
$\{\neg S\}$ $\{\neg P, \neg Q, S, \neg T\}$ $\{\neg R, \neg S\}$
$\{\neg Q, \neg R\}$ $\{\neg P, Q, \neg R, S, \neg T\}$ $\{R, \neg T\}$

e. $\{P, \neg S\}$ $\{\neg R, T\}$ $\{\neg P\}$
$\{\neg R, \neg T\}$ $\{\neg P, Q, \neg S\}$ $\{P, \neg Q\}$
$\{\neg P, Q, \neg R, \neg S\}$ $\{\neg R, S\}$ $\{\neg Q\}$

2.12 GRAPH CLAUSES

The DPP was programmed in the 1950s, but the results were disappointing. Computers were fast but not fast enough, it seemed. The first deep results were obtained by Tseitin in 1968. He proved that the DPP was doomed to being slow on certain sets of clauses that he obtained from graphs. Let us look at his construction.

Given a finite graph $\mathbf{G}$ with set of vertices V and with edge set E, we label each vertex v with a 0 or 1. This number is called the *charge* of the vertex, $charge(v)$. The *total charge* is $\sum_{v\in V} charge(v)$ modulo 2; hence the total charge will be either 0 or 1.

Next, we label each edge with a propositional variable such that distinct edges get distinct variables. Now we have our labeled graph $\widehat{\mathbf{G}}$. For each vertex v let *Var(v)* be the set of propositional variables on the edges attached to v. For each vertex v we construct the set of clauses $Clauses(v)$ by requiring that $\mathsf{C} \in Clauses(v)$ iff

- The propositional variables occurring in C are precisely those in *Var(v)*.
- The number of negative literals in C, plus $charge(v)$, is congruent to 1 modulo 2.

Then we let the set of clauses associated with the labeled graph $\widehat{\mathbf{G}}$ be given by $Clauses(\widehat{\mathbf{G}}) = \bigcup_{v\in V} Clauses(v)$, i.e., the clauses of $\widehat{\mathbf{G}}$ are the clauses attached to the various vertices.

Theorem 2.12.1 [Tseitin, 1968] $Clauses(\widehat{\mathbf{G}})$ is satisfiable iff the total charge is zero.

Let us illustrate his construction.

Example 2.12.2 In Figure 2.12.3 we give a graph $\mathbf{G}$ on four vertices a, b, c, d in the left diagram, assign charges to these vertices in the middle diagram, and assign propositional variables to the edges in the last diagram, which finally gives us a labeled graph $\widehat{\mathbf{G}}$.

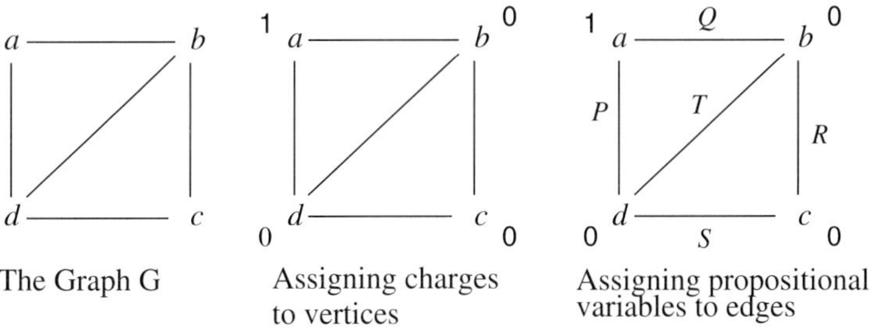

Figure 2.12.3 Example of a labeled graph

Starting with vertex a, which has charge 1, we write down the following sets of clauses:

$$
\begin{aligned}
Clauses(a) &= \{\{P, Q\},\ \{\neg P, \neg Q\}\}.\\
Clauses(b) &= \{\{\neg Q, R, T\},\ \{Q, \neg R, T\},\ \{Q, R, \neg T\},\ \{\neg Q, \neg R, \neg T\}\}.\\
Clauses(c) &= \{\{\neg R, S\},\ \{R, \neg S\}\}.\\
Clauses(d) &= \{\{\neg P, S, T\},\ \{P, \neg S, T\},\ \{P, S, \neg T\},\ \{\neg P, \neg S, \neg T\}\}.
\end{aligned}
$$

Because the total charge is 1, by Tseitin's theorem the set of all 12 clauses is not satisfiable. Note that in Example 2.10.16 (p. 105) we have the details of how to derive the empty clause from these 12 clauses using the DPP.

Exercises

2.12.1 Write out the graph clauses associated with each of the following graphs, then determine if the set of clauses is satisfiable. If not, give a resolution derivation of the empty clause.

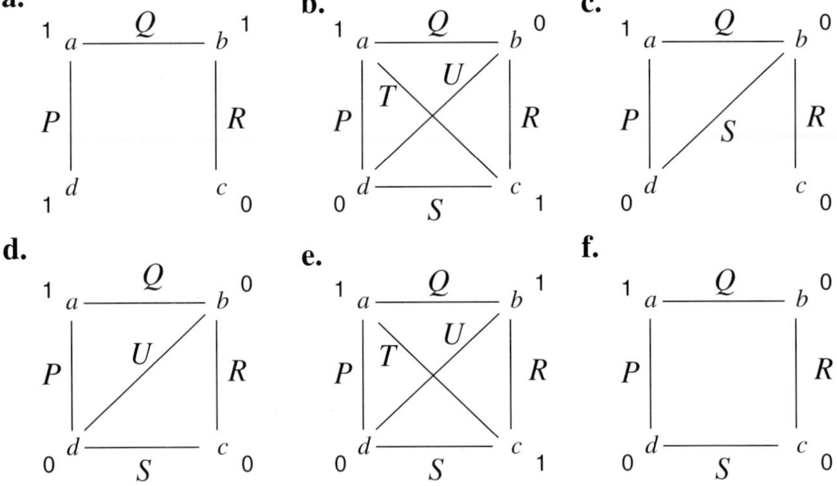

2.13 PIGEONHOLE CLAUSES

In 1974 Cook and Rechkow suggested that the set of clauses expressing a pigeonhole principle would be difficult to prove unsatisfiable by resolution.

The *pigeonhole principle* $\mathbb{P}_n$ says that one cannot put $n+1$ objects into n slots with distinct objects going into distinct slots. Let us see how we can formulate this as a set of clauses.

We choose propositional variables P_{ij} for $1 \le i \le n+1$ and $1 \le j \le n$. Our intended interpretation of P_{ij} is that the ith object goes into the jth slot. So we write down the following $n+1+\binom{n+1}{2}\cdot n$ clauses:

1. $\{P_{i1}, \cdots, P_{in}\}$ for $1 \le i \le n+1$.
2. $\{\neg P_{ik}, \neg P_{jk}\}$ for $1 \le i < j \le n+1$ and $1 \le k \le n$.

Item 1 says each i for $1 \le i \le n+1$ goes into some slot k, where $1 \le k \le n$. Item 2 says that distinct i and j between 1 and $n+1$ cannot go into the same slot. Any truth evaluation that satisfies the conditions will map $n+1$ objects one–to–one into n slots. Of course, this cannot be done, so the clauses must be unsatisfiable. However, if we throw away any one clause, the remaining collection of clauses is satisfiable!

Example 2.13.1 $\mathbb{P}_2$ is the set of nine clauses in six variables:

$$\begin{array}{lll} \{P_{11}, P_{12}\} & \{P_{21}, P_{22}\} & \{P_{31}, P_{32}\} \\ \{\neg P_{11}, \neg P_{21}\} & \{\neg P_{12}, \neg P_{22}\} & \{\neg P_{11}, \neg P_{31}\} \\ \{\neg P_{12}, \neg P_{32}\} & \{\neg P_{21}, \neg P_{31}\} & \{\neg P_{22}, \neg P_{32}\}. \end{array}$$

Exercises

2.13.1 Give a resolution derivation of the empty clause for the pigeonhole principle $\mathbb{P}_2$.

2.14 HISTORICAL REMARKS

The history of propositional logic, or statement logic, is difficult to present in isolation. From near its beginnings propositional logic was often introduced not as a separate subject but as part of a more ambitious logical system. So we will take the liberty to range over a wider realm of topics in an effort to discover how propositional logic evolved.

2.14.1 The Beginnings

A symbolic approach to the logic of propositions was introduced by MacColl in 1877. Within two years Frege presented his great work, *Concept Script*, a full–fledged higher–order logic, as a foundation for the natural numbers. His system contained explicit axioms and rules of inference, and one was to be restricted to those. This was a very modern proof system (except for the two-dimensional notation). The first part of his proof system was devoted to handling the connectives $\rightarrow$ and $\neg$. If we restrict Frege's system to these connectives we have the following:

Connectives $\neg, \rightarrow$
Rules of inference substitution[6] + modus ponens
Axioms

1. $P \rightarrow (Q \rightarrow P)$
2. $(P \rightarrow (Q \rightarrow R)) \rightarrow ((P \rightarrow Q) \rightarrow (P \rightarrow R))$
3. $(P \rightarrow (Q \rightarrow R)) \rightarrow (Q \rightarrow (P \rightarrow R))$
4. $(P \rightarrow Q) \rightarrow (\neg Q \rightarrow \neg P)$
5. $\neg\neg P \rightarrow P$
6. $P \rightarrow \neg\neg P$

This system went essentially unnoticed for half a century.

2.14.2 Statement Logic and the Algebra of Logic

In 1880 Schröder showed that the standard connectives could be defined in terms of the connective $\curlywedge$ (see p. 53), i.e., that $\curlywedge$ is adequate for propositional logic. However, he did not then go on to found a propositional logic based on his connective. In the *Algebra of Logic* he presented his version of the *statement calculus* in the second volume.

In the early 1880s Peirce became involved with a symbolic logic for statements. Like MacColl, he took the notion of "implies" as primitive, and defined "and" and "or" from them. There are of course striking similarities between the logic of classes and the logic

---

[6] We have not used the axiom of substitution in PC. What it says is rather obvious, namely, from any formula $\mathsf{F}(P_1, \cdots, P_n)$ one can infer any substitution instance $\mathsf{F}(\mathsf{F}_1, \cdots, \mathsf{F}_n)$ of it. This axiom is fine as long as one is deriving tautologies but requires some caution when carrying out derivations with premisses. One does not want to substitute into the premisses.

of statements, e.g., we have the following parallels:

$$
\begin{array}{l|l}
A \subseteq B & A \to B \\
A = B & A \leftrightarrow B \\
A \cup B = B \cup A & A \vee B \leftrightarrow B \vee A \\
A \cap B = B \cap A & A \wedge B \leftrightarrow B \wedge A.
\end{array}
$$

This led Peirce to use a single notation for both, identifying $\to$ with $\subseteq$, etc.

For the *Algebra of Logic* Schröder adopted Peirce's collapsed notation convention, opting for the notation he had used with the logic of classes as the dual–purpose notation. As Schröder observed, assertions true in the logic of statements need not be true in the logic of classes. The primary example is

$$(A \cup B = 1) = (A = 1) \cup (B = 1), \tag{22}$$

a statement that does not generally hold in the logic of classes, but translated into propositional logic it gives the true assertion

$$(A \vee B \leftrightarrow 1) \leftrightarrow (A \leftrightarrow 1) \vee (B \leftrightarrow 1). \tag{23}$$

Schröder's approach to the logic of statements was rather cautious. His *primary* statements were the ones about classes, e.g., $A \cup B = 1$, and the *secondary* statements were about the primary statements. Thus the outermost $=$ in the statement (22) was an iff between primary statements, and the innermost $=$'s were the equality of classes. He also recognized that one could have *tertiary* statements about the secondary statements, etc., but he said he would limit himself to the the lower levels of statements.

His program was thus to develop the statement calculus, a formal logic for the statements in the calculus of classes. The calculus of classes had of course been introduced to offer a mathematical translation of the logic of philosophy for the purpose of determining the validity of arguments by doing algebraic manipulations.

The validity of assertion (22) makes it clear that Schröder saw the logic of statements as a two–valued logic. A good part of Vol. II is concerned with sorting out which basic assertions are true in both the logic of statements and the logic of classes. Schröder came to the conclusion that the logic of statements is nothing more than the logic of classes restricted to the two values 0 and 1.

One of the interesting results he presents is the following theorem from the logic of statements:

$$(A \cap B \subseteq C) \subseteq (A \subseteq C) \cup (B \subseteq C). \tag{24}$$

He notes that this assertion makes no sense for the logic of classes, but for the logic of statements it is true. After translating (24) into words, and noting that it might be a bit difficult to comprehend, Schröder sets out to convince the reader as follows (Volume II, pp. 260–261): Since we are talking about statements, they can take on only the values 0 or 1. He lists all the possibilities in a table, with some asterisks:

A,	B,	C
0	0	0*
0	0	1*
0	1	0*
1	0	0*
0	1	1*
1	0	1*
1	1	0
1	1	1*

The asterisks, he explains, mark the rows that satisfy $A \cap B \subseteq C$, the left–hand side of (24). He then notes that the seven rows that satisfy the right–hand side of (24) are the same as those with an asterisk. Thus (24) is satisfied by the rows with an asterisk. For the single row without an asterisk he notes that both the left and right sides of (24) are 0, and thus (24) is also true for this row. Consequently, assertion (24) is a theorem of the statement calculus.

This is the closest Schröder came to using modern truth tables. And indeed this is very close. However, he did not put any special emphasis on this method and sketched it only in a couple of other, simpler places. In particular he did not make explicit the connection between valid arguments and truth tables, nor the truth table method of verifying that a (primary) equation is true in the logic of classes.

2.14.3 Frege's Work Ignored

It is interesting that the scholarly work of Schröder makes no mention of Frege's axioms for handling the connectives $\rightarrow$ and $\neg$, nor any of the technical details of the rest of his work. One factor in this omission was likely Frege's combative personality.

Frege was particularly concerned about the foundational treatment the natural numbers received in mathematics, or rather, the lack of any respectable treatment. He disliked, for example, the explanation of the number 2 in which one holds up two pieces of chalk and asks the listener to "abstract". He wanted a definition of a number that was symbolical in nature, one that could be used in a formal logical system. In 1884 he wrote a philosophical book devoted to a criticism of the existing foundations of natural numbers.

The *only* comment Schröder makes about Frege's work in his *Algebra of Logic* is in the *bibliography* of his first volume. There Schröder says that Frege's 1884 book has many critical remarks about the content of one of Schröder's earlier works, an introductory text on numbers and algebra for school teachers and students.

So it is quite possible that Frege's work was ignored not only because he wrote in a new two–dimensional notation but also because he was short on influential friends. For example, Hilbert briefly exchanged a letter or two with him on foundations. Hilbert's reply to Frege's arguments was that he wanted to think it over. And that was the end of the correspondence. Frege's work received uniformly poor reviews, including ones by Cantor and Schröder.

2.14.4 Bertrand Russell Rescues Frege's Logic

Evidently Peano was not put off by Frege's rather blunt style, in spite of the fact that Frege had insisted that Peano publicly retract his published claims about a system of his being superior to one of Frege's. Peano saw that Frege's objections were well founded, made the retraction, and published Frege's comments. Peano's good will was no doubt Frege's good fortune, for when Peano met the young Bertrand Russell in 1900, at the International Congress of Mathematicians in Paris, he suggested that he read the work of Frege.

Frege had planned a final masterwork on the foundations of numbers, the first volume of which was published in 1893. This is the volume that Russell would read with the keenest of interest. At one point in the book Frege points out that he has introduced as an axiom a new principle, one that he has not seen in the literature, but surely one that everyone believes: Namely, two properties are the same iff they hold of the same class of objects, i.e., the identification of intension with extension.

This new principle is the only point in the book that Frege feels anyone can find fault with insofar as the logical foundations of his work are concerned, and he challenges the reader to either show the fault or to join him in accepting the principle.

By 1901 Russell realized that this axiom, along with the others in the volume, easily led to the derivation of a contradiction. This is the famous Russell paradox about the class that has as it members all classes that do not belong to themselves. Unfortunately, this class belongs to itself iff it does not belong to itself. Thus Frege's powerful logical system was inconsistent, i.e., too powerful. Every statement could be proved in it!

Russell communicated this to Frege in a letter in 1902. Frege was just going over the final proofs of his second volume when the letter arrived. (The first volume had not been a financial success for the publisher, and Frege was paying for the publication of the second volume out of his own pocket.)

One can imagine the feeling of being told that a lifetime of work has collapsed under a contradiction. Some claim that Frege was destroyed by this revelation, and others claim he quickly adapted to the situation. Indeed, Frege promptly added an appendix that he thought patched the problem.[7] Russell had the highest regard for Frege, and Frege lived to see the publication of *Principia Mathematica* (1910–1913) with its simple dedication in the preface:

> In all questions of logical analysis, our chief debt is to Frege.

Given the poor reception Frege met during his most creative period, from 1879 to 1900, one can easily imagine that he was pleased to have someone finally take him seriously, even if they found an error. His neglected work would be, after 30 years, revived in the three volumes of *Principia Mathematica* that dominated mathematical logic for many years.

2.14.5 The Influence of *Principia*

With the publication of *Principia Mathematica* the subject of mathematical logic set out on a different course. First there was the shedding of the arithmetical notation and analogs to algebra (we quote from the preface of *Principia*):

[7] The appendix was eventually shown to reduce the universe to one element.

> The general method which guides our handling of logical symbols is due to Peano. His great merit consists not so much in his definite logical discoveries nor in the details of his notations (excellent as both are), as in the fact that he first showed how symbolic logic was to be freed from its undue obsession with the forms of ordinary algebra, and thereby made it a suitable instrument for research $\cdots$.

Then Russell and Whitehead also limited their work to a straightforward presentation of a symbolic logic:

> We have avoided both controversy and general philosophy, and made our statements dogmatic in form $\cdots$.

For those who did not enjoy pondering such philosophical matters as Frege's distinctions between *sense* and *reference*, this must have been a relief.

In *Principia,* Russell and Whitehead adopted the following set of connectives, axioms, and rules of inference:

Connectives $\neg, \vee$
Rules of inference substitution + modus ponens
Axioms (where "$P \rightarrow Q$" means "$\neg P \vee Q$")

1. $(P \vee P) \rightarrow P$
2. $Q \rightarrow (P \vee Q)$
3. $(P \vee Q) \rightarrow (Q \vee P)$
4. $(P \vee (Q \vee R)) \rightarrow (Q \vee (P \vee R))$
5. $(Q \rightarrow R) \rightarrow ((P \vee Q) \rightarrow (P \vee R))$.

In 1913 H. M. Sheffer showed that $|$, the Sheffer stroke, was the only connective one needed. Russell and Whitehead were quite impressed by this and added several pages to a later edition of *Principia* to show how to adapt *Principia* to this single connective.

In 1917 J. Nicod found the following simplification of the propositional system in *Principia*:

Connectives $\neg, \vee$
Rules of inference substitution + modus ponens
Axioms (where "$P \rightarrow Q$" means "$\neg P \vee Q$")

1. $(P \vee P) \rightarrow P$
2. $P \rightarrow (P \vee Q)$
3. $(P \vee (Q \vee R)) \rightarrow (Q \vee (P \vee R))$
4. $(Q \rightarrow R) \rightarrow ((P \vee Q) \rightarrow (P \vee R))$.

In the same year he showed that using the Sheffer stroke, one could get by with a single axiom and a single rule of inference in addition to substitution:

Connective $|$

Rules of inference substitution + $\dfrac{\mathsf{F},\ \mathsf{F}|(\mathsf{G}|\mathsf{H})}{\mathsf{H}}$

Axiom $(P|(Q|R))|((S|(S|S))|((U|Q)|((P|U)|(P|U))))$.

2.14.6 The Emergence of Truth Tables, Completeness

It is rather surprising that the notion of a truth table emerged so slowly. In spite of Schröder's near truth tables, mentioned in Sec. 2.14.2, the concept of a truth table did not come into clear focus until 1920. According to Kneale [**30**], it seems to have been a concept that crystallized in the minds of several logicians at about the same time, in 1920. Two of the most notable among the enlightened were Wittgenstein[8] and Post. Still, in Hilbert and Ackermann's book of 1928 one does not find truth tables.

Another surprising fact was the lack of a question about the soundness and completeness of the logics, until 1920. Frege suggested that he had found all the principles of logic that one needed to develop mathematics. This idea carried over to Whitehead and Russell, who with proof by example actually carried out portions of mathematics in *Principia,* presumably with the idea that once one sees how to do it, one will know how to continue with other parts of mathematics.

In 1920 Post considered the question of whether or not the propositional logic portion of *Principia* was complete. He answered this in his paper of 1921, in the affirmative, making use of truth tables.

2.14.7 The Hilbert School of Logic

Principia stole the show from the algebra of logic. Hilbert had never shown any particular interest in the latter, but by 1917 he was lecturing in Göttingen on foundations. This interest continued for the next two decades. He was aided by his assistants Ackermann and

[8]Wittgenstein had contacted Frege regarding studying logic and was referred by Frege to Bertrand Russell.

Bernays. During the 1920s Hilbert conceived a two–book project on logic.[9] The first was to be an elementary introduction, posing the important problems of completeness, consistency, and decidability. This text was authored by Hilbert and Ackermann and used the propositional proof system of *Principia* with Axiom 4 deleted (a simplification due to Bernays) and with a slight modification[10] of Axiom 2. Thus Hilbert and Ackermann's system (1928) was:

Connectives $\neg, \vee$
Rules of inference substitution + modus ponens
Axioms (where "$P \to Q$" means "$\neg P \vee Q$")

1. $(P \vee P) \to P$
2. $P \to (P \vee Q)$
3. $(P \vee Q) \to (Q \vee P)$
4. $(Q \to R) \to ((P \vee Q) \to (P \vee R))$.

2.14.8 The Polish School of Logic

During the 1920s there was a particular concentration in the areas of mathematical logic in Poland. Łukasiewicz ran a seminar on logic in Warsaw that enjoyed the participation of future legends like Alfred Tarski. Łukasiewicz was very interested in propositional proof systems, particularly in methods to prove completeness and independence of the axioms.

We have already looked at completeness. The axioms of a propositional proof system are *independent* if one cannot derive any of the axioms from the remaining ones. Likewise, one can formulate the notion of *irredundant* rules of inference. If one can derive the same formulas without using a given rule, then the rule is redundant.

We need to say a few words about axiom schemata. In 1927 von Neumann suggested that one could replace the use of the substitution rule with substitution limited to axioms. That is, if one has an axiom schema like $P \to (Q \to P)$, then one is really saying that one has an infinite set of axioms, namely, all formulas of the form $\mathsf{F} \to (\mathsf{G} \to \mathsf{F})$.

---

[9]The second book, by Hilbert and Bernays, was to appear in the early 1930s. But Gödel's results so transformed the subject of mathematical logic that the project was delayed several years, and became two volumes.

[10]The 1928 book of Hilbert and Ackermann says that their axioms, plus one more, give the axioms used in *Principia Mathematica*. This is not quite correct, as Whitehead and Russell use $Q \to (P \vee Q)$ instead of $P \to (P \vee Q)$. This point was not cleared up in subsequent editions of the Hilbert and Ackermann book.

Then one can naturally formulate what it means for a collection of axiom schemata to be independent.

Our PC has an independent axiom schema, since there is only one, but a highly redundant set of rules.

One of the most popular of the propositional proof systems was found by Łukasiewicz. In 1928 he simplified the system of Frege to the following:

Connectives $\neg, \rightarrow$
Rules of inference substitution + modus ponens
Axioms

1. $P \rightarrow (Q \rightarrow P)$
2. $(P \rightarrow (Q \rightarrow R)) \rightarrow ((P \rightarrow Q) \rightarrow (P \rightarrow R))$
3. $(\neg P \rightarrow \neg Q) \rightarrow (Q \rightarrow P)$.

When Łukasiewicz published this result, he inserted a footnote saying that the system of Frege did not seem to be known, not even in his own country, Germany. In Appendix D the reader will find all details of a completeness proof for this system (which we call FŁ) in worksheet style—the reader is asked to fill in suitable references to axioms or lemmas (where indicated) to finish the proof.

2.14.9 Other Propositional Proof Systems

The fascination with simple propositional proof systems continued. In 1947 E. Götlind and H. Rasiowa simplified the propositional axioms of *Principia* as follows:

Connectives $\neg, \vee$
Rules of inference substitution + modus ponens
Axioms (where "$P \rightarrow Q$" means "$\neg P \vee Q$")

1. $(P \vee P) \rightarrow P$
2. $P \rightarrow (P \vee Q)$
3. $(Q \rightarrow R) \rightarrow ((P \vee Q) \rightarrow (R \vee P))$.

And then, in 1953, C. A. Meredith came up with the following show piece, a sound and complete propositional proof system that has a single rule of inference, the familiar modus ponens, and a single, not so familiar, axiom schema:

Connectives $\neg, \rightarrow$
Rule of inference modus ponens
Axiom schema $((((P \rightarrow Q) \rightarrow (\neg R \rightarrow \neg S)) \rightarrow R) \rightarrow U) \rightarrow ((U \rightarrow P) \rightarrow (S \rightarrow P))$.

2.14.10 Problems with Algorithms

At a mathematical meeting in the United States in 1946 Tarski rekindled interest in the questions Łukasiewicz had studied, namely, how to determine if a propositional proof system was complete, how to determine if the axioms were independent, and how to determine the formulas that could be derived in a given proof system. For the independence question Łukasiewicz (and Bernays) had developed a method popularly known as the method of *matrices*. To show an axiom was independent of the others one created artificial "truth" tables, i.e, matrices, perhaps with more than two truth values, and showed that all the formulas one could derive from the diminished set of axioms had a certain property related to the matrices, but the deleted axiom did not.

For example if we look at the three axioms of Łukasiewicz on the previous page, then to show that the third one does not follow from the first two, we simply use the usual truth table for $\rightarrow$ but change the table for $\neg$ to that of the identity function, i.e., we use

P	$\neg P$
1	1
0	0

Now we define an *M–tautology* to be a formula that has only the value 1 when evaluated at any truth evaluation, using these new tables for the connectives. Then we see that (1) modus ponens preserves M–tautologies, and (2) all formulas proved by using the first two axioms are M–tautologies, but the third axiom is not an M–tautology. Thus we cannot derive the third axiom from the first two. By similar means we can show that the other two axioms are also independent, and thus the set of three axioms is independent. If we delete any one, the system will no longer be complete.

Although Łukasiewicz was able to analyze many propositional proof systems with such methods, he failed to find a general method—and with good reason. In 1949 Linial and Post proved that there do not exist algorithms to determine if a set of axioms is complete or independent. This does not just mean that no one has found an algorithm, but rather that no one will ever find an algorithm. They also exhibited a propositional proof system with the property that there is no algorithm to determine what its theorems are. Numerous papers followed on this theme of undecidability results for propositional proof systems.

2.14.11 Reduction to Propositional Logic

Another direction connected with propositional logic developed when Löwenheim essentially showed, in 1915, that if a first–order statement is true, then it must be possible to prove it by analyzing propositional formulas that can be generated from the formula. This principle was not enunciated in his proof, since this was not the goal of his paper, and remained obscured until its presentations by Skolem (1929) and Herbrand (1930). This is the method we will present in Chapter 5, after discussing skolemization.

Precisely this idea was picked up in the 1950s by mathematicians interested in the possibility of using a computer to prove mathematical theorems. But it turned out that the propositional formulas were too difficult to analyze, even for computers. When it came to determining if a formula was a tautology, nothing better than truth tables was really known.

One of the methods used to analyze propositional formulas was to work with clauses and resolution. After it was found that this method could be slow in practice, mathematicians tried to prove that for some problems it *had* to be slow. The first success was Tseitin's work in 1968, in which he showed that the DPP was slow for graph clauses (see Sec. 2.12). However, the question of whether or not (full) resolution would be slow remained open until 1985, when A. Haken proved that the pigeonhole clauses were difficult for resolution. Then, in 1987, A. Urquhart proved that Tseitin's graph clauses were also difficult for resolution, and in 1988 Chvátal and Szemerédi proved that randomly generated sparse families of clauses usually require exponential time to refute by resolution.

2.14.12 Testing for Satisfiability

The apparent difficulty of the problem of testing a propositional formula to see whether it is a tautology, or satisfiable, became clearer with the work of S. Cook and R. Karp in 1971, when they showed that if one could answer the satisfiability question reasonably fast (i.e., in polynomial time), then one could answer many other apparently hard questions, like testing graphs for isomorphism, reasonably fast. The question of whether or not there is a fast algorithm (Does P = NP?) remains open.

Chapter 3

Equational Logic

SYMBOLS	$\approx$
	function symbols
	constant symbols
	variables

Equational logic studies, of course, equations. It is a *fragment* of first–order logic. To start we need to choose

- a set $\mathcal{F}$ of *function symbols* $f, g, h, \cdots$,
- a set $\mathcal{C}$ of *constant symbols* $c, d, e, \cdots$, and
- a set X of *variables* $x, y, z \cdots$.

The set $\mathcal{L} = \mathcal{F} \cup \mathcal{C}$ is called a *language* (of *algebras*, or *algebraic structures*). (Either $\mathcal{F}$ or $\mathcal{C}$, or both, can be empty.)

We want each function symbol to have a positive integer, called its *arity*, assigned to it. If the number is n, we say f is n*–ary*. For small n we have special names, i.e., *unary* ($n = 1$), *binary* ($n = 2$), and *ternary* ($n = 3$).

Example 3.0.1 One of the examples used in this section to illustrate the ideas is *Boolean algebras*. The language $\mathcal{L}_{BA}$ of Boolean algebras has $\mathcal{F} = \{\vee, \wedge, '\}$ and $\mathcal{C} = \{0, 1\}$, where $\vee$ and $\wedge$ are binary function symbols, and $'$ is a unary function symbol. They have the following standard names:

$\vee$	*join*
$\wedge$	*meet*
$'$	*complement*

The constants are just called by the usual names for these symbols, *zero* and *one*.

3.1 INTERPRETATIONS AND ALGEBRAS

Now we turn to the "meaning" of our symbols. The obvious interpretation of a function symbol is as a function on a set.

Definition 3.1.1 If A is a set and n is a positive integer, then an *n–ary* function f on A is a mapping from A^n to A, i.e., f maps each *n–tuple* $(a_1, \ldots, a_n)$ of elements of A to an element of A, denoted $f(a_1, \ldots, a_n)$. Special cases are *unary* ($n = 1$), *binary* ($n = 2$), and *ternary* ($n = 3$) functions.

Example 3.1.2 We can describe a single unary function on a small set either with a table, e.g.,

	f
0	2
1	1
2	3
3	0

or with a directed graph representation:

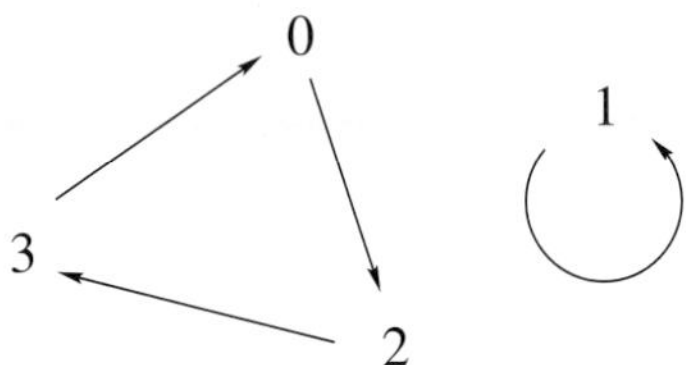

Figure 3.1.3 Picture of a single unary function

Example 3.1.4 We can also describe small binary functions on a set A using a table, called a *Cayley table*. For example, to describe the integers mod 4, with the binary operation of addition mod 4, we have the Cayley table:

+	0	1	2	3
0	0	1	2	3
1	1	2	3	0
2	2	3	0	1
3	3	0	1	2

Example 3.1.5 We can also describe functions on a small set A using a table that is similar to the truth tables used to describe

the connectives, e.g., to describe the ternary function $f(x, y, z) = xy + yz + xz$ on the integers modulo 2 we could use

x	y	z	f
0	0	0	0
0	0	1	0
0	1	0	0
0	1	1	1
1	0	0	0
1	0	1	1
1	1	0	1
1	1	1	1

To describe the binary function $f(x, y) = xy + 1$ on the integers modulo 3 we have the table

x	y	f
0	0	1
0	1	1
0	2	1
1	0	1
1	1	2
1	2	0
2	0	1
2	1	0
2	2	2

Now we give the technical definition of an *interpretation* of a first–order language on a set, followed by notational conventions to make it easier to use, and examples of how an interpretation works.

Definition 3.1.6 An *interpretation* I of the language $\mathcal{L}$ on a nonempty set A is a mapping that assigns to each symbol from $\mathcal{L}$ a function or constant as follows:

- $I(c)$ is an element of A for each constant symbol c in $\mathcal{C}$.
- $I(f)$ is an n–ary function on A for each n–ary function symbol f in $\mathcal{F}$.

An $\mathcal{L}$*-algebra* (or $\mathcal{L}$*-structure*) $\mathbf{A}$ is a pair (A, I) where I is an interpretation of $\mathcal{L}$ on A.

Given an algebra $\mathbf{A}$, the interpretations of the function symbols are sometimes called the *fundamental operations* of the algebra.

We can visualize an interpretation I on the set A as follows:

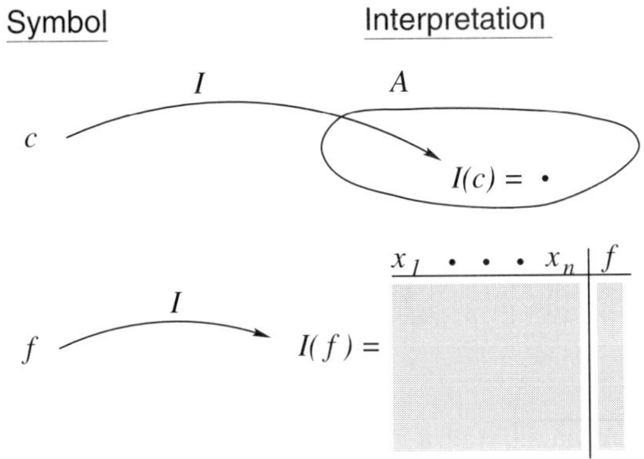

Figure 3.1.7 An interpretation of $\mathcal{L}$ on A

Notational Convention 3.1.8 Given an $\mathcal{L}$–structure $\mathbf{A} = (A, I)$, we prefer to write $c^{\mathbf{A}}$ for $I(c)$, and $f^{\mathbf{A}}$ for $I(f)$ since this immediately reminds us that we are working with the algebra $\mathbf{A}$.

When we feel confident that the context makes the meaning clear (and this is usually the case), we will simply write c for $c^{\mathbf{A}}$, f for $f^{\mathbf{A}}$.

Also, rather than writing (A, I) we often prefer to write $(A, \mathcal{F}, \mathcal{C})$ or simply to list the symbols in $\mathcal{F}$ and $\mathcal{C}$, e.g., the algebra $(R, +, \cdot, 0, 1)$, which is the set of real numbers with addition, multiplication, and two specified constants. Working backward in this example we find $\mathcal{F} = \{+, \cdot\}$ and $\mathcal{C} = \{0, 1\}$.

(We accept the ambiguity of using the same notation for a symbol and its interpretation, and we hope you do too.)

Example 3.1.9 The most familiar algebras are those based on the numbers systems, e.g.,

- $(N, +, \cdot, 0, 1)$
  the natural numbers with $\mathcal{L} = \{+, \cdot, 0, 1\}$;
- $(Z, +, \cdot, -, 0, 1)$
  the integers with $\mathcal{L} = \{+, \cdot, -, 0, 1\}$;
- $(Z_n, +, \cdot, -, 0, 1)$
  the integers modulo n with $\mathcal{L} = \{+, \cdot, -, 0, 1\}$;
- $(Q, +, \cdot, -, 0, 1)$
  the rational numbers with $\mathcal{L} = \{+, \cdot, -, 0, 1\}$;
- $(R, +, \cdot, -, 0, 1)$
  the real numbers with $\mathcal{L} = \{+, \cdot, -, 0, 1\}$;
- $(C, +, \cdot, -, 0, 1)$
  the complex numbers with $\mathcal{L} = \{+, \cdot, -, 0, 1\}$.

Note that the language we are using for N does not have the minus symbol $(-)$, since the natural numbers are not closed under the operation minus. Also, we do not use the division symbol $(\div)$ in any of the cases because in algebraic first–order structures, i.e., algebras, the n–ary function symbols must be interpreted as functions defined on *all* n–tuples. Because $1 \div 0$ is not defined, i.e., $\div$ is not defined on the pair $(1, 0)$, we exclude this operation.

The other number systems given here are closed under minus. One can take minus to be either *subtraction* or *negative of*. The usual convention is to choose the latter, i.e., $-$ is treated as a *unary* operation symbol.

Example 3.1.10 The next most familiar algebras are those learned in linear algebra, namely, algebras of matrices. We use $\mathbf{M}_{m\times n}(R)$ to denote the collection of $m \times n$ matrices with real–number entries.

- $(\mathbf{M}_{m\times n}(R), +, -, O)$
  $m \times n$ matrices with $\mathcal{L} = \{+, -, 0\}$;
- $(\mathbf{M}_{n\times n}(R), +, \cdot, -, I, O)$
  $n \times n$ matrices with $\mathcal{L} = \{+, \cdot, -, 0, 1\}$.

O stands for the zero matrix, and I for the identity matrix. There is no identity matrix in the first example because identity matrices are always square matrices, i.e., we need $m = n$. Likewise, we do not include $\cdot$ in the first example because one cannot multiply two $m \times n$ matrices unless $m = n$.

Example 3.1.11 Let $\mathcal{L} = \mathcal{L}_{BA}$. Let $Su(U)$ be the collection of all subsets of a given set U (U is called the *universe*), and interpret join as *union* $(\cup)$, meet as *intersection* $(\cap)$, and complement as the *complement* $(')$ in U. Interpret 0 as the *empty set* $(\emptyset)$, and 1 as the *universe* (U). Then

$$\mathbf{Su}(U) = (Su(U), \cup, \cap, ', \emptyset, U),$$

the Boolean algebra of subsets of U, is the most basic example of a Boolean algebra.

Boolean algebras were introduced, in their modern form, in the early twentieth century, although the inspiration and basic examples go back to Boole's mid–nineteenth century work with the calculus of classes.

Example 3.1.12 Let $\mathcal{L} = \mathcal{L}_{BA}$. Let $A = \{0, 1\}$, and let the function symbols be interpreted as follows:

$\vee$	0	1
0	0	1
1	1	1

$\wedge$	0	1
0	0	0
1	0	1

	$'$
0	1
1	0

and $0, 1$ are translated in the obvious manner (to $0, 1$). This is the best known of all the Boolean algebras.

It can take a while to realize the enormous diversity of $\mathcal{L}$–algebras — one indication is their sheer numbers. Let us illustrate with some examples.

Example 3.1.13 If $\mathcal{L} = \{f\}$, where f is a unary function symbol, then the number of $\mathcal{L}$–algebras $\mathbf{A}$ (called *monounary algebras*) on a given n–element set $A = \{a_1, \ldots, a_n\}$ is n^n, since for each a_i we can replace the box in $f^{\mathbf{A}}(a_i) = \Box$ in n different ways. This might become more clear if we use the table representation of a function. That is, we arbitrarily replace the n $\Box$'s in the following table using the elements $a_1, \ldots, a_n$:

	f
a_1	$\Box$
$\vdots$	$\vdots$
a_n	$\Box$

Example 3.1.14 If $\mathcal{L} = \{\cdot\}$, where $\cdot$ is a binary function symbol, then the number of $\mathcal{L}$–algebras (called *binary algebras*[1]) on a given n–element set $A = \{a_1, \ldots, a_n\}$ is n^{n^2}, since for each pair (a_i, a_j) we can replace the box in $a_i \cdot a_j = \Box$ in n different ways. Again, using the table representation of a function, we arbitrarily replace the n^2 $\Box$'s in the following table using the elements $a_1, \ldots, a_n$:

$\cdot$	a_1	$\cdots$	a_n
a_1	$\Box$	$\vdots$	$\Box$
$\vdots$	$\vdots$	$\vdots$	
a_n	$\Box$	$\vdots$	$\Box$

Thus we have a staggering $4^{16} = 4,294,967,296$ binary algebras on a four–element set.

[1]Such algebras are called *groupoids* by universal algebraists. Because the word "groupoid" is also used by other mathematicians, with a different meaning, we have selected the name *binary algebra*.

Exercises

3.1.1 Explicitly write out the four different possibilities for a unary function on the two–element set $\{a, b\}$, where each function is to be obtained by filling in a table like the following, replacing each $\square$ with one of a or b:

	f
a	$\square$
b	$\square$

3.1.2 Explicitly write out the 27 different possibilities for a unary function on the three–element set $\{a, b, c\}$, where each function is to be obtained by filling in a table like the following, replacing each $\square$ with one of a or b or c:

	f
a	$\square$
b	$\square$
c	$\square$

3.1.3 Explicitly write out the 16 different possibilities for a binary function on the two–element set $\{a, b\}$, where each function is to be obtained by filling in a table like the following, replacing each $\square$ with one of a or b:

$\cdot$	a	b
a	$\square$	$\square$
b	$\square$	$\square$

3.1.4 How many $\mathcal{L}$ structures can one find on the set $\{a, b\}$ when $\mathcal{L} = \{+, \cdot, -, 0, 1\}$, i.e., how many ways can one fill the $\square$'s in the following, using only a's and b's?

$+$	a	b
a	$\square$	$\square$
b	$\square$	$\square$

$\cdot$	a	b
a	$\square$	$\square$
b	$\square$	$\square$

	$-$
a	$\square$
b	$\square$

0 is $\square$ 1 is $\square$

3.1.5 How many $\mathcal{L}$ structures can one find on the set $\{a, b\}$ when $\mathcal{L} = \{\vee, \wedge, ', 0, 1\}$, i.e., how many ways can one fill the $\square$'s in the following, using only a's and b's?

$\vee$	a	b
a	$\square$	$\square$
b	$\square$	$\square$

$\wedge$	a	b
a	$\square$	$\square$
b	$\square$	$\square$

	$'$
a	$\square$
b	$\square$

0 is $\square$ 1 is $\square$

3.1.6 How many $\mathcal{L}$ structures can one find on the set $\{a, b, c\}$ when $\mathcal{L} = \{\vee, \wedge, ', 0, 1\}$, i.e., how many ways can one fill the $\square$'s in the following, using only a's and b's and c's?

$\vee$	a	b	c
a	$\square$	$\square$	$\square$
b	$\square$	$\square$	$\square$
c	$\square$	$\square$	$\square$

$\wedge$	a	b	c
a	$\square$	$\square$	$\square$
b	$\square$	$\square$	$\square$
c	$\square$	$\square$	$\square$

	$'$
a	$\square$
b	$\square$
c	$\square$

0 is $\square$ 1 is $\square$

3.2 TERMS

Definition 3.2.1 Given a language $\mathcal{L}$ for algebras, the *$\mathcal{L}$-terms* over X are (inductively) defined by the following:

- A variable x in X is an $\mathcal{L}$–term.
- A constant symbol c in $\mathcal{C}$ is an $\mathcal{L}$–term.
- If $t_1, \ldots, t_n$ are $\mathcal{L}$–terms and f is an n–ary function symbol in $\mathcal{F}$, then $ft_1 \cdots t_n$ is an $\mathcal{L}$–term.

$T(X)$ denotes the set of $\mathcal{L}$–terms.

As with the inductive definition of propositional formulas (on p. 38), the preceding definition is to be understood as initially describing some strings of symbols that are terms and then giving a procedure to make new terms from ones that we already have. It is understood that only the strings that can be achieved in this manner will be called terms.

Example 3.2.2 Thus if f is a unary function symbol in $\mathcal{F}$ and x is a variable, then

$$x, \ fx, \ ffx, \text{ etc.},$$

are $\mathcal{L}$–terms, and if also c is a constant symbol in $\mathcal{C}$, then

$$c, \ fc, \ ffc, \text{ etc.},$$

are $\mathcal{L}$–terms.

If also g is a binary function symbol and $x, y, z \in X$, then

$$gxx, \ gxy, \ ggcfxx, \ gxgfzx, \ gfgxzgcy, \text{ etc.},$$

are $\mathcal{L}$–terms.

If also h is a ternary function symbol in $\mathcal{F}$ and $x, y, z, w \in X$, then

$$hxxx, \ hzcy, \ hhxzyygxc, \ hgxxgcyggxfcx, \text{ etc.},$$

are $\mathcal{L}$–terms.

Example 3.2.3 It can be rather difficult to parse the preceding terms to see how they are built up according to the inductive definition of terms. However, there is a simple numerical algorithm to determine if a string of function symbols, variable symbols, and constant symbols is a term. Given a term, the algorithm can be used to find the subterm starting with a given function symbol in the term.

Suppose we are given a language $\mathcal{L}$ of algebras, and $s = fs_1 \cdots s_n$ is a string with f a function symbol, and each s_i is either a function symbol or a constant symbol belonging to this language, or a variable. We define the integer–valued function γ_s on the numbers 0 to n inductively by

$$\gamma_s(0) = 0.$$

$$\gamma_s(i+1) = \begin{cases} \gamma_s(i)+1 & \text{if } s_{i+1} \text{ is a variable;} \\ \gamma_s(i)+1 & \text{if } s_{i+1} \text{ is a constant symbol;} \\ \gamma_s(i)+1-\text{arity}(s_{i+1}) & \text{if } s_{i+1} \text{ is a function symbol.} \end{cases}$$

For $s = fs_1 \cdots s_n$ and γ_s as in the preceding definition, s is a term iff

a. $\gamma_s(i) < \text{arity}(f)$ *for* $i < n$, *and*
b. $\gamma_s(n) = \text{arity}(f)$.

If s is a term, then s must be of the form $ft_1 \cdots t_k$, where $k =$ $\text{arity}(f)$, for some choice of terms $t_1, \ldots, t_k$. One can easily read off the t_j, since the first i such that $\gamma_s(i) = j$ gives the end of the term t_j.

Let us apply this algorithm to the following situation. Let $\mathcal{L} = \{f, g, h, c\}$ be a language of algebras with f unary, g binary, h ternary and c a constant symbol. We want to determine if $s = hgxxgcyggxfcx$ is a term and if so, to find the three subterms t_1, t_2, t_3 such that $ht_1t_2t_3 = hgxxgcyggxfcx$. The following table shows the calculations involved:

i	0	1	2	3	4	5	6	7	8	9	10	11	12
s_i	h	g	x	x	g	c	y	g	g	x	f	c	x
$\gamma(i)$	0	−1	0	1	0	1	2	1	0	1	1	2	3

Thus conditions (a) and (b) of the theorem are fulfilled, as the arity of h is 3, and the value of $\gamma(i)$ is below 3 until the last step, where it is 3. Then t_1 ends at the fourth symbol of s, because that gives the first place where $\gamma(i)$ is 1, t_2 ends at the seventh symbol of s, as that gives the first place where $\gamma(i)$ is 2, and t_3 ends at the last symbol of s. Thus $t_1 = gxx$, $t_2 = gcy$, $t_3 = ggxfcx$.

Remark 3.2.4 One of the important features of prefix notation, and indeed of any good notation, is the *unique readability* of a term. For example, the term $hgxxgcyggxfcx$ from above is to be parsed as $h\ gxx\ gcy\ ggxfcx$, setting off the three subterms that follow h. The proof that prefix notation has unique readability is given later, in Theorem 3.12.17 on p. 197.

Notational Convention 3.2.5 With our favorite binary function symbols, like $+$ and $\cdot$, we usually prefer infix notation, with conventions for dropping parentheses, e.g., we write $x + y$ instead of $+xy$, and $x \cdot (y + z)$ instead of $\cdot x + yz$.

Let us look at some examples using this familiar notation.

Example 3.2.6 Let $\mathcal{L}$ be the language $\{+, \cdot, -, 0, 1\}$ for the integers. Then the following are examples of terms:

$$\begin{array}{llll} 0 & x & x+y & (x+y)\cdot z \\ 0+(-1) & 1+x & x+(x\cdot y) & -(x+y)+(x\cdot z) \\ 0+(1\cdot 0) & (1+x)\cdot(x+y) & (x\cdot y)+(x\cdot z). & \end{array}$$

Example 3.2.7 Let $\mathcal{L}$ be the language $\mathcal{L}_{BA}$. Then the following are examples of terms:

$$\begin{array}{llll} 0 & x & x\vee y & (x\vee y)\wedge z \\ 0\vee 0 & 1\vee x & x\vee(x\wedge y) & (x\vee y)'\vee(x\wedge z) \\ 0\vee(1\wedge 0) & (1\vee x)\wedge(x\vee y) & (x\wedge y)\vee(x\wedge z). & \end{array}$$

It is convenient to stop referring to $\mathcal{L}$ in any situation where it is fixed and just to say "terms". The rules for forming terms are remarkably similar to the rules used for forming propositional formulas. Indeed, if we think of $\neg$ as a unary function symbol, and if we think of the binary connectives as binary function symbols, then indeed *the propositional formulas are just terms in this language*. Now we inductively define $\mathsf{Tree}(t)$ for terms:

$\mathsf{Tree}(x)$ is x

$\mathsf{Tree}(c)$ is c

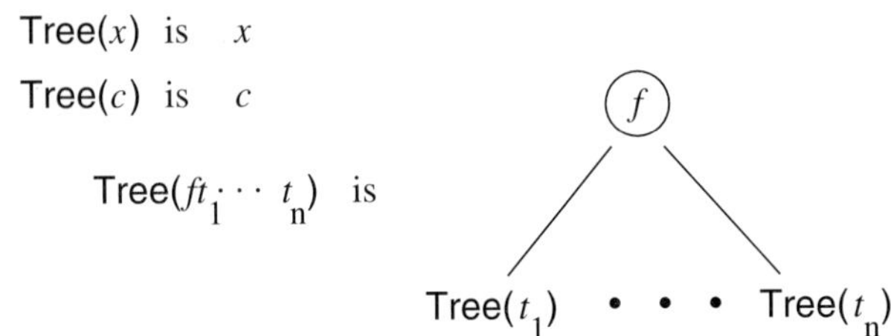

Figure 3.2.8 Inductive definition of $\mathsf{Tree}(t)$

Then, for example, $\mathsf{Tree}(fxgxyhx)$ is (with f binary, g ternary, h unary):

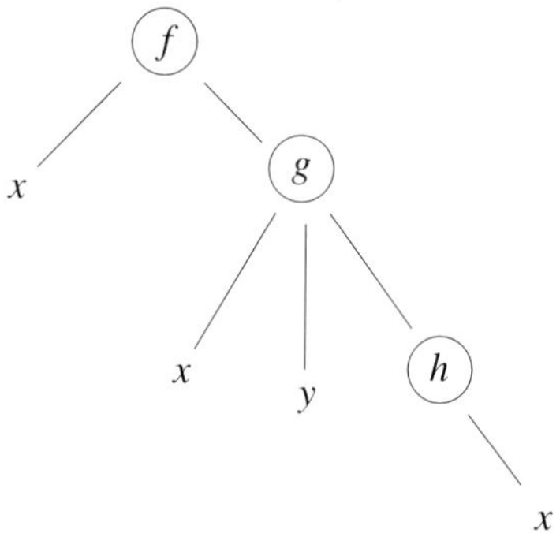

Figure 3.2.9 Example of a term tree

Just as propositional formulas have subformulas, terms have subterms.

Definition 3.2.10 The *subterms* of a term t are defined inductively by the following:

- The only subterm of a variable x is the variable x itself.
- The only subterm of a constant symbol c is the symbol c itself.
- The subterms of $ft_1 \cdots t_n$ are $ft_1 \cdots t_n$ itself and all the subterms of the t_i, for $1 \leq i \leq n$.

Example 3.2.11 If f is binary, g ternary, and h unary, then the subterms of $fxgxyhx$ are $fxgxyhx$, $gxyhx$, hx, x, y.

In our more familiar notation we have the following:

Example 3.2.12 The subterms of

$$(1 + (x \cdot (-y))) \cdot ((x \cdot x) + (y \cdot y))$$

are given by

$$\begin{array}{l} (1 + (x \cdot (-y))) \cdot ((x \cdot x) + (y \cdot y)) \\ 1 + (x \cdot (-y)) \\ 1 \\ x \cdot (-y) \\ x \\ -y \\ y \end{array}$$

$$(x \cdot x) + (y \cdot y)$$
$$x \cdot x$$
$$y \cdot y.$$

Example 3.2.13 The subterms of

$$(1 \vee (x \wedge y')) \wedge ((x \wedge x) \vee (y \wedge y))$$

are given by

$$(1 \vee (x \wedge y')) \wedge ((x \wedge x) \vee (y \wedge y))$$
$$1 \vee (x \wedge y')$$
$$1$$
$$x \wedge y'$$
$$x$$
$$y'$$
$$y$$
$$(x \wedge x) \vee (y \wedge y)$$
$$x \wedge x$$
$$y \wedge y.$$

Notational Convention 3.2.14 If t is a term, then we use the notation $t(x_1, \dots, x_n)$ to indicate that the variables occurring in t are *among* $x_1, \dots, x_n$.

Exercises

3.2.1 Draw the trees of the following terms, and find all subterms.

a. $(x + (y \cdot y)) \cdot (y + (x \cdot x))$
b. $1 + (x \cdot ((x + z) + ((-y) \cdot y)))$
c. $(((x \cdot x) + (y \cdot y)) + (z \cdot z)) + (w \cdot w)$

3.2.2 Draw the trees of the following terms, and find all subterms.

a. $(x \vee (y \wedge y)') \wedge (y' \vee (x \wedge x))$
b. $1' \vee (x \wedge ((x \vee z)' \vee (y' \wedge y)))$
c. $x \vee (y \wedge (z \vee (w \wedge x)')')'$

3.2.3 In the following, f_n is an n–ary function symbol, for $n = 1, 2, 3$.

Determine which of the following are terms. Draw the tree of any terms, and find all subterms.

a. $f_1 f_1 f_1 v$
b. $f_1 f_1 f_2 v y$

c. $f_3uf_1f_2yvyz$
d. $f_3f_2f_1f_2ywzyf_2xf_3uuw$
e. $f_2f_3wyf_3wf_3f_3f_2f_2yyuuxzzz$
f. $f_2f_3wvf_3f_1zf_2zuyf_3xwf_3f_2f_1f_1ywxf_2yu$

3.3 TERM FUNCTIONS

Now we focus our attention on the *meanings* of terms in a given algebra $\mathbf{A}$, namely,

terms $t(x_1, \ldots, x_n)$ define *functions* $t^{\mathbf{A}} : A^n \to A$

Given an $\mathcal{L}$–algebra $\mathbf{A}$ we can associate with each term $t(x_1, \cdots, x_n)$ an n–ary function $t^{\mathbf{A}} : A^n \to A$ as follows.

Definition 3.3.1 The *term functions* $t^{\mathbf{A}}$ on $\mathbf{A}$ are defined inductively by the following:

- If $t(x_1, \ldots, x_n)$ is x_i then $t^{\mathbf{A}}(a_1, \ldots, a_n) = a_i$.
- If $t(x_1, \ldots, x_n)$ is $c \in \mathcal{C}$ then $t^{\mathbf{A}}(a_1, \ldots, a_n) = c^{\mathbf{A}}$.
- If $t = ft_1 \cdots t_k$ then $t^{\mathbf{A}} = f^{\mathbf{A}}(t_1^{\mathbf{A}}, \ldots, t_k^{\mathbf{A}})$.

Example 3.3.2 Using the usual language for the natural numbers, we have a term $t(x, y, z)$ given by $x \cdot (y + (x + 1) \cdot z)$. The corresponding term function $t^{\mathbf{N}} : N^3 \to N$ maps the triple $(1, 3, 2)$ to 7 as $t^{\mathbf{N}}(1, 3, 2) = 1 \cdot (3 + (1 + 1) \cdot 2) = 7$.

3.3.1 Evaluation Tables

When examining term functions on small algebras it is very handy to have *evaluation tables*. They correspond to the truth tables we used extensively in Chapter 2.

Example 3.3.3 Let $\mathbf{A}$ be the two–element Boolean algebra as defined in Example 3.1.12 (p. 138), and let $t(x, y, z)$ be the term $x \wedge (y \vee z')$. Then the evaluation table for the term t is as follows, where $\boxed{t}$ locates the term t:

x	y	z	t
1	1	1	1
1	1	0	1
1	0	1	0
1	0	0	1
0	1	1	0
0	1	0	0
0	0	1	0
0	0	0	0

or

x	y	z	z'	$y \vee z'$	$\boxed{t}$ $x \wedge (y \vee z')$
1	1	1	0	1	1
1	1	0	1	1	1
1	0	1	0	0	0
1	0	0	1	1	1
0	1	1	0	1	0
0	1	0	1	1	0
0	0	1	0	0	0
0	0	0	1	1	0

Example 3.3.4 Let $\mathbf{A} = \{A, f\}$ be the algebra on $A = \{a, b\}$, where f is the ternary function defined by

x	y	z	f
a	a	a	a
a	a	b	a
a	b	a	b
a	b	b	b
b	a	a	b
b	a	b	a
b	b	a	a
b	b	b	a

Then the evaluation table for the term $t(x, y) = f(x, f(x, y, x), y)$ is given by the following:

x	y	$f(x, y, x)$	$\boxed{t}$ $f(x, f(x, y, x), y)$
a	a	a	a
a	b	b	b
b	a	a	b
b	b	a	a

Exercises

3.3.1 Let $\mathbf{A} = \{A, \vee, \wedge, ', 0, 1\}$ be the two–element Boolean algebra defined in Example 3.1.12 (p. 138). Give evaluation tables for the following terms:

a. $t(x, y) = (x \wedge y) \vee (x' \wedge y')$
b. $t(x, y, z) = (x \wedge y) \vee (y \wedge z) \vee (x \wedge z)$

c. $t(x, y, z) = ((x \wedge y) \wedge z') \vee ((x' \wedge y') \wedge z)$
d. $t(x, y, z) = ((x \vee y) \vee z') \wedge ((x' \vee y') \vee z)$
e. $t_0(x, y) = (x \wedge y') \vee (x' \wedge y)$
f. $t(x, y, z) = t_0(t_0(x, y), z)$
g. $t(x, y, z) = t_0(x, t_0(y, z))$.

3.3.2 Let $\mathbf{A} = (A, +, \cdot, -, 0, 1)$ be the algebra obtained by interpreting the language of the integers on the two–element set $A = \{a, b\}$ as follows:

$+$	a	b
a	a	b
b	b	a

$\cdot$	a	b
a	a	a
b	a	b

	$-$
a	b
b	a

0 is a 1 is b

Find evaluation tables for the following terms:

a. $t(x) = 1 + x$
b. $t(x, y) = (x + y) + (x \cdot y)$
c. $t(x, y, z) = x + (y + z)$
d. $t(x, y, z) = (x + y) + z$
e. $t(x, y, z) = (x \cdot y) + ((x \cdot z) + (y \cdot z))$.

3.3.3 Assume our language has a ternary function symbol f, and we have a three–element algebra on $\{0, 1, 2\}$, where f is interpreted as $f(a, b, c) = a - b + c$ modulo 3.

Find the values of the term functions for the following terms, evaluated at the indicated arguments in this algebra. (You need to find only one line of the evaluation table.)

a. $t(x) = ffxxxfxxx$ at 1
b. $t(x, y) = fxfxyxy$ at $(0, 1)$
c. $t(x, y, z) = fxfyzfxyzz$ at $(0, 1, 0)$
d. $t(x, y, z, u) = ffyzfuxyzfuxu$ at $(0, 1, 0, 1)$

3.3.4 Assume our language has a ternary function symbol f, and we have a two–element algebra on $\{0, 1\}$, where f is interpreted as follows:

x	y	z	f
1	1	1	0
1	1	0	0
1	0	1	1
1	0	0	1
0	1	1	0
0	1	0	0
0	0	1	0
0	0	0	1

Find the values of the term functions for the following terms, evaluated at the indicated arguments in this algebra. (You need to find only one line of the evaluation table.)

a. $t(x) = ffxxxxfxxx$ at 1
b. $t(x,y) = fxfxyxy$ at $(0,1)$
c. $t(x,y,z) = fxfyzfxyzz$ at $(0,1,0)$
d. $t(x,y,z,u) = ffyzfuxyzfuxu$ at $(0,1,0,1)$

3.3.5 Assume our language has a unary function symbol f, a binary function symbol g, and a ternary function symbol h, and we have a two–element algebra on $\{a,b\}$, where f,g,h are interpreted as follows:

x	f
a	b
b	a

x	y	g
a	a	a
a	b	b
b	a	b
b	b	b

x	y	z	h
a	a	a	a
a	a	b	a
a	b	a	b
a	b	b	b
b	a	a	a
b	a	b	a
b	b	a	a
b	b	b	b

Find the values of the term functions for the following terms, evaluated at the indicated arguments in this algebra. (You need to find only one line of the evaluation table.)

a. $t(x) = hhxxfxxhxgxxx$ at b
b. $t(x,y) = hfgxyhxygyxy$ at (a,b)
c. $t(x,y,z) = fhfxhgxzzhxygxyz$ at (a,b,a)
d. $t(x,y,z,u) = hfhfygzyhugzuyzhguyxu$ at (a,b,a,b)

3.4 EQUATIONS

Definition 3.4.1 An $\mathcal{L}$-*equation* is an expression of the form $s \approx t$, where s and t are $\mathcal{L}$–terms.

As mentioned in the preface, we will be using two symbols for equality, starting with this section. The usual symbol $=$ is to be understood simply as "identical to," whereas $\approx$ is used in the formulas of our logics to express equality.

Our basic objective is to gain some understanding of when one equation follows from other equations, e.g., if we assume $x + y \approx y + x$, then we can conclude $x + (y + z) \approx (y + z) + x$ but not $x+x \approx y+y$. We want to look at correct arguments about equations from two vantage points:

- (*semantic*) a notion of when an equation is true in an algebra;
- (*syntactic*) axioms and rules of inference for deriving equational consequences of equations.

Once we develop these components for equational logic we will show that they are equivalent, i.e., we will prove the soundness and completeness of the system. Let us start with the semantic side of equational logic.

3.4.1 The Semantics of Equations

Definition 3.4.2 Suppose $\mathcal{L}$ is a language of algebras.

- For $\mathbf{A}$ an $\mathcal{L}$–structure and $s \approx t$ an $\mathcal{L}$–equation we write
$$\mathbf{A} \models s \approx t$$
(read: $\mathbf{A}$ *satisfies* $s \approx t$ or $s \approx t$ *holds* in $\mathbf{A}$) if $s^{\mathbf{A}} = t^{\mathbf{A}}$, i.e., s and t define the same term function on A.
- If $\mathbf{A}$ is an $\mathcal{L}$–structure and $\mathcal{S}$ is a set of $\mathcal{L}$–equations, then
$$\mathbf{A} \models \mathcal{S}$$
(read: $\mathbf{A}$ *satisfies* $\mathcal{S}$ or $\mathcal{S}$ *holds* in $\mathbf{A}$) if $\mathbf{A} \models s \approx t$ for *every* equation $s \approx t$ in $\mathcal{S}$.
- If $\mathcal{S}$ is a set of $\mathcal{L}$–equations and $s \approx t$ is an $\mathcal{L}$–equation, then we write
$$\mathcal{S} \models s \approx t$$
(read: $s \approx t$ is a *consequence* of $\mathcal{S}$ or *follows from* $\mathcal{S}$) if $\mathbf{A} \models s \approx t$ whenever $\mathbf{A} \models \mathcal{S}$.

For *satisfies* we also use the word *models*, and we say **A** is a *model* of an equation or a set of equations.

Definition 3.4.3 A binary algebra **A** is *associative* if it satisfies the associative law $x \cdot (y \cdot z) \approx (x \cdot y) \cdot z$. In such case it is also called a *semigroup*. It is *commutative* if it satisfies the commutative law $x \cdot y \approx y \cdot x$. And it is *idempotent* if it satisfies the idempotent law $x \cdot x \approx x$.

The idempotent law is easy to check, since one looks down the main diagonal of the table of $\cdot$ to see that $x \cdot x$ always has the value x.

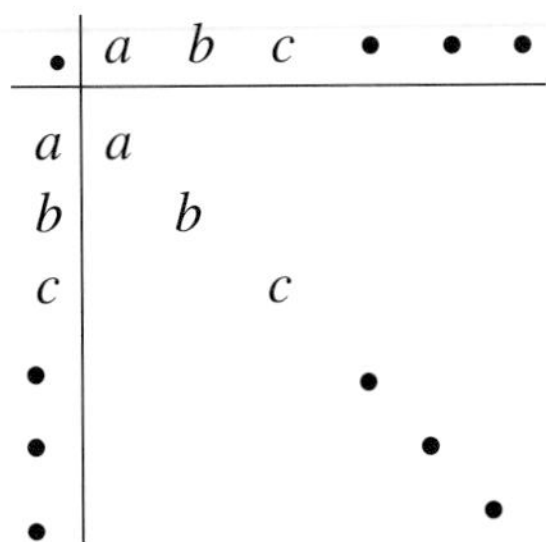

Figure 3.4.4 Checking the idempotent law

The commutative law is also easy to check, as one looks at the table for $\cdot$ to see if it is *symmetric* about the main diagonal.

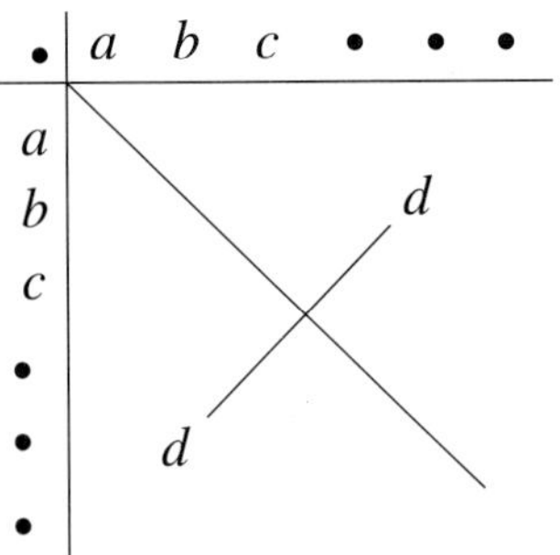

Figure 3.4.5 Checking the commutative law

Let us show how one would check associativity using an evaluation table.

Example 3.4.6 The binary algebra $\mathbf{A}$ given by

$\cdot$	a	b
a	a	b
b	b	b

is idempotent, commutative, and associative, i.e., we have

$$\mathbf{A} \models x \cdot x \approx x$$
$$\mathbf{A} \models x \cdot y \approx y \cdot x$$
$$\mathbf{A} \models x \cdot (y \cdot z) \approx (x \cdot y) \cdot z.$$

The first two properties follow by the preceding remarks and diagrams.

To check the associative law we construct the following evaluation table for the terms $s(x, y, z) = x \cdot (y \cdot z)$ and $t(x, y, z) = (x \cdot y) \cdot z$ and show that the term functions are equal, i.e., $s^{\mathbf{A}} = t^{\mathbf{A}}$:

					$\boxed{s}$		$\boxed{t}$
	x	y	z	$y \cdot z$	$x \cdot (y \cdot z)$	$x \cdot y$	$(x \cdot y) \cdot z$
1.	a	a	a	a	a	a	a
2.	a	a	b	b	b	a	b
3.	a	b	a	b	b	b	b
4.	a	b	b	b	b	b	b
5.	b	a	a	a	b	b	b
6.	b	a	b	b	b	b	b
7.	b	b	a	b	b	b	b
8.	b	b	b	b	b	b	b

Since the columns for $x \cdot (y \cdot z)$ and $(x \cdot y) \cdot z$ are the same, the associative law holds.

Actually, we could argue that the associative law holds without making the evaluation table for the associative law, in this example. Namely, we see that if we evaluate a term involving x, y, z at any row where one of x, y, z has the value b, then the value of the term will be b. Otherwise the value is a.

Example 3.4.7 The binary algebra $\mathbf{A}$ given by

$\cdot$	a	b
a	b	b
b	a	b

is not idempotent nor commutative nor associative. The failure of the idempotent law is clear as $a \cdot a = b$. The failure of the commutative law also is easy to spot, as the table is not symmetric. In particular, $a \cdot b \neq b \cdot a$.

The failure of the associative law is not something easily seen in the table of this basic operation. So let us use an evaluation table:

					$\boxed{s}$		$\boxed{t}$
	x	y	z	$y \cdot z$	$x \cdot (y \cdot z)$	$x \cdot y$	$(x \cdot y) \cdot z$
1.	a	a	a	b	b	b	a
2.	a	a	b	b	b	b	b
3.	a	b	a	a	b	b	a
4.	a	b	b	b	b	b	b
5.	b	a	a	b	b	a	b
6.	b	a	b	b	b	a	b
7.	b	b	a	a	a	b	a
8.	b	b	b	b	b	b	b

Indeed, associativity fails precisely on lines 1 and 3.

However, this binary algebra does satisfy an interesting equation, namely,

$$\mathbf{A} \models x \cdot (y \cdot z) \approx y \cdot (x \cdot z).$$

We can see this by finding the term functions for $s(x, y, z) = x \cdot (y \cdot z)$ and $t(x, y, z) = y \cdot (x \cdot z)$ and showing that they are equal, using an evaluation table:

				$\boxed{s}$		$\boxed{t}$
x	y	z	$y \cdot z$	$x \cdot (y \cdot z)$	$x \cdot z$	$y \cdot (x \cdot z)$
a	a	a	b	b	b	b
a	a	b	b	b	b	b
a	b	a	a	b	b	b
a	b	b	b	b	b	b
b	a	a	b	b	a	b
b	a	b	b	b	b	b
b	b	a	a	a	a	a
b	b	b	b	b	b	b

Example 3.4.8 Now let us suppose our language has two unary function symbols, f and g. What can we say about f and g if the algebra $\mathbf{A}$ satisfies $fgx \approx x$?

If we choose any elements a, b from A, then letting $c = g(a)$ and $d = g(b)$, we must have $f(c) = a$ and $f(d) = b$. Thus if $a \neq b$, then we must have $c \neq d$. But this just says that the function g maps A *one–to–one* to A, i.e., $a \neq b$ implies $g(a) \neq g(b)$.

Also, if we choose any a from A, it is clear that by letting $b = g(a)$ we have $f(b) = a$. This just says that f maps A *onto* A.

Thus our conclusion is: The equation $fgx \approx x$ implies that "f is onto" and that "g is one–to–one."

Example 3.4.9 [WILKIE'S IDENTITY] In 1980 Wilkie created a bit of a stir when he found that the following equation was true of the positive integers P with the usual operations $+, \cdot, \uparrow$ and 1 ($\uparrow$ is exponentiation) but was not a consequence of the equations (involving just the operations $+, \cdot, \uparrow$ and 1) that one learns in high school:

$$\begin{aligned} & \left((1+x)^y + (1+x+x^2)^y\right)^x \cdot \left((1+x^3)^x + (1+x^2+x^4)^x\right)^y \\ \approx\ & \left((1+x)^x + (1+x+x^2)^x\right)^y \cdot \left((1+x^3)^y + (1+x^2+x^4)^y\right)^x. \end{aligned} \tag{25}$$

To see that this equation is actually true on the positive integers we need only observe that we can factor some of the polynomials:

$$\begin{aligned} 1+x^3 &\approx (1+x)\cdot(1-x+x^2) \\ 1+x^2+x^4 &\approx (1+x+x^2)\cdot(1-x+x^2). \end{aligned}$$

So if we let

$$\begin{aligned} A &= 1+x \\ B &= 1+x+x^2 \\ C &= 1-x+x^2, \end{aligned}$$

then (25) becomes

$$\begin{aligned} & (A^y + B^y)^x \cdot \left((A\cdot C)^x + (B\cdot C)^x\right)^y \\ \approx\ & (A^x + B^x)^y \cdot \left((A\cdot C)^y + (B\cdot C)^y\right)^x, \end{aligned} \tag{26}$$

and this is true in the positive integers.

Note that we proved Wilkie's identity by using elementary facts, but we had to bring in the operation "$-$" to do the job. It is not so easy to show that Wilkie's identity cannot be derived from the equations (involving only addition, multiplication, exponentiation and 1) that one learns in high school (see [**6**]).

3.4.2 Classes of Algebras Defined by Equations

Some of the most interesting classes of algebras in mathematics are defined by equations. We will introduce some of them in this section, mainly because we want to use the equations presented here, and the examples of algebras, in the examples and exercises in later sections. So treat this section as a repository of basic examples for our work.

No previous study of abstract algebra is assumed as background for this book. In this section we will give equational axioms for some of the most popular classes of algebras, namely, rings, semigroups, monoids, groups, and Boolean algebras. We will also give a few basic examples of algebras from each class. The examples of algebras given in this section are the only ones (from these classes) that are needed for this book.

Example 3.4.10 [RINGS] $\mathcal{R}$ is the following set of equations (in the language $\mathcal{L}_{\mathcal{R}} = \{+, \cdot, -, 0, 1\}$, where $+, \cdot$ are binary, $-$ is unary, and $0, 1$ are constants) defining *rings:*

R1.	$x + 0$	$\approx$	x	additive identity
R2.	$x + (-x)$	$\approx$	0	additive inverse
R3.	$x + y$	$\approx$	$y + x$	$+$ is commutative
R4.	$x + (y + z)$	$\approx$	$(x + y) + z$	$+$ is associative
R5.	$x \cdot 1$	$\approx$	x	right multiplicative identity
R6.	$1 \cdot x$	$\approx$	x	left multiplicative identity
R7.	$x \cdot (y \cdot z)$	$\approx$	$(x \cdot y) \cdot z$	$\cdot$ is associative
R8.	$x \cdot (y + z)$	$\approx$	$(x \cdot y) + (x \cdot z)$	left distributive
R9.	$(x + y) \cdot z$	$\approx$	$(x \cdot z) + (y \cdot z)$	right distributive.

All algebras $\mathbf{R} = (R, +, \cdot, -, 0, 1)$ that satisfy $\mathcal{R}$ are called *rings*. $\mathcal{R}$ is called a *set of axioms* or a *set of defining equations* for rings. The equations defining rings are obviously what we encounter with our usual number systems.

The most obvious examples of rings come from number systems and matrices.

a. (*Infinite number systems*)
$(Z, +, \cdot, -, 0, 1)$, $(Q, +, \cdot, -, 0, 1)$, $(R, +, \cdot, -, 0, 1)$, $(C, +, \cdot, -, 0, 1)$;

b. (*Finite number systems*)
$(Z_n, +, \cdot, -, 0, 1)$, the integers modulo n;

c. (*Other number systems*)
$(Z[i], +, \cdot, -, 0, 1)$, where $Z[i] = \{m + in : m, n \in Z\}$, $i = \sqrt{-1}$
$(Z[\sqrt{2}], +, \cdot, -, 0, 1)$, where $Z[\sqrt{2}] = \{m + \sqrt{2}n : m, n \in Z\}$;

d. $(M_{n \times n}(R), +, \cdot, -, O, I)$, the $n \times n$ matrices with real entries.

Example 3.4.11 [BOOLEAN ALGEBRAS] $\mathcal{BA}$ is a set of equations (in the language $\mathcal{L}_{\mathcal{BA}}$ from Example 3.0.1, p. 133) that defines

Boolean algebras, namely, we choose $\mathcal{BA}$ to be

B1.	$x \vee y$	$\approx$	$y \vee x$	commutative
B2.	$x \wedge y$	$\approx$	$y \wedge x$	commutative
B3.	$x \vee (y \vee z)$	$\approx$	$(x \vee y) \vee z$	associative
B4.	$x \wedge (y \wedge z)$	$\approx$	$(x \wedge y) \wedge z$	associative
B5.	$x \wedge (x \vee y)$	$\approx$	x	absorption
B6.	$x \vee (x \wedge y)$	$\approx$	x	absorption
B7.	$x \wedge (y \vee z)$	$\approx$	$(x \wedge y) \vee (x \wedge z)$	distributive
B8.	$x \vee x'$	$\approx$	1	
B9.	$x \wedge x'$	$\approx$	0	
B10.	$x \vee 1$	$\approx$	1	
B11.	$x \wedge 0$	$\approx$	$0.$	

Note that we have made a selection of the equations from the list of equations for the logic of classes that appears on p. 12. Derivations for some of the other equations are given in Exercise 3.8.8 on p. 181.

All algebras $\mathbf{B} = (B, \vee, \wedge, ', 0, 1)$ that satisfy $\mathcal{BA}$ are called *Boolean algebras*. $\mathcal{BA}$ is called a *set of axioms* or a *set of defining equations* for Boolean algebra.

One of the interesting theorems about finite Boolean algebras is that they appear precisely in sizes that are a power of 2, i.e., they can appear in sizes 1,2,4,8, etc., but a Boolean algebra cannot have 3 elements, or 12 elements, etc.

Here are tables for a two–element Boolean algebra $\mathbf{B}_2$ (that we saw in Example 3.1.12, p. 138), and a four–element Boolean algebra $\mathbf{B}_4$:

$\mathbf{B}_2$:

$\vee$	0	1
0	0	1
1	1	1

$\wedge$	0	1
0	0	0
1	0	1

	$'$
0	1
1	0

$\mathbf{B}_4$:

$\vee$	0	1	a	b
0	0	1	a	b
1	1	1	1	1
a	a	1	a	1
b	b	1	1	b

$\wedge$	0	1	a	b
0	0	0	0	0
1	0	1	a	b
a	0	a	a	0
b	0	b	0	b

	$'$
0	1
1	0
a	b
b	a

We also have the Boolean algebra $\mathbf{Su}(U) = (Su(U), \cup, \cap, ', \emptyset, U)$, as described in Example 3.1.11 on p. 137. Note that if U is a finite set with n elements, then $\mathbf{Su}(U)$ is a finite Boolean algebra with 2^n elements.

Example 3.4.12 [SEMIGROUPS] $\mathcal{SG}$ is the following set consisting of one equation (in the language $\mathcal{L}_{\mathcal{SG}} = \{\cdot\}$ consisting of a single binary operation) that defines *semigroups*, namely,

$$\mathcal{SG} = \{(x \cdot y) \cdot z \approx x \cdot (y \cdot z)\}.$$

Models of $\mathcal{SG}$ are called *semigroups*, and $\mathcal{SG}$ *axiomatizes* or *defines* the class of semigroups. Semigroups come up in computer science in the study of languages, or in any subject that is based on the study of *strings* of symbols.

First there are the obvious semigroups from the classical number systems with addition or multiplication, e.g., $(N, +)$, the nonnegative integers with addition, $(N, \cdot)$, the nonnegative integers with multiplication, $(P, +)$, the positive integers with addition, $(P, \cdot)$, the positive integers with multiplication, $(Z, +)$, the integers with addition, $(Z, \cdot)$, the integers with multiplication, $(2Z, +)$, the even integers with addition, $(2Z, \cdot)$, the even integers with multiplication, and so forth.

Also there are a huge number of semigroups of small size. On a two–element set $\{a, b\}$ we find that 8 of the 16 possible tables are semigroups. Perhaps you will recognize some of these from the study of truth tables (i.e., $\wedge$ is associative, etc.):

$\cdot$	a	b
a	a	a
b	a	a

$\cdot$	a	b
a	b	b
b	b	b

$\cdot$	a	b
a	a	b
b	a	b

$\cdot$	a	b
a	a	a
b	b	b

$\cdot$	a	b
a	a	b
b	b	a

$\cdot$	a	b
a	b	a
b	a	b

$\cdot$	a	b
a	a	b
b	b	b

$\cdot$	a	b
a	a	a
b	a	b

For X a set let $X^\star$ denote the set of *strings* on X, including the *empty* string. (X^+ is the usual notation for the set of *nonempty strings* on X.) X is called an *alphabet*. Thus if $a, b, c \in X$, then we have the string *abbaca* in $X^\star$. We define multiplication on the strings to be *concatenation*, i.e., the product of *abba* and *cab* is *abbacab*. This gives a semigroup $(X^\star, \cdot)$.

Example 3.4.13 [MONOIDS] $\mathcal{M}$ is the following set of three equations (in the language $\mathcal{L}_{\mathcal{M}} = \{\cdot, 1\}$, where $\cdot$ is binary and 1 is a constant symbol) that defines *monoids*, namely,

$$\mathcal{M} = \{(x \cdot y) \cdot z \approx x \cdot (y \cdot z), \quad x \cdot 1 \approx x, \quad 1 \cdot x \approx x\}.$$

Any algebra $\mathbf{A}$ that satisfies $\mathcal{M}$ is called a *monoid*, and $\mathcal{M}$ is a set of *axioms* or *defining equations* for monoids.

Monoids are very close to semigroups. Every monoid can be viewed as a semigroup (just drop the constant symbol), and every semigroup with an element that behaves like a multiplicative identity can be viewed as a monoid by interpreting 1 to be such an element. If a semigroup does not have an element that behaves like 1, then one can simply attach a new element 1 and define $1 \cdot a$ and $a \cdot 1$ to be a for every a in the semigroup, and presto, one has a monoid.

The number of two–element monoids is just 2 once we specify the element 1, namely, there are only the following possibilities on the set $\{1, b\}$:

$\cdot$	1	b
1	1	b
b	b	b

and

$\cdot$	1	b
1	1	b
b	b	1

When we look at the set of strings $X^\star$ on the alphabet X, we have a monoid by interpreting 1 as the empty string, and the multiplication symbol as concatenation.

Example 3.4.14 [GROUPS] $\mathcal{G}$ is a set of equations (in the language $\mathcal{L}_{\mathcal{G}} = \{\cdot, ^{-1}, 1\}$, where $\cdot$ is binary, $^{-1}$ is unary, and 1 is a constant symbol) defining *groups*, namely, we choose

$$\begin{array}{lrcl} \text{G1}: & x \cdot 1 & \approx & x \\ \text{G2}: & x \cdot x^{-1} & \approx & 1 \\ \text{G3}: & (x \cdot y) \cdot z & \approx & x \cdot (y \cdot z). \end{array}$$

G1 says we have a *right* multiplicative identity element, G2 says every element has a *right* inverse, and G3 is of course the associative law.

There is also an *additive* notation for groups, namely, the language is $\{+, -, 0\}$ and is usually reserved for groups that are *commutative*, i.e., when the equation $x + y \approx y + x$ holds. In this language the axioms for groups become:

$$\begin{array}{lrcl} \text{G1}': & x + 0 & \approx & x \\ \text{G2}': & x + (-x) & \approx & 0 \\ \text{G3}': & (x + y) + z & \approx & x + (y + z). \end{array}$$

Here are some standard examples of groups, first looking at ones that come from numbers and matrices, and then at some of the small finite groups. The notation $A \setminus B$ means the set of elements in A but not in B.

a. (*Infinite number systems under addition*)
$(Z, +, -, 0)$, $(Q, +, -, 0)$, $(R, +, -, 0)$, $(C, +, -, 0)$;

b. (*Finite number systems under addition*)
$(Z_n, +, -, 0)$, the integers modulo n;

c. (*Numbers under multiplication*)
$(Q \setminus \{0\}, \cdot, {}^{-1}, 1)$, $(R \setminus \{0\}, \cdot, {}^{-1}, 1)$, $(C \setminus \{0\}, \cdot, {}^{-1}, 1)$;

d. (*Finite number systems under multiplication*)
$(Z_p \setminus \{0\}, \cdot, {}^{-1}, 1)$, the nonzero integers modulo a prime p;

e. $(M_{m\times n}(R), +, -, O)$, the $m \times n$ matrices with addition;

f. $(GL_n(R), \cdot, {}^{-1}, I)$, the invertible $n \times n$ matrices with multiplication.

g. Here are some examples of small finite groups, where we have selected the interpretation of the constant. The commutative examples are given in both multiplicative and additive notation:

$\mathbf{G}_{2\times}$

$\cdot$	1	b
1	1	b
b	b	1

	$^{-1}$
1	1
b	b

$\mathbf{G}_{2+}$

$+$	0	b
0	0	b
b	b	0

	$-$
0	0
b	b

$\mathbf{G}_{3\times}$

$\cdot$	1	b	c
1	1	b	c
b	b	c	1
c	c	1	b

	$^{-1}$
1	1
b	c
c	b

$\mathbf{G}_{3+}$

$+$	0	b	c
0	0	b	c
b	b	c	0
c	c	0	b

	$-$
0	0
b	c
c	b

$\mathbf{G}_{6\times}$

$\cdot$	1	b	c	d	e	f
1	1	b	c	d	e	f
b	b	c	1	e	f	d
c	c	1	b	f	d	e
d	d	f	e	1	c	b
e	e	d	f	b	1	c
f	f	e	d	c	b	1

	$^{-1}$
1	1
b	c
c	b
d	d
e	e
f	f

3.4.3 Three Very Basic Properties of Equations

Our first observation regarding $\approx$ is that it behaves like an equivalence relation.

Lemma 3.4.15 Let $\mathbf{A}$ be an $\mathcal{L}$–structure. Then for any $\mathcal{L}$–terms s, s_1, s_2, s_3 we have

- $\mathbf{A} \models s \approx s$
- $\mathbf{A} \models s_1 \approx s_2$ implies $\mathbf{A} \models s_2 \approx s_1$
- $\mathbf{A} \models s_1 \approx s_2$ and $\mathbf{A} \models s_2 \approx s_3$ implies $\mathbf{A} \models s_1 \approx s_3$.

Proof. Use the fact that $\mathbf{A} \models s \approx t$ iff $s^{\mathbf{A}} = t^{\mathbf{A}}$, and that ordinary equality, $=$, is reflexive, symmetric, and transitive. ☐

Theorem 3.4.16 Let $\mathcal{S}$ be a set of $\mathcal{L}$–equations. Then for any $\mathcal{L}$–terms s, s_1, s_2, s_3 we have

- $\mathcal{S} \models s \approx s$
- $\mathcal{S} \models s_1 \approx s_2$ implies $\mathcal{S} \models s_2 \approx s_1$
- $\mathcal{S} \models s_1 \approx s_2$ and $\mathcal{S} \models s_2 \approx s_3$ implies $\mathcal{S} \models s_1 \approx s_3$.

Proof. This follows from the definition of $\mathcal{S} \models s \approx t$ and the previous lemma. ☐

Exercises

3.4.1 Determine if the following equations are true in the monounary algebra (A, f) described by the table

	f
0	1
1	2
2	2

a. $fx \approx ffx$
b. $ffx \approx fffx$
c. $ffx \approx ffy$

3.4.2 Determine if the following equations are true in the biunary algebra (A, f, g) described by the tables:

	f
0	1
1	2
2	2

	g
0	1
1	2
2	0

a. $fgx \approx gfx$
b. $gffx \approx gfffx$
c. $fgfgx \approx fgx$
d. $gfgx \approx gfgy$

3.4.3 Determine if the following equations are true in the algebra $(A, +, \cdot, -, 0, 1)$ on $A = \{a, b\}$ as defined by the following tables:

$+$	a	b
a	a	b
b	b	a

$\cdot$	a	b
a	a	a
b	a	b

	$-$
a	b
b	a

0 is a 1 is b

a. $x + x \approx 0$
b. $x \cdot x \approx 1$
c. $x + y \approx y + x$
d. $x \cdot y \approx y \cdot x$
e. $x + (y + z) \approx (x + y) + z$
f. $x \cdot (y \cdot z) \approx (x \cdot y) \cdot z$

3.4.4 Given the binary algebra

$\cdot$	a	b
a	b	b
b	a	b

determine if the following equations hold:

a. $x \cdot x \approx x \cdot (y \cdot x)$
b. $y \cdot x \approx x \cdot (y \cdot x)$
c. $(x \cdot y) \cdot (x \cdot y) \approx (x \cdot y)$
d. $x \cdot (y \cdot z) \approx (x \cdot y) \cdot (x \cdot z)$
e. $x \cdot (y \cdot z) \approx z \cdot (y \cdot x)$
f. $x \cdot (y \cdot x) \approx y \cdot (x \cdot y)$

3.4.5 Determine if the following equations are true in the binary algebra $(A, \cdot)$ described by the table

$\cdot$	0	1	2
0	0	0	0
1	0	1	1
2	0	1	2

a. $x \cdot x \approx x$
b. $x \cdot y \approx y \cdot x$
c. $x \cdot (y \cdot z) \approx (x \cdot y) \cdot z$
d. $x \cdot (y \cdot z) \approx x \cdot (y \cdot w)$

3.4.6 (The language is $\{f, g, 0\}$, where f, g are unary and 0 is a constant.) Does

$$\{fgx \approx x,\ gfx \approx 0\}$$

have a nontrivial model? [A *nontrivial model* is a model on a set having *more* than one element. Note that a one–element algebra satisfies *every* identity!]

3.4.7 (The language is $\{f, g, h, k, c\}$, where f, g, h, k are unary and c is a constant.) Does $\{gfx \approx x,\ khx \approx x,\ ghx \approx c\}$ have a nontrivial model?

3.4.8 Show that the natural numbers satisfy Wilkie's identity by verifying that (26) holds.

3.5 VALID ARGUMENTS

Definition 3.5.1 Given an algebra **A**, an argument

$$s_1 \approx t_1, \cdots, s_n \approx t_n \qquad \therefore s \approx t,$$

also written

$$\begin{array}{l} s_1 \approx t_1 \\ \quad\vdots \\ s_n \approx t_n \\ \therefore s \approx t, \end{array}$$

is *valid in* **A** (or *correct in* **A**) provided either some equation $s_i \approx t_i$ does not hold in **A**, or $s \approx t$ holds in **A**, i.e., if all the premisses hold in **A**, then so does the conclusion.

When **A** is a finite algebra, then we can use a *combined* evaluation table to determine if an argument is valid. This is similar to our use of truth tables in the propositional logic. However, there is a *major* difference, namely, we do not use the "one row at a time" analysis that worked for analyzing arguments in propositional logic.

We start off with a simple argument, namely, if $x \approx y$ and $y \approx z$, then $x \approx z$ in a given algebra.

Example 3.5.2 Let **A** be a one–element binary algebra on the set $A = \{a\}$. It must have the following table for the fundamental operation:

$\cdot$	a
a	a

To check the validity of the argument

$$x \approx y,\ y \approx z \qquad \therefore x \approx z$$

let us form a combined evaluation table, where

$$s_1 = x \quad t_1 = y \quad s_2 = y \quad t_2 = z \quad s = x \quad t = z,$$

as follows:

			$\boxed{s_1}$	$\boxed{t_1}$	$\boxed{s_2}$	$\boxed{t_2}$	$\boxed{s}$	$\boxed{t}$
x	y	z	x	y	y	z	x	z
a	a	a	a	a	a	a	a	a

Examining the table we see that both premisses and the conclusion are *true*. Thus the argument is valid in **A**.

Example 3.5.3 Let **A** be the two–element Boolean algebra given in Example 3.1.12 (p. 138). To check the validity of the argument

$$x \approx y, \ y \approx z \qquad \therefore x \approx z$$

we form an evaluation table, where

$$s_1 = x, \quad t_1 = y, \quad s_2 = y, \quad t_2 = z, \quad s = x, \quad t = z,$$

as follows:

			s_1	t_1	s_2	t_2	s	t
x	y	z	x	y	y	z	x	z
0	0	0	0	0	0	0	0	0
0	0	1	0	0	0	1	0	1
0	1	0	0	1	1	0	0	0
0	1	1	0	1	1	1	0	1
1	0	0	1	0	0	0	1	0
1	0	1	1	0	0	1	1	1
1	1	0	1	1	1	0	1	0
1	1	1	1	1	1	1	1	1

Looking at the evaluation table we see that the premisses $s_1 \approx t_1$ and $s_2 \approx t_2$ are both *false* in **A** (the appropriate *columns* are not the same), and the conclusion $s \approx t$ is also false (the last two columns do not agree).

Thus this is a valid argument in **A**.

Indeed, following these two examples, we can show that the argument

$$x \approx y, \ y \approx z \qquad \therefore x \approx z$$

is valid in any algebra **A**. The premisses do not hold in any algebra with more than one element.

Let us emphasize the fact that when we say an equation $s \approx t$ is true in **A**, we are requiring that it be true for *all* evaluations of the variables of s and t. Thus given the equations $x \approx y$ and $y \approx z$, we are not inferring that the y of the first equation represents the same element as the y of the second equation.

To refer to *particular* elements x, y, z in *equational logic* we need to use *constants* instead of variables. The argument

$$a \approx b, \ b \approx c \qquad \therefore a \approx c$$

is valid in any algebra that has a language with the constant symbols a, b, c, and the symbol b refers to the same element in both premisses.

Example 3.5.4 Let **A** be the two–element Boolean algebra given in Example 3.1.12 (p. 138). To check the validity of the argument

$$x \wedge y \approx y \wedge x,\ x \wedge (y \vee x) \approx x \qquad \therefore x \wedge x' \approx x \vee x'$$

we form an evaluation table, where $s_1 = x \wedge y,\ t_1 = y \wedge x,\ s_2 = x \wedge (y \vee x)\ t_2 = x\ \ s = x \wedge x'\ \ t = x \vee x'$, as follows:

		s_1	t_1	s_2	t_2	s	t
x	y	$x \wedge y$	$y \wedge x$	$x \wedge (y \vee x)$	x	$x \wedge x'$	$x \vee x'$
0	0	0	0	0	0	0	1
0	1	0	0	0	0	0	1
1	0	0	0	1	1	0	1
1	1	1	1	1	1	0	1

Looking at the evaluation table we see that the premisses $s_1 \approx t_1$ and $s_2 \approx t_2$ are true in **A** (the appropriate columns are equal), but the conclusion $s \approx t$ is false (the last two columns do not agree).

Thus this in not a valid argument in **A**.

Having looked at valid arguments in a particular algebra **A**, we now generalize to look at arguments that are valid in *all* algebras **A**.

Definition 3.5.5 An argument

$$s_1 \approx t_1, \cdots, s_n \approx t_n \qquad \therefore s \approx t,$$

also written

$$\begin{aligned} &s_1 \approx t_1 \\ &\quad\vdots \\ &s_n \approx t_n \\ &\therefore s \approx t, \end{aligned}$$

is *valid* (or *correct*) provided it is valid in *every* algebra **A**, i.e., for every algebra **A**, either some equation $s_i \approx t_i$ does not hold in **A**, or $s \approx t$ holds in **A**.

Of course this becomes impossible to check using evaluation tables because there are an infinite number of algebras to examine. So how do we ever verify that an equational argument is valid? There are two ways. One is to use the proof system in Sec. 3.8.1. The other is to study abstract algebra courses to learn special methods that can aid in a semantic analysis of validity.

At this point one direction that we can handle is that of *refuting* an equational argument. An equational argument

$$s_1 \approx t_1, \cdots, s_n \approx t_n \qquad \therefore s \approx t$$

is *not* valid iff one can find an algebra **A** such that all the premisses hold in **A**, but the conclusion does not hold. Such an **A** is called a *counterexample* to the argument.

It can be very difficult to find counterexamples, in general, to invalid arguments. The smallest counterexample known that shows "Wilkie's identity (see p. 153) is not a consequence of what one learns in high school" has 14 elements and was found by Marcel Jackson in 1995. To keep the material manageable, all the counterexamples needed in this chapter will be either very small finite algebras (two or three elements should do in most cases) or well–known infinite algebras (e.g., the number systems).

Remark 3.5.6 Note that we can *never* use a one–element algebra to refute an equational argument because in a one–element algebra *all* equations are true (there is only one value for the term functions to take).

Example 3.5.7 Show that the argument

$$x \cdot y \approx y \cdot x \qquad \therefore x \cdot x \approx x$$

is not valid.

We need to find a binary algebra that is commutative but not idempotent. The following two–element binary algebra does the job:

$\cdot$	a	b
a	b	b
b	b	b

Example 3.5.8 Show that the argument

$$fffx \approx ffx \qquad \therefore ffx \approx fx$$

is not valid.

The following three–element monounary algebra provides a counterexample:

	f
a	b
b	c
c	c

Such examples we usually discover by drawing pictures such as the following:

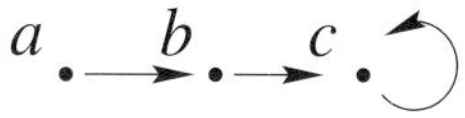

Figure 3.5.9 A monounary counterexample

Example 3.5.10 Show that the argument

$$fgx \approx x \qquad \therefore gfx \approx x$$

is not valid.

Referring to Example 3.4.8 (p. 152) we see that if $\mathbf{A}$ is a *finite* algebra, and the premiss $fgx \approx x$ holds on $\mathbf{A}$, then f is a *bijection* (since it is onto and A is finite) and g is also a *bijection* (since it is one–to–one on a finite set). But then f and g are inverses, and $gfx \approx x$ also holds on $\mathbf{A}$. Thus *the argument is valid for finite algebras.*

To find a counterexample we are forced to go to infinite algebras. We need f to be onto and g to be one–to–one, in order to satisfy $fgx \approx x$. But neither can be a bijection, otherwise they will be inverses and the conclusion will also hold. So we want f to be one–to–one, but not onto, and g to be onto but not one–to–one.

The following gives a simple counterexample on the natural numbers N:

$$\begin{aligned} g(n) &= n+1 \\ f(n+1) &= n \\ f(0) &= 0. \end{aligned}$$

A picture for this algebra would be as follows:

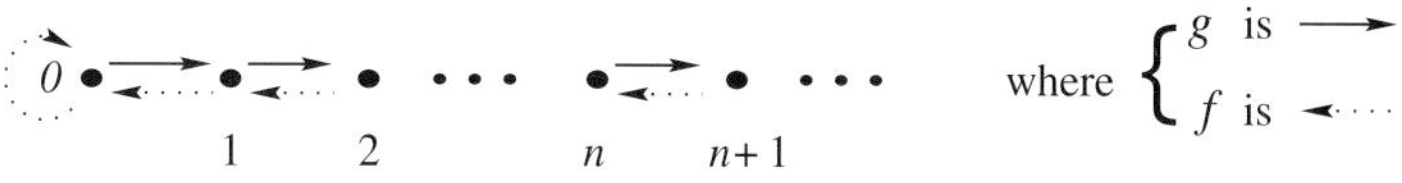

Figure 3.5.11 A biunary counterexample

Exercises

3.5.1 Determine if the following arguments are valid for the monounary algebra $\mathbf{A} = (A, f)$, where $A = \{a, b, c\}$ and f is defined by

	f
a	b
b	c
c	b

a. $fffx \approx fx \quad \therefore ffx \approx x$
b. $ffffx \approx fx \quad \therefore ffx \approx fx$
c. $fffx \approx fffy \quad \therefore fffx \approx fx$
d. $fffx \approx fy \quad \therefore ffx \approx ffffx$

3.5.2 Determine if the following arguments are valid for the biunary algebra $\mathbf{A} = (A, f, g)$, where $A = \{a, b, c\}$ and f and g are defined by

	f
a	b
b	c
c	b

	g
a	b
b	c
c	a

a. $fgfx \approx gfgx \quad \therefore fgx \approx gfx$
b. $gggx \approx x \quad \therefore fffx \approx fx$
c. $fgfgx \approx gfgfx \quad \therefore fgx \approx fgy$
d. $gfx \approx fgx \quad \therefore ggggx \approx gx$

3.5.3 Determine if the following arguments are valid for the binary algebra $\mathbf{A} = (A, \cdot)$, where $A = \{a, b\}$ and $\cdot$ is given by

$\cdot$	a	b
a	a	b
b	b	a

a. $x \cdot y \approx y \cdot x \quad \therefore x \cdot (y \cdot y) \approx x \cdot y$
b. $x \cdot (y \cdot z) \approx x \cdot z \quad \therefore x \cdot y \approx x$
c. $x \cdot (y \cdot z) \approx (x \cdot y) \cdot z \quad \therefore x \cdot x \approx x$
d. $x \cdot y \approx y \cdot x \quad \therefore x \cdot (y \cdot x) \approx x \cdot y$

3.5.4 Determine if the following arguments are valid for the algebra $(A, +, \cdot, -, 0, 1)$ on $A = \{a, b\}$ as defined by the following tables:

$+$	a	b
a	a	b
b	b	a

$\cdot$	a	b
a	a	a
b	a	b

	$-$
a	b
b	a

0 is a 1 is b

a. $x + x \approx 0,\ x \cdot y \approx y \cdot x \quad \therefore x + y \approx y + x$
b. $x \cdot x \approx x,\ x + y \approx y + x \quad \therefore x \cdot (y + z) \approx x \cdot y$
c. $x \cdot x \approx 1,\ x \cdot (x \cdot x) \approx x \quad \therefore x \cdot y \approx x$

d. $x \cdot (y + z) \approx (x \cdot y) + (x \cdot z),\ x + (y \cdot y) \approx x + y$
$\therefore (x + x) + x \approx 0$

3.5.5 Determine if the following arguments are valid for the group $\mathbf{G}_{3\times}$ defined on p. 158.

a. $x \cdot x \approx 1 \qquad \therefore x \cdot y \approx y \cdot x$
b. $x \cdot (x \cdot x) \approx 1 \qquad \therefore x \cdot x \approx 1 \cdot x$
c. $x \cdot (x \cdot x) \approx 1 \qquad \therefore (x \cdot y)^{-1} \approx x \cdot y$
d. $x \cdot (x \cdot x) \approx 1 \qquad \therefore (x \cdot y)^{-1} \approx (x \cdot y) \cdot (x \cdot y)$

3.5.6 Find counterexamples to the following arguments.

a. $fx \approx ffx$
$\therefore\ fx \approx fy$
b. $fffx \approx x$
$ffffx \approx fx$
$\therefore\ fx \approx x$
c. $x \cdot y \approx y \cdot x$
$\therefore\ x \cdot (y \cdot z) \approx (x \cdot y) \cdot z$
d. $x \cdot (y \cdot z) \approx (x \cdot y) \cdot z$
$\therefore\ x \cdot y \approx y \cdot x$
e. $x \cdot (y + z) \approx (x \cdot y) + (x \cdot z)$
$\therefore\ (x + y) \cdot z \approx (x \cdot z) + (y \cdot z)$
f. $f(x, y, z) \approx f(x, z, y)$
$\therefore\ f(x, y, z) \approx f(z, x, y)$

3.5.7 Find counterexamples to the following arguments involving the sets of equations $\mathcal{R}$, $\mathcal{BA}$, and $\mathcal{G}$ introduced in Sec. 3.4.2.

a. $\mathcal{R} \quad \therefore x \cdot x \approx x$
b. $\mathcal{R} \quad \therefore x + x \approx 0$
c. $\mathcal{R} \quad \therefore x \cdot y \approx y \cdot x$
d. $\mathcal{BA} \quad \therefore x \vee y' \approx (x \vee y)'$
e. $\mathcal{BA} \quad \therefore (x \vee y)' \approx x' \vee y'$
f. $\mathcal{BA} \quad \therefore x \vee (y \wedge z)' \approx (x \vee y') \wedge (x \vee z')$
g. $\mathcal{SG} \quad \therefore (x \cdot x) \cdot x \approx x$
h. $\mathcal{SG} \quad \therefore (x \cdot x) \cdot x \approx x \cdot x$
i. $\mathcal{SG} \quad \therefore (x \cdot y) \cdot x \approx (x \cdot x) \cdot y$
j. $\mathcal{G} \quad \therefore x \cdot x \approx 1$
k. $\mathcal{G} \quad \therefore (x \cdot x) \cdot x \approx x$
l. $\mathcal{G} \quad \therefore x \cdot y \approx y \cdot x$
m. $\mathcal{G} \quad \therefore (x \cdot y) \cdot x^{-1} \approx y \cdot y$
n. $\mathcal{G} \quad \therefore (x \cdot y) \cdot (x^{-1} \cdot y^{-1}) \approx 1$

3.6 SUBSTITUTION

Definition 3.6.1 Given a term $s(x_1, \ldots, x_n)$ and terms $s_1, \ldots, s_n$, the expression $s(s_1, \ldots, s_n)$ denotes the result of (simultaneously) substituting s_i for x_i in s. The notation we use for the *substitution* is

$$\begin{pmatrix} x_1 \leftarrow s_1 \\ \vdots \\ x_n \leftarrow s_n \end{pmatrix}.$$

We can illustrate the substitution procedure with a picture, where we use a simple triangle[2] to represent the tree of a term.

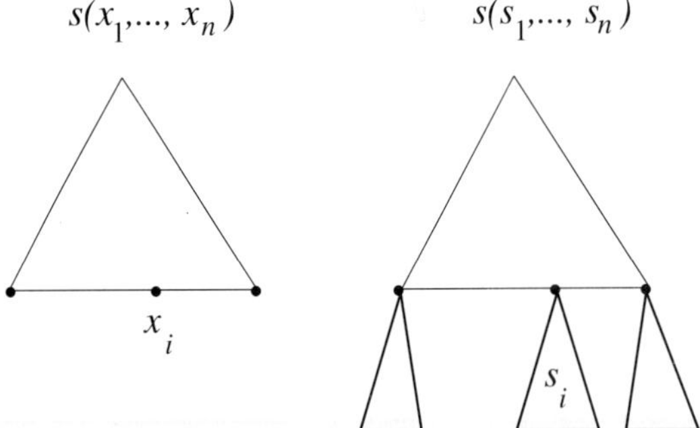

Figure 3.6.2 Visualizing substitution

At the bottom of the tree of the term $s(x_1, \ldots, x_n)$ we have the variables, the *leaves* of the tree. These are simultaneously replaced by the trees for the s_i to give the tree of $s(s_1, \ldots, s_n)$.

The importance of properly carrying out a *simultaneous* substitution is shown by the following example.

Example 3.6.3 Let $t(x, y) = (x + y) \cdot x$, and consider the substitution

$$\begin{pmatrix} x \leftarrow s_1 \\ y \leftarrow s_2 \end{pmatrix}$$

where $s_1 = x + y$, $s_2 = x \cdot y$. Then we have

$$t(s_1, s_2) = ((x + y) + (x \cdot y)) \cdot (x + y).$$

[2]We first encountered these abstract trees in the work of Nachum Dershowitz.

Now, if instead of *simultaneous* substitution we do *sequential* substitution, say, substitute for x first, and then for y, then we have the two steps:

$((x+y)+y)\cdot(x+y)$	applying just $(x \leftarrow x+y)$;
$((x+(x\cdot y))+(x\cdot y))\cdot(x+(x\cdot y))$	then applying $(y \leftarrow x\cdot y)$.

This is quite a different answer from that obtained by simultaneous substitution.

Now we come to one of the main properties, that satisfaction is preserved by substitutions.

Theorem 3.6.4 If

$$\mathbf{A} \models t(x_1,\ldots,x_n) \approx t'(x_1,\ldots,x_n)$$

and $s_1,\cdots,s_n$ are terms, then

$$\mathbf{A} \models t(s_1,\ldots,s_n) \approx t'(s_1,\ldots,s_n).$$

Proof. Since the term functions $t^{\mathbf{A}}$ and $t'^{\mathbf{A}}$ are equal on all values from $\mathbf{A}$, it follows that they are equal on the values that come from the $s_i^{\mathbf{A}}$. □

Example 3.6.5 Since the system of integers $\mathbf{Z}$ satisfies $x+y \approx y+x$, by substitution it also satisfies $(x\cdot x)+(y\cdot z) \approx (y\cdot z)+(x\cdot x)$.

Example 3.6.6 The binary algebra in Example 3.4.7, p. 151, satisfies the equation

$$(x\cdot y)\cdot((u\cdot v)\cdot(w\cdot z)) \approx (u\cdot v)\cdot((x\cdot y)\cdot(w\cdot z))$$

by substitution into $x\cdot(y\cdot z) \approx y\cdot(x\cdot z)$.

Theorem 3.6.7 [SUBSTITUTION THEOREM] If $\mathcal{S} \models s \approx t$ and $s' \approx t'$ is a substitution instance of $s \approx t$, then $\mathcal{S} \models s' \approx t'$.

Proof. This follows immediately from Theorem 3.6.4. □

Example 3.6.8 $\mathcal{R} \models x\cdot(y+z) \approx (x\cdot y)+(x\cdot z)$, since the left distributive law is actually in $\mathcal{R}$. Thus, by substitution

$$\mathcal{R} \models (u+v)\cdot(y+z) \approx ((u+v)\cdot y)+((u+v)\cdot z),$$

i.e., this equation is true of every ring.

Example 3.6.9 $\mathcal{BA} \models x\wedge(x\vee y) \approx x$, since this absorption law is in $\mathcal{BA}$. Thus, by substitution

$$\mathcal{BA} \models (u\vee w)'\wedge((u\vee w)'\vee(u\wedge w')) \approx (u\vee w)',$$

i.e., this equation is true of every Boolean algebra.

Example 3.6.10 $\mathcal{G} \models x \cdot x^{-1} \approx 1$, since this law is in $\mathcal{G}$. Thus, by substitution

$$\mathcal{G} \models (x \cdot y^{-1})^{-1} \cdot ((x \cdot y^{-1})^{-1})^{-1} \approx 1,$$

i.e., this equation is true of every group.

Exercises

3.6.1 Carry out the substitution

$$\begin{pmatrix} x \leftarrow 1 + (y \cdot x) \\ y \leftarrow (x + z) \cdot 0 \\ z \leftarrow (x \cdot y) + (-z) \end{pmatrix}$$

on the following terms:

a. $x + y$
b. $(x \cdot y) + x$
c. $(x + (y \cdot (z + y))) \cdot y$
d. $((x \cdot y) + (z \cdot x)) \cdot ((x + y) \cdot (y + z))$.

3.6.2 Carry out the substitution

$$\begin{pmatrix} x \leftarrow (y \wedge 1) \vee x' \\ y \leftarrow ((x \vee 0) \wedge y)' \\ z \leftarrow (x \vee y) \wedge z \end{pmatrix}$$

on the following terms:

a. $x \vee y$
b. $(x \vee y) \wedge x$
c. $(z \wedge (y \vee (z' \wedge y))) \vee y'$
d. $((x' \wedge y) \vee (z \wedge x)') \wedge ((x \vee y) \wedge (y \vee z))'$.

3.6.3 Carry out the substitution

$$\begin{pmatrix} x \leftarrow y \cdot x^{-1} \\ y \leftarrow (1 \cdot y)^{-1} \\ z \leftarrow (x \cdot z)^{-1} \end{pmatrix}$$

on the following terms:

a. $x \cdot y$
b. $(x \cdot y) \cdot x$
c. $x \cdot (y \cdot x)$
d. $(x \cdot (1 \cdot (x \cdot z)^{-1})^{-1})^{-1} \cdot y$
e. $((x \cdot 1) \cdot (z \cdot x)) \cdot ((z \cdot 1^{-1}) \cdot (y \cdot x))$.

3.6.4 In the following we have a unary function symbol f, a binary function symbol g, and a ternary function symbol h. Carry out the substitution

$$\begin{pmatrix} x \leftarrow fgxy \\ y \leftarrow fgxhxyz \\ z \leftarrow hxfygzx \end{pmatrix}$$

on the following terms:

a. $fffx$
b. $ffgzy$
c. $hxfgyxz$
d. $hgfgyxzygxhzxz$.

In the following problems use the substitution theorem to justify the claims.

3.6.5 Show that the following equations $s \approx t$ are true of rings, i.e., that $\mathcal{R} \models s \approx t$.

a. $(1 + (x \cdot y)) + (x \cdot z) \approx (x \cdot z) + (1 + (x \cdot y))$
b. $(1+x)\cdot((1+y)+(1+z)) \approx ((1+x)\cdot(1+y))+((1+x)\cdot(1+z))$
c. $(x\cdot(1+y))\cdot((x\cdot y)\cdot((x\cdot x)\cdot y)) \approx ((x\cdot(1+y))\cdot(x\cdot y))\cdot((x\cdot x)\cdot y)$

3.6.6 Show that the following equations $s \approx t$ are true of Boolean algebras, i.e., that $\mathcal{BA} \models s \approx t$.

a. $((x \wedge y) \vee z) \wedge ((x \vee y) \vee (y \vee z))$
$\approx (((x \wedge y) \vee z) \wedge (x \vee y)) \vee (((x \wedge y) \vee z) \wedge (y \vee z))$
b. $((x \wedge y)' \vee z) \wedge (((x \wedge y)' \vee z) \vee (x \vee y)') \approx (x \wedge y)' \vee z$

3.6.7 Show that the following equations $s \approx t$ are true of groups, i.e., that $\mathcal{G} \models s \approx t$.

a. $(x \cdot y) \cdot (x^{-1} \cdot y^{-1}) \approx ((x \cdot y) \cdot x^{-1}) \cdot y^{-1}$
b. $(x \cdot y) \cdot ((x^{-1} \cdot y^{-1}) \cdot (x^{-1} \cdot y^{-1})^{-1})$
$\approx ((x \cdot y) \cdot (x^{-1} \cdot y^{-1})) \cdot (x^{-1} \cdot y^{-1})^{-1}$

3.7 REPLACEMENT

Replacement is essentially a nonuniform version of substitution. Starting with a term $t(\cdots s \cdots)$ with an occurrence of a subterm s in it, we *replace* s with another term s'.

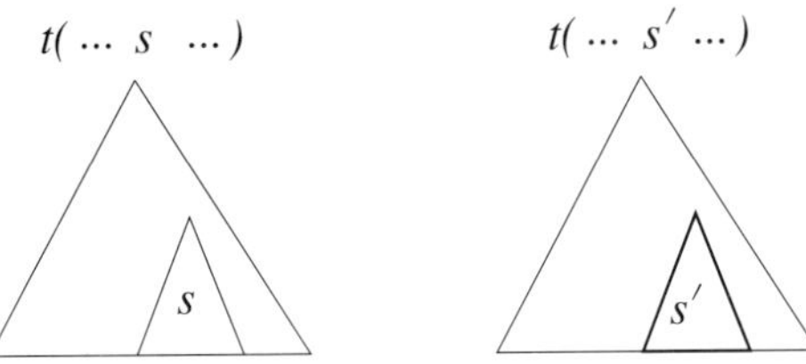

Figure 3.7.1 Visualizing replacement

Example 3.7.2 Replacing the second occurrence (from the left) of $x + (y \cdot z)$ in

$$((x + (y \cdot z)) \cdot z) + ((y + (x + (y \cdot z))) \cdot z)$$

with the term $(u + v) \cdot w$ gives the term

$$((x + (y \cdot z)) \cdot z) + ((y + ((u + v) \cdot w)) \cdot z).$$

Example 3.7.3 Replacing the second occurrence (from the left) of $x \vee (y \wedge z)$ in

$$((x \vee (y \wedge z)) \wedge (x \vee y)) \vee (((x \wedge y) \vee (x \vee (y \wedge z))) \wedge (x \vee (y \wedge z)))$$

with the term $(x \vee y) \wedge (x \vee z)$ gives the term

$$((x \vee (y \wedge z)) \wedge (x \vee y)) \vee (((x \wedge y) \vee ((x \vee y) \wedge (x \vee z))) \wedge (x \vee (y \wedge z))).$$

Now we show that replacements "near the top" of a term preserve satisfaction. The following figure shows what it means to carry out such a replacement.

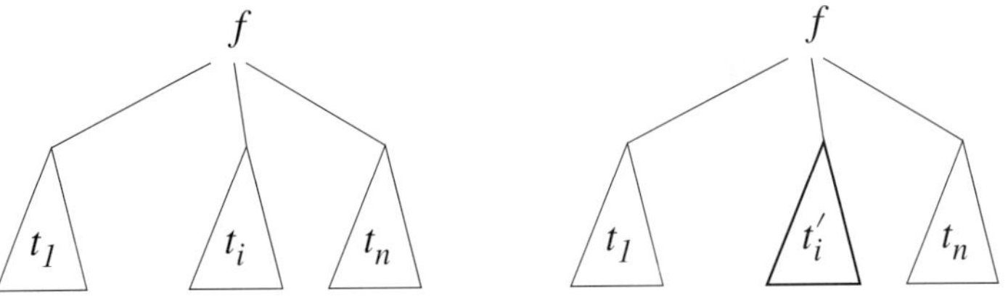

Figure 3.7.4 Visualizing replacement "near the top"

Lemma 3.7.5 Let a language $\mathcal{L}$ of algebras be given and let $\mathbf{A}$ be an $\mathcal{L}$–structure. Let f be an n–ary function symbol in $\mathcal{F}$, and let $t_1, \ldots, t_n$ be $\mathcal{L}$–terms. Suppose t_i' is an $\mathcal{L}$–term with $\mathbf{A} \models t_i \approx t_i'$. Then $\mathbf{A} \models ft_1 \cdots t_i \cdots t_n \approx ft_1 \cdots t_i' \cdots t_n$.

Proof. Observe that

$$\begin{aligned}(ft_1 \cdots t_i \cdots t_n)^{\mathbf{A}} &= f^{\mathbf{A}}(t_1^{\mathbf{A}}, \cdots, t_i^{\mathbf{A}}, \cdots, t_n^{\mathbf{A}}) \\ &= f^{\mathbf{A}}(t_1^{\mathbf{A}}, \cdots, t_i'^{\mathbf{A}}, \cdots, t_n^{\mathbf{A}}) \\ &= (ft_1 \cdots t_i' \cdots t_n)^{\mathbf{A}}.\end{aligned}$$

□

Example 3.7.6 The equation

$$(x \cdot (y+z)) + (x \cdot y) \approx ((x \cdot y) + (x \cdot z)) + (x \cdot y)$$

is a consequence of $\mathcal{R}$, since it is obtained by a "near the top" replacement of (the only occurrence of) $x \cdot (y+z)$ on the left with $(x \cdot y) + (x \cdot z)$, and $x \cdot (y+z) \approx (x \cdot y) + (x \cdot z)$ is in $\mathcal{R}$.

Thus this equation is true in all rings, since they, by definition, satisfy $\mathcal{R}$.

Next, we want to show that replacement, like substitution, preserves equational consequences.

Theorem 3.7.7 Let s, t, t' be $\mathcal{L}$–terms. If $\mathbf{A}$ is an $\mathcal{L}$–structure and $\mathbf{A} \models t \approx t'$, and if s' is the result of replacing an occurrence of t in s with t', then $\mathbf{A} \models s \approx s'$.

Proof. We essentially repeat the proof of Theorem 2.4.5 (p. 49), that replacement preserves truth equivalence for propositional formulas. Let us say that an $\mathcal{L}$–term s has the *replacement property* if our Theorem 3.7.7 holds for s. We will use proof by induction over terms in $T(X)$ to show that every s in $T(X)$ has the replacement property.

a. $s = x$ for some $x \in X$.

If t occurs as a subterm of s, then clearly $t = s$, so $s' = t'$. Thus $\mathbf{A} \models s \approx s'$.

b. $s = c$ for $c \in \mathcal{C}$.

If t occurs as a subterm of s, then $t = c$, so $s' = t'$. Thus $\mathbf{A} \models s \approx s'$.

c. $s = fs_1 \cdots s_n$ and each of $s_1, \ldots, s_n$ has the replacement property.

Suppose we have an occurrence of t as a subterm of s. If $t = s$, then $t' = s'$, so $\mathbf{A} \models s \approx s'$. So now we suppose $t \neq s$. Then the occurrence of t in s is actually an occurrence of t in one of $s_1, \ldots, s_n$, say in s_i. Let s_i' be the result of replacing that occurrence of t with t'. Clearly, $s' = fs_1 \cdots s_i' \ldots s_n$. By our induction hypothesis, $\mathbf{A} \models s_i \approx s_i'$. And then, by Lemma 3.7.5, $\mathbf{A} \models fs_1 \cdots s_i \cdots s_n \approx fs_1 \cdots s_i' \cdots s_n$, so $\mathbf{A} \models s \approx s'$.

This finishes the proof by induction on terms. □

Example 3.7.8 The binary algebra in Example 3.4.7 (p. 151), satisfies the equation

$$v \cdot ((x \cdot (y \cdot z)) \cdot u) \approx v \cdot ((y \cdot (x \cdot z)) \cdot u)$$

by replacement, using $x \cdot (y \cdot z) \approx y \cdot (x \cdot z)$.

Theorem 3.7.9 [REPLACEMENT THEOREM] Let s, t, t' be $\mathcal{L}$–terms. If $\mathcal{S}$ is a set of $\mathcal{L}$–equations and $\mathcal{S} \models t \approx t'$, and if s' is the result of replacing an occurrence of t in s with t', then $\mathcal{S} \models s \approx s'$.

Proof. This is a simple application of Theorem 3.7.7. □

Example 3.7.10 The equation

$$\begin{aligned} & (x \cdot (y \cdot (z \cdot (u \cdot (v + w))))) + (x \cdot y) \\ \approx \quad & (x \cdot (y \cdot (z \cdot ((u \cdot v) + (u \cdot w))))) + (x \cdot y) \end{aligned}$$

follows from $\mathcal{R}$, since we have applied a replacement using the left distributive law, and the left distributive law is in $\mathcal{R}$.

Thus all rings satisfy this equation.

Exercises

3.7.1 In the following, replace the second occurrence (from the left) of the term $x \cdot y$ with $(y + z) \cdot (z + x)$:

a. $(x + (x \cdot y)) \cdot (x + (x \cdot y))$
b. $((x \cdot z) + (x \cdot y)) + (z + (x + (x \cdot y)))$
c. $(((x + y) \cdot (y + x)) \cdot (x \cdot y)) + ((x + z) \cdot (x + (x \cdot y)))$.

3.7.2 In the following, replace the last occurrence of the term $(x \vee y) \wedge z$ with $(x \wedge z) \vee (y \wedge z)$:

a. $(x \vee ((x \vee y) \wedge z)) \wedge (x \vee (x \wedge y))$
b. $((x \wedge z) \vee (x \wedge y)) \vee (((x \vee y) \wedge z) \vee (((x \vee y) \wedge z) \vee (x \wedge y)))$
c. $(((x \vee y) \wedge z) \vee (y \vee x)) \wedge (((x \vee y) \wedge z) \vee ((x \vee z) \wedge (x \vee (x \wedge y))))$.

3.7.3 In the following we have a unary function symbol f, a binary function symbol g, and a ternary function symbol h. Replace the last occurrence of $gffyz$ in each of the following with $hxfygyz$:

a. $ffgffyz$
b. $hxfgffyzz$
c. $hgfggffyzxzyggffyzhzxz$
d. $ghgffyzyhyhhggyyxxgffyzzxzy$.

3.7.4 Use the substitution and replacement theorems to explain why the following equations hold in rings.

a. $(x + y) \cdot (u + v) \approx ((x + y) \cdot u) + ((x + y) \cdot v)$
b. $1 + ((x + y) \cdot (x + y)) \approx 1 + (((x + y) \cdot x) + ((x + y) \cdot y))$
c. $(1 + ((x + y) \cdot (x + y))) \cdot x \approx (1 + (((x + y) \cdot x) + ((x + y) \cdot y))) \cdot x$

3.7.5 Use the substitution and replacement theorems to explain why the following equations hold in Boolean algebras.

a. $(x \vee y) \wedge (u \vee v) \approx ((x \vee y) \wedge u) \vee ((x \vee y) \wedge v)$
b. $x \vee ((x \vee y) \wedge (x \vee y)) \approx x \vee (((x \vee y) \wedge x) \vee ((x \vee y) \wedge y))$
c. $(x \vee ((x \vee y) \wedge (x \vee y))) \wedge x \approx (x \vee (((x \vee y) \wedge x) \vee ((x \vee y) \wedge y))) \wedge x$

3.8 A PROOF SYSTEM FOR EQUATIONAL LOGIC

In the previous sections we saw ways to obtain new consequences from given ones. Now we collect these together to form a proof system for equations.

3.8.1 Birkhoff's Rules

On the syntactic side of equational logic, Birkhoff studied the following five rules.

BIRKHOFF'S RULES OF INFERENCE FOR EQUATIONAL LOGIC

RULE	NAME	EXAMPLE
$\dfrac{}{s \approx s}$	Reflexive	$\dfrac{}{x^2 \approx x^2}$
$\dfrac{s \approx t}{t \approx s}$	Symmetric	$\dfrac{x \approx x^2}{x^2 \approx x}$
$\dfrac{r \approx s,\ s \approx t}{r \approx t}$	Transitive	$\dfrac{x \approx x^2,\ x^2 \approx yz}{x \approx yz}$
$\dfrac{r(\vec{x}) \approx s(\vec{x})}{r(\vec{t}) \approx s(\vec{t})}$	Substitution	$\dfrac{xy \approx yx}{(x+y)z \approx z(x+y)}$
$\dfrac{s \approx t}{r(\cdots s \cdots) \approx r(\cdots t \cdots)}$	Replacement	$\dfrac{xy \approx yx}{x^2 + xy \approx x^2 + yx}$

Definition 3.8.1 If $\mathcal{S}$ is a set of $\mathcal{L}$–equations, then a *derivation* of an equation $s \approx t$ from $\mathcal{S}$ is a sequence $s_1 \approx t_1, \ldots, s_n \approx t_n$ of equations such that each one is either a member of $\mathcal{S}$ or is the result of applying one of the five rules of inference to previous members of the sequence, and the last equation is $s \approx t$.

We write $\mathcal{S} \vdash s \approx t$ if such a derivation exists.

Remark 3.8.2 The first rule of inference simply allows us to use $s \approx s$ for any step of a derivation.

Example 3.8.3 If $\vdash s \approx t$, then $s = t$, i.e., s and t are exactly the same term. To see this, note that if $s_1 \approx t_1, \ldots, s_n \approx t_n$ is a derivation witnessing $\vdash s \approx t$, then the first step must be an application of the reflexive rule, since this is the only rule that applies when there are no premisses or previous steps in the derivation. Thus $s_1 = t_1$. Now we can show that any of the rules applied to obtain $s_2 \approx t_2$ will have $s_2 = t_2$, and so on. Thus we finally have $s_n = t_n$, i.e., $s = t$.

In the following examples we will use the abbreviations *refl, symm, trans, subs,* and *repl* for the five rules. The notation "subs 3" means "use substitution on line 3", and so forth.

Example 3.8.4 Give a derivation to witness $fffx \approx fy \vdash fx \approx fy$.

1.	$fffx \approx fy$	given
2.	$fffx \approx fx$	subs 1
3.	$fx \approx fffx$	symm 2
4.	$fx \approx fy$	trans 1,3

Example 3.8.5 Show $fgx \approx x,\ ggx \approx x \vdash fx \approx gx$.

1.	$fgx \approx x$	given
2.	$ggx \approx x$	given
3.	$fggx \approx gx$	subs 1
4.	$fggx \approx fx$	repl 2
5.	$fx \approx fggx$	symm 4
6.	$fx \approx gx$	trans 3,5

Example 3.8.6 Show $\mathcal{R} \vdash x \cdot 0 \approx 0$.

First let us give a thumbnail sketch, namely, the kind of argument we use everyday. Since $0 + 0 \approx 0$, we have $x \cdot (0 + 0) \approx x \cdot 0$, so we have $x \cdot 0 + x \cdot 0 \approx x \cdot 0$. Subtracting $x \cdot 0$ from both sides gives $x \cdot 0 \approx 0$.

In terms of Birkhoff's rules, much was hidden in this brief argument. Now we give a detailed derivation. However, to keep the lines within the margins we are going to use the convention that $\cdot$ takes precedence over $+$, e.g., $a \cdot b + c \cdot d$ is shorthand for $(a \cdot b) + (c \cdot d)$.

1.	$x + 0$	$\approx$	x	given (R1)
2.	$0 + 0$	$\approx$	0	subs 1
3.	$x \cdot (0 + 0)$	$\approx$	$x \cdot 0$	repl 2
4.	$x \cdot (y + z)$	$\approx$	$x \cdot y + x \cdot z$	given (R8)
5.	$x \cdot (0 + 0)$	$\approx$	$x \cdot 0 + x \cdot 0$	subs 4
6.	$x \cdot 0 + x \cdot 0$	$\approx$	$x \cdot (0 + 0)$	symm 5
7.	$x \cdot 0 + x \cdot 0$	$\approx$	$x \cdot 0$	trans 3,6
8.	$(x \cdot 0 + x \cdot 0) + (-(x \cdot 0))$	$\approx$	$x \cdot 0 + (-(x \cdot 0))$	repl 7
9.	$x + (-x)$	$\approx$	0	given (R2)
10.	$x \cdot 0 + (-(x \cdot 0))$	$\approx$	0	subs 9
11.	$(x \cdot 0 + x \cdot 0) + (-(x \cdot 0))$	$\approx$	0	trans 8,10
12.	$x + (y + z)$	$\approx$	$(x + y) + z$	given (R4)
13.	$x \cdot 0 + (x \cdot 0 + (-(x \cdot 0)))$	$\approx$	$(x \cdot 0 + x \cdot 0) + (-(x \cdot 0))$	subs 12
14.	$x \cdot 0 + (x \cdot 0 + (-(x \cdot 0)))$	$\approx$	0	trans 11,13
15.	$x \cdot 0 + (-(x \cdot 0))$	$\approx$	0	subs 9
16.	$x \cdot 0 + (x \cdot 0 + (-(x \cdot 0)))$	$\approx$	$(x \cdot 0) + 0$	repl 15
17.	$x \cdot 0 + 0$	$\approx$	$x \cdot 0$	subs 1
18.	$x \cdot 0 + (x \cdot 0 + (-(x \cdot 0)))$	$\approx$	$x \cdot 0$	trans 16,17
19.	$x \cdot 0$	$\approx$	$x \cdot 0 + (x \cdot 0 + (-(x \cdot 0)))$	symm 18
20.	$x \cdot 0$	$\approx$	0	trans 14,19

Example 3.8.7 We will give a derivation that witnesses $\mathcal{G} \vdash 1 \cdot x \approx x$. Unfortunately we do not have a helpful thumbnail sketch for this example.

1.	$x \cdot 1$	$\approx$	x	given (G1)
2.	$x \cdot x^{-1}$	$\approx$	1	given (G2)
3.	$(x \cdot y) \cdot z$	$\approx$	$x \cdot (y \cdot z)$	given (G3)
4.	$(y \cdot x) \cdot x^{-1}$	$\approx$	$y \cdot (x \cdot x^{-1})$	subs 3
5.	$y \cdot (x \cdot x^{-1})$	$\approx$	$y \cdot 1$	repl 2
6.	$(y \cdot x) \cdot x^{-1}$	$\approx$	$y \cdot 1$	trans 4,5
7.	$y \cdot 1$	$\approx$	y	subs 1
8.	$(y \cdot x) \cdot x^{-1}$	$\approx$	y	trans 6,7
9.	y	$\approx$	$(y \cdot x) \cdot x^{-1}$	symm 8
10.	$y \cdot x$	$\approx$	$((y \cdot x) \cdot x^{-1}) \cdot (x^{-1})^{-1}$	subs 9
11.	$((y \cdot x) \cdot x^{-1}) \cdot (x^{-1})^{-1}$	$\approx$	$y \cdot (x^{-1})^{-1}$	repl 8
12.	$y \cdot x$	$\approx$	$y \cdot (x^{-1})^{-1}$	trans 10,11
13.	x	$\approx$	$(x \cdot x^{-1}) \cdot (x^{-1})^{-1}$	subs 9
14.	$(x \cdot x^{-1}) \cdot (x^{-1})^{-1}$	$\approx$	$1 \cdot (x^{-1})^{-1}$	repl 2
15.	x	$\approx$	$1 \cdot (x^{-1})^{-1}$	trans 13,14
16.	$1 \cdot (x^{-1})^{-1}$	$\approx$	x	symm 15
17.	$1 \cdot x$	$\approx$	$1 \cdot (x^{-1})^{-1}$	subs 12
18.	$1 \cdot x$	$\approx$	x	trans 17,16

This derivation, the shortest that we know of to witness $\mathcal{G} \vdash 1 \cdot x \approx x$ using Birkhoff's rules, was discovered by William McCune of Argonne National Laboratory in the summer of 1992, using the automated theorem proving program OTTER (see p. 259). In ordinary mathematical practice we would write the proof out in about 10 steps by (1) not writing $\mathcal{S}$ as steps 1–3, (2) omitting the parentheses to take care of the associative law, and (3) combining the symmetry steps with other steps.

3.8.2 Is There a Strategy for Finding Equational Derivations?

Perhaps you will be impressed by the nontriviality of equational derivations, especially the last example, on groups. There are many more such clever derivations that we could present. Unlike the propositional logic, where we have *guaranteed*, if sometimes slow, methods, e.g., truth tables, DNFs, DPP, etc., for analyzing arguments, the equational logic has *no such algorithms*. This does not mean that we have not found algorithms yet, but rather that they simply do not exist. The way to become more efficient with equational derivations is (1), to *practice* and (2), to have a touch of *cleverness*.

The difficulty in dealing with equations, even fairly simple looking ones, has led to a number of special strategies in the attempts at automated theorem proving. We will look at one of them, *term rewrite systems*, later in this chapter.

Exercises

3.8.1 Find derivations for the following, where f is a unary function symbol.

a. $fx \approx fy \vdash ffx \approx fx$
b. $fffx \approx x, ffx \approx fx \vdash fx \approx x$
c. $ffx \approx y \vdash x \approx y$
d. $fx \approx ffffx, fffffx \approx fffx \vdash ffx \approx fffx$

3.8.2 Find derivations for the following, where f, g are unary function symbols.

a. $fx \approx gy \vdash fx \approx fy$
b. $gx \approx gy \vdash gfx \approx gfy$
c. $fx = x, fgx \approx fgy \vdash gx \approx gy$
d. $fgx \approx x, gfx \approx gfy \vdash fx \approx fy$

3.8.3 Find derivations for the following, where $\cdot$ is a binary function symbol.

a. $x \cdot y \approx y \vdash x \cdot (y \cdot z) \approx (x \cdot y) \cdot z$
b. $x \cdot y \approx u \cdot v \vdash x \cdot (y \cdot z) \approx (x \cdot y) \cdot z$
c. $x \cdot y \approx u \cdot v \vdash x \cdot y \approx y \cdot x$
d. $x \cdot y \approx u \cdot v \vdash x \cdot x \approx y \cdot y$

3.8.4 From $(x \cdot y) \cdot z \approx x \cdot (y \cdot z)$ deduce $(((x \cdot y) \cdot z) \cdot u) \cdot v \approx x \cdot (y \cdot (z \cdot (u \cdot v)))$ by finding a derivation.

3.8.5

a. Find a derivation to justify the claim that $\mathcal{R} \vdash 0 \cdot x \approx 0$.
b. Fill in the reasons for the steps in the following derivation that shows $\mathcal{R} \vdash (x \cdot y) + (x \cdot (-y)) \approx x \cdot 0$.

1.	$x \cdot (y + z) \approx (x \cdot y) + (x \cdot z)$	given
2.	$x \cdot (y + (-y)) \approx (x \cdot y) + (x \cdot (-y))$	?
3.	$(x \cdot y) + (x \cdot (-y)) \approx x \cdot (y + (-y))$	?
4.	$x + (-x) = 0$	given
5.	$y + (-y) = 0$	?
6.	$x \cdot (y + (-y)) \approx x \cdot 0$	?
7.	$(x \cdot y) + (x \cdot (-y)) \approx x \cdot 0$	?

3.8.6

a. Fill in the reasons for the steps in the following derivation that shows $\mathcal{G} \vdash 1 \cdot (x^{-1})^{-1} \approx x$.

1.	$x \cdot 1 \approx x$	given
2.	$x \cdot x^{-1} \approx 1$	given
3.	$(x \cdot y) \cdot z \approx x \cdot (y \cdot z)$	given
4.	$1 \approx x \cdot x^{-1}$	?
5.	$1 \cdot (x^{-1})^{-1} \approx (x \cdot x^{-1}) \cdot (x^{-1})^{-1}$	?
6.	$(x \cdot x^{-1}) \cdot (x^{-1})^{-1} \approx x \cdot (x^{-1} \cdot (x^{-1})^{-1})$	?
7.	$1 \cdot (x^{-1})^{-1} \approx x \cdot (x^{-1} \cdot (x^{-1})^{-1})$	?
8.	$x^{-1} \cdot (x^{-1})^{-1} \approx 1$	?
9.	$x \cdot (x^{-1} \cdot (x^{-1})^{-1}) \approx x \cdot 1$	?
10.	$x \cdot (x^{-1} \cdot (x^{-1})^{-1}) \approx x$	?
11.	$1 \cdot (x^{-1})^{-1} \approx x$	?

b. Fill in the reasons for the steps in the following derivation that shows $\mathcal{G} \vdash (x \cdot y) \cdot (y^{-1} \cdot x^{-1}) \approx 1$.

1.	$x \cdot 1 \approx x$	given
2.	$x \cdot x^{-1} \approx 1$	given
3.	$(x \cdot y) \cdot z \approx x \cdot (y \cdot z)$	given

4.	$(x \cdot y) \cdot (y^{-1} \cdot x^{-1}) \approx x \cdot (y \cdot (y^{-1} \cdot x^{-1}))$	?
5.	$(y \cdot y^{-1}) \cdot x^{-1} \approx y \cdot (y^{-1} \cdot x^{-1})$	?
6.	$y \cdot (y^{-1} \cdot x^{-1}) \approx (y \cdot y^{-1}) \cdot x^{-1}$	?
7.	$y \cdot y^{-1} \approx 1$	?
8.	$(y \cdot y^{-1}) \cdot x^{-1} \approx 1 \cdot x^{-1}$	?
9.	$y \cdot (y^{-1} \cdot x^{-1}) \approx 1 \cdot x^{-1}$	?
10.	$x \cdot ((y \cdot y^{-1}) \cdot x^{-1}) \approx x \cdot (1 \cdot x^{-1})$	?
11.	$(x \cdot y) \cdot (y^{-1} \cdot x^{-1}) \approx x \cdot (1 \cdot x^{-1})$	?
12.	$(x \cdot 1) \cdot x^{-1} \approx x \cdot (1 \cdot x^{-1})$	?
13.	$x \cdot (1 \cdot x^{-1}) \approx (x \cdot 1) \cdot x^{-1}$	?
14.	$(x \cdot 1) \cdot x^{-1} \approx x \cdot x^{-1}$	?
15.	$(x \cdot 1) \cdot x^{-1} \approx 1$	?
16.	$x \cdot (1 \cdot x^{-1}) \approx 1$	?
17.	$(x \cdot y) \cdot (y^{-1} \cdot x^{-1}) \approx 1$	?

3.8.7 In the following, $a, b, c, 1$ are constants. From

$$\begin{aligned} x \cdot 1 &\approx x \\ x \cdot x^{-1} &\approx 1 \\ (x \cdot y) \cdot z &\approx x \cdot (y \cdot z) \\ a \cdot c &\approx b \cdot c \end{aligned}$$

deduce $a \approx b$ (this is the *cancellation law*) by filling in the reasons for the steps of the following derivation.

1.	$x \cdot 1 \approx x$	given
2.	$x \cdot x^{-1} \approx 1$	given
3.	$(x \cdot y) \cdot z \approx x \cdot (y \cdot z)$	given
4.	$a \cdot c \approx b \cdot c$	given
5.	$(a \cdot c) \cdot c^{-1} \approx (b \cdot c) \cdot c^{-1}$	?
6.	$(a \cdot c) \cdot c^{-1} \approx a \cdot (c \cdot c^{-1})$	?
7.	$c \cdot c^{-1} \approx 1$	?
8.	$a \cdot (c \cdot c^{-1}) \approx a \cdot 1$	?
9.	$a \cdot 1 \approx a$	?
10.	$a \cdot (c \cdot c^{-1}) \approx a$	?
11.	$a \approx a \cdot (c \cdot c^{-1})$	?
12.	$a \cdot (c \cdot c^{-1}) \approx (a \cdot c) \cdot c^{-1}$	?
13.	$a \approx (a \cdot c) \cdot c^{-1}$	?
14.	$a \approx (b \cdot c) \cdot c^{-1}$	?
15.	$(b \cdot c) \cdot c^{-1} \approx b \cdot (c \cdot c^{-1})$	?
16.	$a \approx b \cdot (c \cdot c^{-1})$	?
17.	$b \cdot (c \cdot c^{-1}) \approx b \cdot 1$	?

18. $a \approx b \cdot 1$?
19. $b \cdot 1 \approx b$?
20. $a \approx b$?

3.8.8

a. Fill in the reasons for the following derivation of the idempotent law $x \wedge x \approx x$ for Boolean algebras, i.e., to show $\mathcal{BA} \vdash x \wedge x \approx x$.

1. $x \vee (x \wedge y) \approx x$ given
2. $x \wedge (x \vee y) \approx x$ given
3. $x \vee (x \wedge x) \approx x$?
4. $x \wedge (x \vee (x \wedge x)) \approx x$?
5. $x \wedge (x \vee (x \wedge x)) \approx x \wedge x$?
6. $x \wedge x \approx x \wedge (x \vee (x \wedge x))$?
7. $x \wedge x \approx x$?

b. Explain how to use the previous part to find a derivation of the other idempotent law, $x \vee x \approx x$, for Boolean algebras.

c. Fill in the reasons for the following derivation to show $\mathcal{BA} \vdash x \wedge 1 \approx x$.

1. $x \vee x' \approx 1$ given
2. $x \wedge (x \vee y) \approx x$ given
3. $1 \approx x \vee x'$?
4. $x \wedge 1 \approx x \wedge (x \vee x')$?
5. $x \wedge (x \vee x') \approx x$?
6. $x \wedge 1 \approx x$?

d. Explain how to use the previous part to find a derivation for $\mathcal{BA} \vdash x \vee 0 \approx x$.

e. Fill in the reasons for the steps in the following derivation of the second distributive law for Boolean algebras, i.e., show $\mathcal{BA} \vdash x \vee (y \wedge z) \approx (x \vee y) \wedge (x \vee z)$.

1. $x \vee (x \wedge y) \approx x$ given
2. $x \wedge (x \vee y) \approx x$ given
3. $x \vee (y \vee z) \approx (x \vee y) \vee z$ given
4. $x \wedge y \approx y \wedge x$ given
5. $x \wedge (y \vee z) \approx (x \wedge y) \vee (x \wedge z)$ given
6. $x \vee (x \wedge z) \approx x$?
7. $(x \vee (x \wedge z)) \vee (y \wedge z) \approx x \vee (y \wedge z)$?
8. $x \vee ((x \wedge z) \vee (y \wedge z)) \approx (x \vee (x \wedge z)) \vee (y \wedge z)$?
9. $x \vee ((x \wedge z) \vee (y \wedge z)) \approx x \vee (y \wedge z)$?

10. $x \vee (y \wedge z) \approx x \vee ((x \wedge z) \vee (y \wedge z))$?
11. $x \wedge z \approx z \wedge x$?
12. $x \vee ((x \wedge z) \vee (y \wedge z)) \approx x \vee ((z \wedge x) \vee (y \wedge z))$?
13. $x \vee (y \wedge z) \approx x \vee ((z \wedge x) \vee (y \wedge z))$?
14. $y \wedge z \approx z \wedge y$?
15. $x \vee ((z \wedge x) \vee (y \wedge z)) \approx x \vee ((z \wedge x) \vee (z \wedge y))$?
16. $x \vee (y \wedge z) \approx x \vee ((z \wedge x) \vee (z \wedge y))$?
17. $z \wedge (x \vee y) \approx (z \wedge x) \vee (z \wedge y)$?
18. $(z \wedge x) \vee (z \wedge y) \approx z \wedge (x \vee y)$?
19. $z \wedge (x \vee y) \approx (x \vee y) \wedge z$?
20. $(z \wedge x) \vee (z \wedge y) \approx (x \vee y) \wedge z$?
21. $x \vee ((z \wedge x) \vee (z \wedge y)) \approx x \vee ((x \vee y) \wedge z)$?
22. $x \vee (y \wedge z) \approx x \vee ((x \vee y) \wedge z)$?
23. $x \approx x \wedge (x \vee y)$?
24. $x \wedge (x \vee y) \approx (x \vee y) \wedge x$?
25. $x \approx (x \vee y) \wedge x$?
26. $x \vee ((x \vee y) \wedge z) \approx ((x \vee y) \wedge x) \vee ((x \vee y) \wedge z)$?
27. $(x \vee y) \wedge (x \vee z) \approx ((x \vee y) \wedge x) \vee ((x \vee y) \wedge z)$?
28. $((x \vee y) \wedge x) \vee ((x \vee y) \wedge z) \approx (x \vee y) \wedge (x \vee z)$?
29. $x \vee ((x \vee y) \wedge z) \approx (x \vee y) \wedge (x \vee z)$?
30. $x \vee (y \wedge z) \approx (x \vee y) \wedge (x \vee z)$?

f. Find a derivation to witness $\mathcal{BA} \vdash x'' \approx x$.

3.8.9 Let $\mathsf{Conseq}(\mathcal{S})$ be the set of all equational consequences of a set of equations $\mathcal{S}$, i.e.,

$$\mathsf{Conseq}(\mathcal{S}) = \{s \approx t : \mathcal{S} \models s \approx t\}.$$

This is a natural notion, for if $\mathcal{S}$ is $\mathcal{R}$, the axioms for rings, then $\mathsf{Conseq}(\mathcal{R})$ is the set of equations true in all rings. Likewise, $\mathsf{Conseq}(\mathcal{BA})$ is the set of equations true in all Boolean algebras, and so on.

Show that Conseq has the following properties:

a. $\mathcal{S} \subseteq \mathsf{Conseq}(\mathcal{S})$.

b. $s \approx s \in \mathsf{Conseq}(\mathcal{S})$.

c. $s_1 \approx s_2 \in \mathsf{Conseq}(\mathcal{S})$ implies $s_2 \approx s_1 \in \mathsf{Conseq}(\mathcal{S})$.

d. $s_1 \approx s_2 \in \mathsf{Conseq}(\mathcal{S})$ and $s_2 \approx s_3 \in \mathsf{Conseq}(\mathcal{S})$ imply $s_1 \approx s_3 \in \mathsf{Conseq}(\mathcal{S})$.

e. $\mathsf{Conseq}(\mathcal{S})$ is closed under *substitution*.

f. $\mathsf{Conseq}(\mathcal{S})$ is closed under *replacement*.

g. Prove that if $\mathcal{S}_1 \subseteq \mathcal{S}_2$ then $\mathsf{Conseq}(\mathcal{S}_1) \subseteq \mathsf{Conseq}(\mathcal{S}_2)$.

h. Prove that for any set of equations $\mathcal{S}$ we have $\mathsf{Conseq}(\mathsf{Conseq}(\mathcal{S})) = \mathsf{Conseq}(\mathcal{S})$.

i. Prove that for any sets of equations $\mathcal{S}_1$ and $\mathcal{S}_2$ we have

$$\mathsf{Conseq}(\mathcal{S}_1) \cup \mathsf{Conseq}(\mathcal{S}_2) \subseteq \mathsf{Conseq}(\mathcal{S}_1 \cup \mathcal{S}_2).$$

j. Prove that for any sets of equations $\mathcal{S}_1$ and $\mathcal{S}_2$ we have

$$\mathsf{Conseq}(\mathcal{S}_1 \cup \mathsf{Conseq}(\mathcal{S}_2)) = \mathsf{Conseq}(\mathcal{S}_1 \cup \mathcal{S}_2).$$

3.9 SOUNDNESS

Theorem 3.9.1 Birkhoff's rules are sound, i.e.,

$$\mathcal{S} \vdash s \approx t \text{ implies } \mathcal{S} \models s \approx t.$$

Proof. Suppose $\mathcal{S} \vdash s \approx t$, and let $s_1 \approx t_1, \ldots, s_n \approx t_n$ be a derivation to witness this fact. Then $\mathcal{S} \models s_1 \approx t_1$, since either $s_1 = t_1$ or $s_1 \approx t_1$ is in $\mathcal{S}$. Now, applying any rule to consequences of $\mathcal{S}$ gives a consequence of $\mathcal{S}$. Thus we see that $\mathcal{S} \models s_2 \approx t_2$, ..., $\mathcal{S} \models s_n \approx t_n$. But then $\mathcal{S} \models s \approx t$. □

The soundness of Birkhoff's rules gives our basic tool for proving that $\mathcal{S} \models s \approx t$, namely, we look for a derivation of $s \approx t$ from $\mathcal{S}$.

3.10 COMPLETENESS

The proof of the completeness of Birkhoff's rules for equational logic is brief but somewhat demanding in its level of sophistication. Not only is this an interesting result, but it is an excellent warm–up for the style of argument needed to prove Gödel's completeness theorem for first–order logic.

3.10.1 The Construction of $\mathbf{Z}_n$

First, we recall the construction of Z_n, the integers modulo n, and how we prove that Z_n is a ring. These ideas will be paralleled in the completeness theorem that follows.

The steps are as follows.

a. Let $\equiv_n$ be the relation of *congruence modulo* n on the integers, namely, $a \equiv_n b$ iff $n \mid a - b$ (n divides $a - b$).

b. Prove that $\equiv_n$ is an *equivalence* relation on Z, i.e., it is reflexive, symmetric and transitive.

c. Define $[a]$ to be the equivalence class of a with respect to $\equiv_n$, i.e.,

$$[a] = \{j \in Z \, : \, j \equiv_n a\}.$$

d. Interpret the ring function symbols and constants as follows:

$$\begin{array}{l} 0 = [0],\ 1 = [1] \\ -[a] = [-a] \\ [a] + [b] = [a+b] \\ [a] \cdot [b] = [a \cdot b]. \end{array}$$

e. Show that this interpretation is *well–defined*, i.e., that we have assigned functions to the function symbols.

f. Verify the ring axioms $\mathcal{R}$.

3.10.2 The Proof of the Completeness Theorem

Theorem 3.10.1 [Birkhoff, 1935] Birkhoff's rules are complete, i.e.,

$$\mathcal{S} \models s \approx t \text{ implies } \mathcal{S} \vdash s \approx t.$$

Proof. To show $\mathcal{S} \models s \approx t$ implies $\mathcal{S} \vdash s \approx t$ we would like to exhibit a derivation of $s \approx t$ from $\mathcal{S}$. But that is not the way we will proceed. Instead, we will look at the contrapositive, $\mathcal{S} \nvdash s \approx t$ implies $\mathcal{S} \not\models s \approx t$.

We consider the binary relation $\sim$ on the set of terms $T(X)$ defined by: $s \sim t$ holds iff $\mathcal{S} \vdash s \approx t$. Then $\sim$ is an equivalence relation on $T(X)$ that is preserved by substitution and replacement. Let $[t]$ be the equivalence class of t under $\sim$.

Now we will do a construction similar to that used to create Z_n, using $T(X)$ instead of Z, and $\sim$ instead of $\equiv_n$.

Let A be the set of equivalence classes of terms modulo the equivalence relation $\sim$, i.e., $A = \{[t] \, : \, t \in T(X)\}$. Next, we define an interpretation I on A as follows:

- $I(c) = [c]$ for $c \in \mathcal{C}$;
- $I(f)([t_1], \ldots, [t_n]) = [ft_1 \cdots t_n]$.

We need to check that I is well defined, namely, if $s_i \sim t_i$, then $[fs_1 \cdots s_n] = [ft_1 \cdots t_n]$, i.e., $fs_1 \cdots s_n \sim ft_1 \cdots t_n$. But this follows from repeated use of replacement and transitivity, e.g., we have

$$\mathcal{S} \vdash fs_1 s_2 \cdots s_n \approx ft_1 s_2 \ldots s_n$$

by replacement. And then

$$\mathcal{S} \vdash ft_1s_2\cdots s_n \approx ft_1t_2\ldots s_n,$$

also by replacement, and so forth.

Let $\mathbf{A} = (A, I)$. Now this is a remarkable algebra.

Claim 1 $\mathbf{A} \models \mathcal{S}$.

Suppose $s(x_1,\ldots,x_n) \approx s'(x_1,\ldots,x_n) \in \mathcal{S}$. Then for any choice of elements $[t_1],\ldots,[t_n] \in A$ we have

$$\mathcal{S} \vdash s(t_1,\ldots,t_n) \approx s'(t_1,\ldots,t_n),$$

so

$$[s(t_1,\ldots,t_n)] = [s'(t_1,\ldots,t_n)],$$

and thus

$$s^{\mathbf{A}}([t_1],\ldots,[t_n])] = s'^{\mathbf{A}}([t_1],\ldots,[t_n])].$$

But then we have

$$\mathbf{A} \models s \approx s'.$$

Claim 2 Suppose $\mathcal{S} \nvdash s(x_1,\ldots,x_n) \approx t(x_1,\ldots,x_n)$. Then $\mathbf{A} \not\models s \approx t$.

We have $[s(x_1,\ldots,x_n)] \neq [t(x_1,\ldots,x_n)]$, so $s^{\mathbf{A}}([x_1],\ldots,[x_n]) \neq t^{\mathbf{A}}([x_1],\ldots,[x_n])$. Thus $\mathbf{A} \not\models s \approx t$.

With these two claims we see that if $\mathcal{S} \nvdash s \approx t$, then we have $\mathbf{A} \models \mathcal{S}$, but $\mathbf{A} \not\models s \approx t$. Thus $\mathcal{S} \not\models s \approx t$. □

3.10.3 Valid Arguments Revisited

With the completeness theorem we now have our arsenal for analyzing arguments:

- An equational argument

$$s_1 \approx t_1, \cdots, s_n \approx t_n \qquad \therefore s \approx t$$

  is valid iff there is a *derivation* of the conclusion from the premisses, i.e., iff

$$s_1 \approx t_1, \cdots, s_n \approx t_n \vdash s \approx t.$$

  Thus to show an argument is valid we can look for a derivation.

- An equational argument

$$s_1 \approx t_1, \cdots, s_n \approx t_n \qquad \therefore s \approx t$$

is invalid iff there is a *counterexample* **A**, i.e., iff some **A** makes the premisses true and the conclusion false.

Thus we can show an argument is not valid by finding a counterexample.

Exercises

For the following problems, show that each argument is either valid by finding a derivation of the conclusion from the premisses, or show it is not valid by finding a counterexample.

3.10.1

a. $x \cdot y \approx x$
$\therefore\ x \cdot y \approx x \cdot z$

b. $x \cdot y \approx x$
$\therefore\ x \cdot y \approx y \cdot x$

c. $x \cdot y \approx x$
$\therefore\ (x \cdot y) \cdot z \approx x \cdot (y \cdot z)$

d. $x \cdot y \approx u \cdot v$
$\therefore\ x \cdot y \approx y \cdot x$

e. $x \cdot y \approx u \cdot v$
$\therefore\ (x \cdot y) \cdot z \approx x \cdot (y \cdot z)$

f. $x \cdot y \approx u \cdot v$
$\therefore\ x \approx x \cdot x$

g. $x \cdot y \approx y \cdot x$
$(x \cdot y) \cdot z \approx x \cdot (y \cdot z)$
$\therefore\ x \cdot (y \cdot z) \approx z \cdot (y \cdot x)$

h. $x \cdot x \approx x$
$x \cdot y \approx y \cdot x$
$(x \cdot y) \cdot z \approx x \cdot (y \cdot z)$
$\therefore\ x \cdot (y \cdot z) \approx x \cdot (y \cdot w)$

i. $fffx \approx fx \qquad \therefore ffx \approx x$

j. $fx \approx fy \qquad \therefore ffx \approx fx$

k. $fffx \approx fffy \qquad \therefore fffx \approx fx$

l. $fx \approx fffx \qquad \therefore fx \approx fffffx$

m. $fgfx \approx gfgx \qquad \therefore fgx \approx gfx$

n. $ffx \approx ggx \qquad \therefore fx \approx gx$

o. $fx \approx gx \qquad \therefore ffx \approx ggx$

A set of equational axioms $\mathcal{A}$ is said to be *independent* if for any $s \approx t \in \mathcal{A}$ we have

$$\mathcal{A} \setminus \{s \approx t\} \not\models s \approx t.$$

3.10.2 Show that the group axioms $\mathcal{G}$ are independent, i.e., that any two of the three axioms can be satisfied by an algebra (two elements will do!) that does not satisfy the other axiom.

3.11 CHAIN DERIVATIONS

Now we look at a briefer format for presenting equational proofs. It is briefer because we collapse several of the preceding steps into one—which is closer to everyday practice when working with equations.

Definition 3.11.1 An *elementary derivation* from a set $\mathcal{S}$ of equations is an equation of the form

$$r(\cdots s' \cdots) \approx r(\cdots t' \cdots), \tag{27}$$

where $s' \approx t'$ is a substitution instance of $s \approx t$, and one of the following holds for $s \approx t$:

- $s = t$
- $s \approx t \in \mathcal{S}$
- $t \approx s \in \mathcal{S}$.

Equation (27) is the result of replacing a single occurrence of s' on the left with t' to obtain the right–hand side.

Clearly, elementary derivations from $\mathcal{S}$ are consequences of $\mathcal{S}$.

Definition 3.11.2 A *chain derivation* of $s \approx t$ from $\mathcal{S}$ is a sequence of elementary derivations from $\mathcal{S}$ of the form $t_1 \approx t_2$, $t_2 \approx t_3$, … , $t_{n-1} \approx t_n$, with $s = t_1$ and $t = t_n$. This sequence is often written in the compact form

$$s = t_1 \approx t_2 \approx \cdots \approx t_n = t.$$

Example 3.11.3 From the single premiss (for *left zero* multiplication)

$$\mathbf{LZ} : x \cdot y \approx x$$

we will derive $x \cdot (y \cdot z) \approx (x \cdot y) \cdot z$ using a chain derivation:

1.	$x \cdot (y \cdot z)$	$\approx$	x	from LZ
2.		$\approx$	$x \cdot y$	from LZ
3.		$\approx$	$(x \cdot y) \cdot z$	from LZ.

Example 3.11.4 For a more challenging situation we give a chain derivation for the result in Example 3.8.7 (p. 177), namely, we give a chain derivation that witnesses $\mathcal{G} \vdash 1 \cdot x \approx x$:

1.	$1 \cdot x$	$\approx$	$(1 \cdot x) \cdot 1$	from G1
2.		$\approx$	$(1 \cdot x) \cdot (x^{-1} \cdot (x^{-1})^{-1})$	from G2
3.		$\approx$	$((1 \cdot x) \cdot x^{-1}) \cdot (x^{-1})^{-1}$	from G3
4.		$\approx$	$(1 \cdot (x \cdot x^{-1})) \cdot (x^{-1})^{-1}$	from G3
5.		$\approx$	$(1 \cdot 1) \cdot (x^{-1})^{-1}$	from G2
6.		$\approx$	$1 \cdot (x^{-1})^{-1}$	from G2
7.		$\approx$	$(x \cdot x^{-1}) \cdot (x^{-1})^{-1}$	from G2
8.		$\approx$	$x \cdot (x^{-1} \cdot (x^{-1})^{-1})$	from G3
9.		$\approx$	$x \cdot 1$	from G2
10.		$\approx$	x	from G1.

It is no accident that we were able to find a chain derivation in the last example, knowing that there was a derivation. Indeed, all consequences of a set of equations $\mathcal{S}$ can be found by chain derivations.

Theorem 3.11.5 $\mathcal{S} \vdash s \approx t$ holds iff there is a chain derivation of $s \approx t$ from $\mathcal{S}$.

Proof. Clearly, if there is a chain derivation of $s \approx t$ from $\mathcal{S}$, then there is a Birkhoff derivation of $s \approx t$ from $\mathcal{S}$. This is because each of the elementary derivations can be obtained by several steps of Birkhoff derivation, and the chaining is just an application of transitivity.

Let $\mathsf{Chain}(\mathcal{S})$ be the set of equations $s \approx t$ that can be obtained from $\mathcal{S}$ by chain derivations. Clearly, $\mathcal{S} \subseteq \mathsf{Chain}(\mathcal{S})$. To show that $\mathsf{Chain}(\mathcal{S})$ has all consequences of $\mathcal{S}$ it suffices to show that $\mathsf{Chain}(\mathcal{S})$ is closed under Birkhoff's rules.

a. $\mathsf{Chain}(\mathcal{S})$ is closed under the reflexive rule:
For any term s, $s \approx s \in \mathsf{Chain}(\mathcal{S})$, as $s \approx s$ is an elementary derivation.

b. $\mathsf{Chain}(\mathcal{S})$ is closed under the symmetric rule:
If $t_1 \approx \cdots \approx t_n$ is a chain derivation from $\mathcal{S}$, then so is $t_n \approx \cdots \approx t_1$.

c. $\mathsf{Chain}(\mathcal{S})$ is closed under the transitive rule:
If $t_1 \approx \cdots \approx t_m$ and $t_m \approx \cdots \approx t_n$ are chain derivations from $\mathcal{S}$, then so is $t_1 \approx \cdots \approx t_m \approx \cdots \approx t_n$.

d. Chain($\mathcal{S}$) is closed under the substitution rule:
To see this, first note that if

$$r(\cdots s' \cdots) \approx r(\cdots t' \cdots)$$

is an elementary derivation from $\mathcal{S}$, then for any substitution into this equation, which we denote by a superscript $\star$, we have

$$(r(\cdots s' \cdots))^\star \approx (r(\cdots t' \cdots))^\star$$

is also an elementary derivation from $\mathcal{S}$; namely, we replace an occurrence of $s'^\star$ in the left–hand side with $t'^\star$ to obtain the right. So if $t_1 \approx \cdots \approx t_n$ is a chain derivation from $\mathcal{S}$, then so is $t_1{}^\star \approx \cdots \approx t_n{}^\star$.

e. Chain($\mathcal{S}$) is closed under the replacement rule:
To see this, first note that if $r'(\cdots s' \cdots) \approx r'(\cdots t' \cdots)$ is an elementary derivation from $\mathcal{S}$, then we also have an elementary derivation $r(\cdots r'(\cdots s' \cdots) \cdots) \approx r(\cdots r'(\cdots t' \cdots) \cdots)$. So if $t_1 \approx \cdots \approx t_n$ is a chain derivation from $\mathcal{S}$, then so is $r(\cdots t_1 \cdots) \approx \cdots \approx r(\cdots t_n \cdots)$. □

Exercises

3.11.1 Find chain derivations for claims b, c, and d in Exercise 3.8.1 on p. 178.

3.11.2 Find chain derivations for claims a, c, and d in Exercise 3.8.2 on p. 178.

3.11.3 Find chain derivations for claim a in Exercise 3.8.3 on p. 179.

3.11.4 Find chain derivations for claim b in Exercise 3.8.5 on p. 179.

3.11.5 Find chain derivations for claim a in Exercise 3.8.6 on p. 179.

3.11.6 Find chain derivations for claims b and c in Exercise 3.8.8 on p. 181.

3.12 UNIFICATION

One of the most popular and powerful tools of automated theorem proving is an algorithm to find most general unifiers.

3.12.1 Unifiers

We assume we are working in a fixed language of algebras.

Definition 3.12.1 Let $s(x_1, \ldots, x_n)$ and $s'(x_1, \ldots, x_n)$ be two terms. A *unifier* of s and s' is a substitution

$$\begin{pmatrix} x_1 \leftarrow t_1 \\ \vdots \\ x_n \leftarrow t_n \end{pmatrix}$$

such that $s(t_1, \ldots, t_n) = s'(t_1, \ldots, t_n)$, i.e, after the substitution has been carried out, the two terms have become the same term.

If a unifier can be found for s and t, we say they *can be unified*, or they *are unifiable*.

Example 3.12.2 Let $s(x, y) = x + y \cdot y$ and let $t(x, y) = x + y \cdot x$. Then the substitution

$$\begin{pmatrix} x \leftarrow y \\ y \leftarrow y \end{pmatrix}$$

is a unifier. After this substitution is applied, both s and t become $y + y \cdot y$. Actually, there are many unifiers, namely, we also have

$$\begin{pmatrix} x \leftarrow 0 \\ y \leftarrow 0 \end{pmatrix}$$

and

$$\begin{pmatrix} x \leftarrow u + v \\ y \leftarrow u + v \end{pmatrix}$$

and

$$\begin{pmatrix} x \leftarrow z \\ y \leftarrow z \end{pmatrix}.$$

Of the four unifiers given, the first one is the "best" because by doing a further substitution we can obtain the other unifiers. If we follow the first substitution by

$$\begin{pmatrix} y \leftarrow 0 \end{pmatrix},$$

we have the second substitution. And if we follow the first by

$$\begin{pmatrix} y \leftarrow u + v \end{pmatrix},$$

we have the third substitution. We say the first unifier is *more general* than the second and third unifiers. Note that the first and fourth unifiers are more general than each other, since we can go from either to the other by a further substitution.

Exercises

3.12.1 For the following substitutions

$$\textbf{1.}\ \begin{pmatrix} x \leftarrow ffx \\ y \leftarrow fffx \end{pmatrix} \qquad \textbf{2.}\ \begin{pmatrix} x \leftarrow ffx \\ y \leftarrow fx \end{pmatrix}$$

determine if they are unifiers of the following pairs of terms:

a. $ffffx$ and $fffy$
b. ffx and fx
c. fgx and $fgfy$.

3.12.2 For the following substitutions

$$\textbf{1.}\ \begin{pmatrix} x \leftarrow u+v \\ y \leftarrow u \cdot v \end{pmatrix} \qquad \textbf{2.}\ \begin{pmatrix} x \leftarrow u+u \\ y \leftarrow u+v \end{pmatrix}$$

determine if they are unifiers of the following pairs of terms:

a. $x + (u \cdot v)$ and $(u + v) + y$
b. $(u + x) \cdot (u + v)$ and $(u + (u + u)) \cdot y$.

3.12.2 A Unification Algorithm

The main result on unification is that if two terms s and t are unifiable, then there is a *most general* unifier μ of the two terms, i.e., if σ is any other unifier, then μ is more general than σ, so $\sigma = \tau\mu$ for some substitution τ. This result says that every unifier of s and t is a special case of μ. Furthermore, there is an algorithm to determine if s and t are unifiable, and if so, it finds the most general unifier. The basic idea and results on unification were known to Herbrand in 1930, but the first detailed proof was published by Robinson in 1965.

Let us look at the algorithm first, and in Sec. 3.12.5 we will prove that it works as claimed.

So let us start with two terms s and t and write them in *prefix* notation. We assume $s \neq t$, otherwise every substitution is a unifier, and the most general unifier is the identity substitution $(x \leftarrow x)$, for all variables x.

We arrange them one above the other, in subdivided rectangles with one square for each symbol. The *shaded region* denotes precisely the initial squares where they are identical.

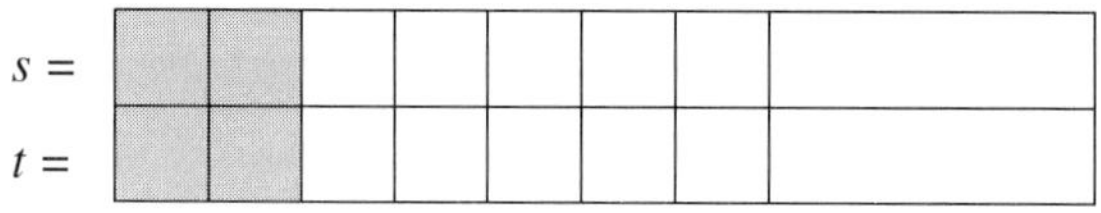

Figure 3.12.3 The grid used for the unification algorithm

Let s' and t' be the subterms of s, respectively t, that start with the first unshaded square. These are called the *critical subterms* of the pair.

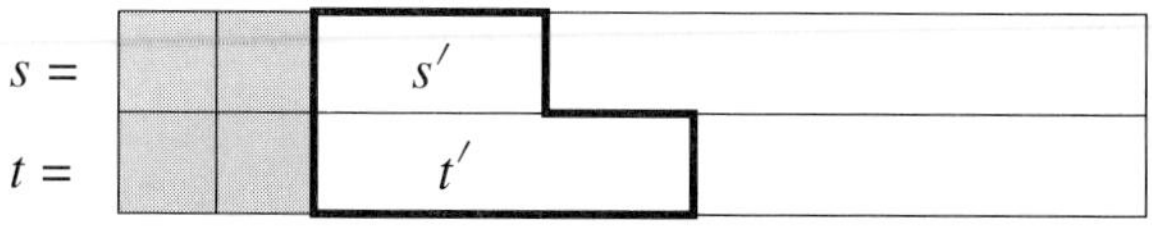

Figure 3.12.4 The critical subterms s' and t'

Example 3.12.5 For the terms

$$\begin{aligned} s &= (x+y)\cdot((x\cdot y)+x) \\ t &= (x+y)\cdot((y+x\cdot y)+y) \end{aligned}$$

the critical subterms are

$$\begin{aligned} s' &= x\cdot y \\ t' &= y+(x\cdot y), \end{aligned}$$

as we have

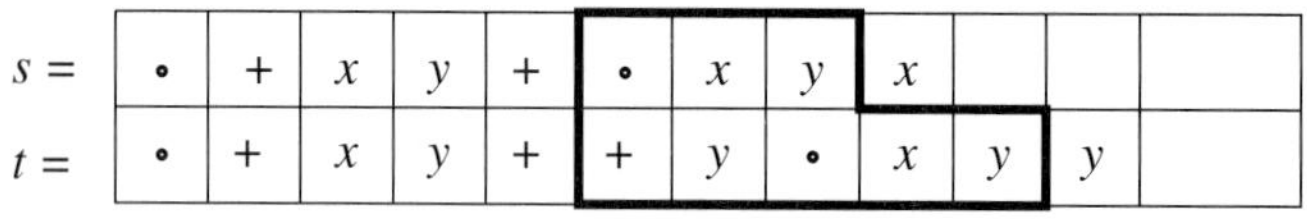

$s =$	•	+	x	y	+	•	x	y	x			
$t =$	•	+	x	y	+	+	y	•	x	y	y	

Figure 3.12.6 The critical subterms of $(x+y)\cdot((x\cdot y)+x)$ and $(x+y)\cdot((y+x\cdot y)+y)$

Definition 3.12.7 The *critical subterm condition* (CSC) is satisfied by s and t if their critical subterms s' and t' consist of a variable, say x, and another term, say r, that has no occurrence of x in it.

Now we can present the unification algorithm:

> let μ be the identity substitution
> WHILE $s \neq t$
>   find the critical subterms s', t'
>   **if** the CSC fails, return NOT UNIFIABLE
>   **else** with $\{s', t'\} = \{x, r\}$
>   apply $(x \leftarrow r)$ to both s and t, i.e.
> $s \leftarrow (x \leftarrow r)s$ and $t \leftarrow (x \leftarrow r)t$
>   apply $(x \leftarrow r)$ to μ, i.e.
> $\mu \leftarrow (x \leftarrow r)\mu$
> ENDWHILE
> return (μ)

Figure 3.12.8 captures the "else" condition inside the WHILE loop, where s and t satisfy the critical subterm condition. We depict the situation where s' is a variable. The substitution $x \leftarrow r$ takes the darkly shaded region in the first figure to the darkly shaded region in the second figure. The lightly shaded region shows what happens to the critical subterms after the substitution, namely, that they both become r. (The new terms may agree farther to the right of the lightly shaded region.)

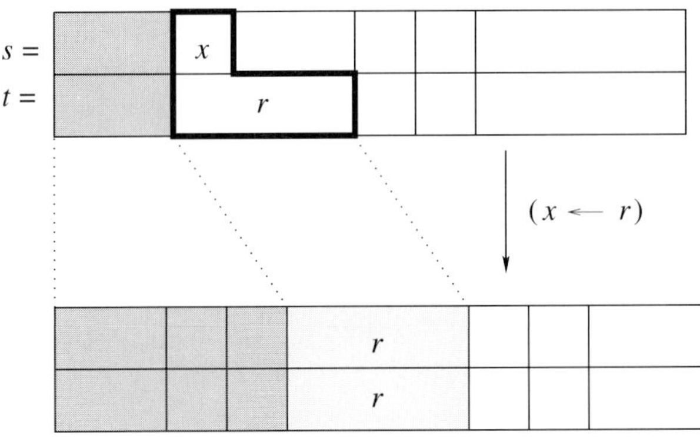

Figure 3.12.8 Applying the substitution $(x \leftarrow r)$

Example 3.12.9 Applying the unification algorithm to the terms $x + y$ and $x \cdot y$:

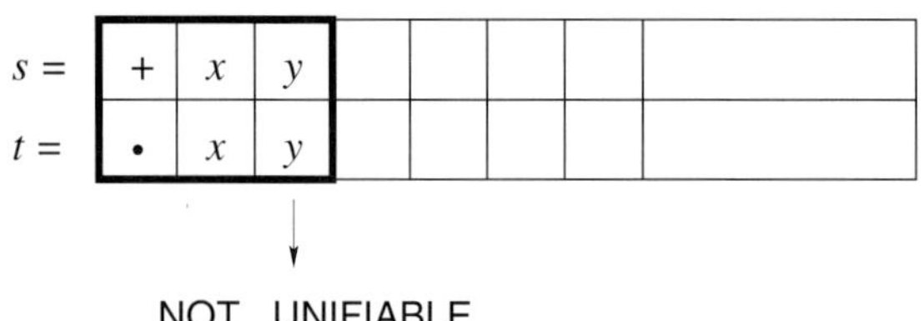

Figure 3.12.10 Applying the unification algorithm to $x + y$ and $x \cdot y$

The CSC fails on the first step. The critical subterms are in the bold outlined boxes. Neither is a variable.

Example 3.12.11 Applying the unification algorithm to $x + (y \cdot y)$ and $(x \cdot x) + y$:

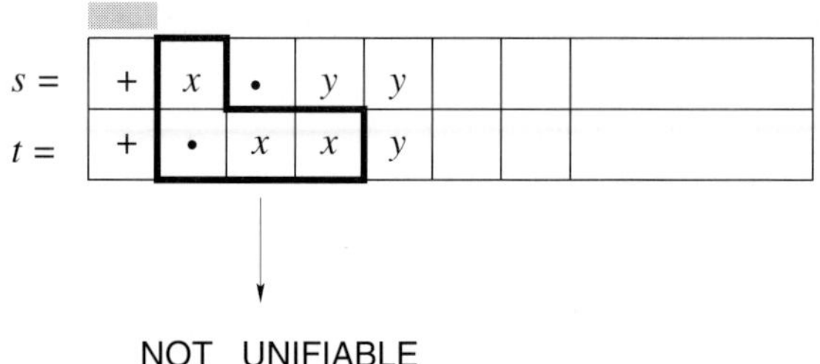

Figure 3.12.12 Applying the unification algorithm to $x + (y \cdot y)$ and $(x \cdot x) + y$

Again the CSC fails on the first step. The identical initial segments are indicated by the *shaded bar* above the grid. One of the critical subterms is a variable, but it also appears in the other critical subterm.

Example 3.12.13 Applying the unification algorithm to the terms $x + x$ and $x + (y \cdot y)$:

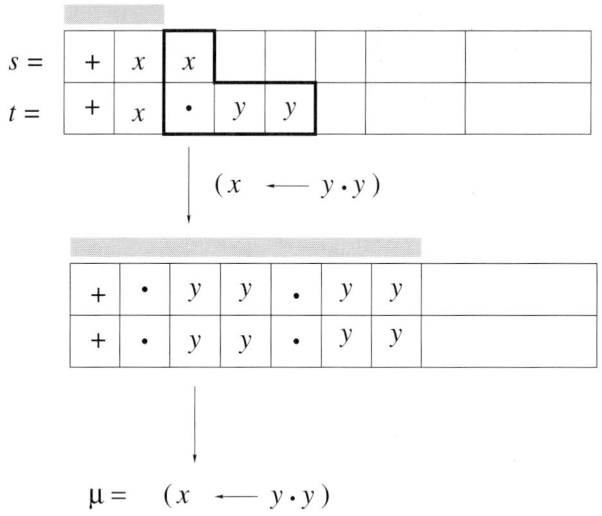

Figure 3.12.14 Applying the unification algorithm to $x + x$ and $x + (y \cdot y)$

These two terms are unifiable, and after one step we have the most general unifier:

$$\mu = (\ x \leftarrow y \cdot y\ ) \,.$$

Example 3.12.15 Applying the unification algorithm to the terms $x + (y \cdot y)$ and $(y \cdot y) + z$:

$s =$	+	x	•	y	y			
$t =$	+	•	y	y	z			

$(x \leftarrow y \cdot y)$

+	•	y	y	•	y	y	
+	•	y	y	z	•		

$(z \leftarrow y \cdot y)$

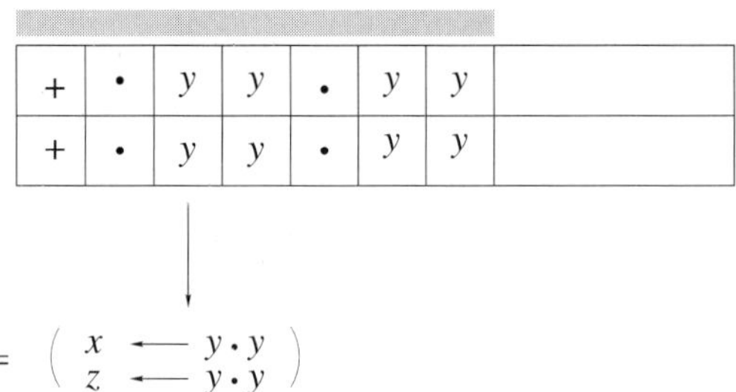

Figure 3.12.16 Applying the unification algorithm to $x + (y \cdot y)$ and $(y \cdot y) + z$

These two terms are unifiable, and after two steps we have the most general unifier:

$$\mu = \begin{pmatrix} x \leftarrow y \cdot y \\ z \leftarrow y \cdot y \end{pmatrix}.$$

Exercises

3.12.3 Find the most general unifier of each of the following pairs, or show that no unifier exists.

- **a.** $((x + y) \cdot fu) + z$ and $v + (y \cdot (fw \cdot (u + u)))$
- **b.** $(x + y) \cdot (x + z)$ and $(x + z) \cdot (y + z)$
- **c.** $x + (y + x)$ and $(x + y) + x$
- **d.** $(x \cdot y) \cdot z$ and $u \cdot u^{-1}$
- **e.** $x \cdot y$ and $u \cdot u^{-1}$
- **f.** $(x \cdot y) \cdot z$ and $u \cdot v$
- **g.** $(x \cdot y) \cdot z$ and $u \cdot (v^{-1} \cdot v)$
- **h.** $(x \cdot y) \cdot z$ and $v \cdot (u \cdot v)^{-1}$
- **i.** $x \cdot y$ and $x \cdot (y \cdot (u \cdot v)^{-1})$
- **j.** $x \cdot y$ and $y \cdot (u \cdot v)^{-1}$
- **k.** $(x \cdot y) \cdot z$ and $v \cdot (u \cdot v)^{-1}$

3.12.3 Properties of Prefix Notation for Terms

Before proving the correctness of the unification algorithm, we need to review some simple properties of our terms, namely, that given a term t it must be either a variable or a constant, or it must be of the form $ft_1 \cdots t_n$. In the latter case the choice of f and the t_i is unique, i.e., we have *unique readability*. Thus, for example, given that f is unary, g is binary, and h is ternary, the term

$$ghwvhfzgzuyhxwhgffywxgyu$$

must be of the form gt_1t_2, and t_1, t_2 are unique, namely,

$$\begin{aligned} t_1 &= hwvhfzgzuy; \\ t_2 &= hxwhgffywxgyu. \end{aligned}$$

We also need the property that if a term s is an *initial segment* of a term t, i.e., s is the first k symbols of t, then we must have $s = t$. For example, consider the term $x + (y \cdot y)$. In prefix notation we have $+x \cdot yy$, and the initial segments are

$$+ \qquad +x \qquad +x\,\cdot \qquad +x \cdot y \qquad +x \cdot yy,$$

and only the last one is a term. The initial segments that are not the whole term are called *proper initial segments*.

In the proof of the following theorem we let $|t|$, the *length* of t, be the number of symbols in t, e.g., the length of $+x \cdot yy$ is 5.

Theorem 3.12.17 Terms are uniquely readable, and proper initial segments of terms cannot be terms.

Proof. We will prove both facts together, using induction on the length of terms. The assertion to be proved is: (i) for every term t either t is a variable or t is a constant symbol or t is uniquely of the form $ft_1 \cdots t_n$, (ii) no proper initial segment of t is a term, and (iii) t is not a proper initial segment of any term.

- **Ground cases** If $|t| = 1$, then t is either a variable or a constant symbol. Then (i)—(iii) are easily seen to hold.
- **Induction hypothesis:** We suppose the theorem is true of terms of length at most ℓ.
- **Induction step:** We want to show that if $|t| = \ell + 1$, then t has properties (i) and (ii).

  First, we write $t = ft_1 \cdots t_n$, for suitable f, n, and $t_1, \dots, t_n$. (Every term that is not a variable or a constant symbol is in this form.) Then we observe that all t_i have length at most ℓ, so the induction hypothesis applies to them.

  For (i) suppose that we also have $t = fs_1 \cdots s_m$. (Of course, the first symbol of t must be f.) Suppose some s_i is not the same as the term t_i. Let k be the first subscript i where $s_i \neq t_i$. Then either s_k is a proper initial segment of t_k, or vice versa. But this contradicts the induction hypothesis regarding t_k satisfying (ii). Thus (i) holds for t.

Next, suppose s is a proper initial segment of t, say $s = fs_1 \cdots s_m$. But then $|s| \leq \ell$, so the induction hypothesis guarantees unique readability. As f is n–ary, there must be n terms following f, so $m = n$. Now each s_i must be the same as t_i, otherwise we look at k, the first i for which they differ, and discover (as before) that t_k violates condition (ii). But then $s = ft_1 \cdots t_n = t$, and we do not have a proper initial segment after all. Thus (ii) holds for t.

Finally, to show (iii) holds for t suppose t is a proper initial segment of a term $s = fs_1 \cdots s_n$. Then some t_i must be either a proper intital segment or a proper extension of some s_i. But this is impossible, as t_i satisfies the induction hypothesis. Thus (iii) holds. □

3.12.4 Notation for Substitutions

To make the notion of "more general" workable it helps to introduce symbolic names for substitutions. We will use lowercase Greek letters for this purpose. Thus σ represents a substitution

$$\begin{pmatrix} x_1 \leftarrow t_1 \\ \vdots \\ x_n \leftarrow t_n \end{pmatrix},$$

and we write $\sigma(s)$, or just σs, when referring to the substitution instance $s(t_1, \ldots, t_n)$ of the term $s(x_1, \ldots, x_n)$. Note that we have the simple property

$$\sigma s(x_1, \ldots, x_n) = s(\sigma x_1, \ldots, \sigma x_n),$$

since $\sigma x_i = t_i$. More generally,

$$\sigma s(s_1, \ldots, s_n) = s(\sigma s_1, \ldots, \sigma s_n).$$

This notation is particularly handy for describing "one substitution after another." For example, suppose σ and σ' are substitutions; say σ is given by

$$\begin{pmatrix} x_1 \leftarrow t_1(x_1, \ldots, x_n) \\ \vdots \\ x_n \leftarrow t_n(x_1, \ldots, x_n) \end{pmatrix},$$

and σ' is given by

$$\begin{pmatrix} x_1 \leftarrow t'_1(x_1, \ldots, x_n) \\ \vdots \\ x_n \leftarrow t'_n(x_1, \ldots, x_n) \end{pmatrix}.$$

Then σ, followed by σ', is given by

$$\begin{pmatrix} x_1 \leftarrow t_1(t'_1, \ldots, t'_n) \\ \vdots \\ x_n \leftarrow t_n(t'_1, \ldots, t'_n) \end{pmatrix},$$

i.e., by

$$\begin{pmatrix} x_1 \leftarrow t_1(\sigma' x_1, \ldots, \sigma' x_n) \\ \vdots \\ x_n \leftarrow t_n(\sigma' x_1, \ldots, \sigma' x_n) \end{pmatrix}.$$

This is

$$\begin{pmatrix} x_1 \leftarrow \sigma' t_1(x_1, \ldots, x_n) \\ \vdots \\ x_n \leftarrow \sigma' t_n(x_1, \ldots, x_n) \end{pmatrix},$$

which reduces to

$$\begin{pmatrix} x_1 \leftarrow \sigma' \sigma x_1 \\ \vdots \\ x_n \leftarrow \sigma' \sigma x_n \end{pmatrix}.$$

Thus doing σ' after σ gives the *composition* of the two substitutions, $\sigma'\sigma$. Doing the substitution σ to a term t and then σ' to the result is the same as doing the single substitution $\sigma'\sigma$ to the term t.

Now we are ready to give our definition of "more general."

Definition 3.12.18 The substitution σ is *more general* than the substitution σ' if there is a substitution τ such that

$$\sigma' = \tau\sigma,$$

i.e., the substitution σ' can be obtained from σ by applying a further substitution, τ.

Exercises

The first three problems refer to the substitutions

$$\sigma_1 = \begin{pmatrix} x \leftarrow y \\ y \leftarrow z \\ z \leftarrow x \end{pmatrix} \qquad \sigma_2 = \begin{pmatrix} x \leftarrow y \\ y \leftarrow z \\ z \leftarrow y \end{pmatrix} \qquad \sigma_3 = \begin{pmatrix} x \leftarrow x + y \\ y \leftarrow y + z \\ z \leftarrow x + z \end{pmatrix}.$$

3.12.4

a. Find $\sigma_1\sigma_2$.
b. Find $\sigma_2\sigma_2$.
c. Find $\sigma_2\sigma_3$.
d. Find $\sigma_1\sigma_2\sigma_3$.

3.12.5

a. Determine if $\sigma_2\sigma_3 = \sigma_3\sigma_2$.
b. Determine if $\sigma_1\sigma_3 = \sigma_3\sigma_1$.

3.12.6

a. Determine if σ_2 is more general than σ_1.
b. Determine if σ_1 is more general than σ_2.
c. Determine if σ_3 is more general than σ_1.
d. Determine if σ_1 is more general than σ_3.

3.12.7 Let

$$\sigma = \begin{pmatrix} x \leftarrow t_1 \\ y \leftarrow t_2 \end{pmatrix}.$$

a. Describe all pairs of terms t_1, t_2 such that $\sigma\sigma$ is the identity substitution.
b. Describe all pairs of terms t_1, t_2 such that $\sigma\sigma = \sigma$.

3.12.8 For variables $x_1, \ldots, x_n$ and for y_i among the x_j consider substitutions of the form

$$\sigma = \begin{pmatrix} x_1 \leftarrow y_1 \\ \vdots \\ x_n \leftarrow y_n \end{pmatrix}.$$

a. How many different substitutions σ are of this form?
b. What condition must hold for such a σ to be more general than the identity substitution?

3.12.5 Verification of the Unification Algorithm

In this section we assume that the unification algorithm (p. 193) is being applied to a pair of terms s, t, where the set of variables appearing in s, t is $\{x_1, \ldots, x_n\}$.

Lemma 3.12.19 The unification algorithm terminates.

Proof. To see this, note that any pass in which the CSC fails leads to immediate termination, and with each pass where the CSC holds, the number of variables in the two terms is reduced by 1 since the substitution $x \leftarrow r$ eliminates the variable x. If at some stage a pass through the WHILE loop finishes with no variables, then (i) either the terms are identical and the algorithm terminates, or (ii) they are distinct and the CSC must fail, so the algorithm terminates. □

We now introduce some notation to describe the history of the finitely many passes of the unification algorithm through the WHILE loop.

- Let $s_0 = s$ and $t_0 = t$.
- Let μ_0 be the identity substitution.
- Let k (≥ 0) be the total number of completed passes through the loop, i.e., passes for which the critical subterm condition is satisfied.
- Let s_{i-1}, t_{i-1} be the terms entering the ith pass through the loop.
- For $1 \leq i \leq k$ let $\{s_i', t_i'\} = \{x_i, r_i(x_{i+1}, \ldots, x_n)\}$ be the critical subterms of s_{i-1}, t_{i-1}. (By appropriately relabeling the variables there is no loss in generality in assuming that the variable x_i is one of the critical subterms in the ith pass.)
- For $1 \leq i \leq k$ let

$$\begin{aligned} \mu_i &= (x_i \leftarrow r_i)\mu_{i-1} \\ s_i &= (x_i \leftarrow r_i)s_{i-1} \\ t_i &= (x_i \leftarrow r_i)t_{i-1}. \end{aligned}$$

Lemma 3.12.20 Either the terms s_k, t_k do not satisfy the CSC, or $s_k = t_k$.

Proof. If $s_k \neq t_k$, then we enter the WHILE loop for the $(k+1)$st pass. But there are only k completed passes. Thus the CSC condition must fail to hold. □

Lemma 3.12.21 The terms s_i, t_i are unifiable iff the terms s_{i-1}, t_{i-1} are unifiable, for $1 \leq i \leq k$. Furthermore, if

$$\sigma = \begin{pmatrix} x_i \leftarrow \sigma x_i \\ \vdots \\ x_n \leftarrow \sigma x_n \end{pmatrix}$$

is a unifier of s_{i-1}, t_{i-1}, then

$$\sigma' = \begin{pmatrix} x_{i+1} & \leftarrow & \sigma x_{i+1} \\ & \vdots & \\ x_n & \leftarrow & \sigma x_n \end{pmatrix}$$

is such that $\sigma = \sigma'(x_i \leftarrow r_i)$.

Proof. If s_i, t_i are unifiable, then for some substitution σ we have $\sigma s_i = \sigma t_i$. Now, using the definition of s_i and t_i, we have

$$\sigma(x_i \leftarrow r_i)s_{i-1} = \sigma(x_i \leftarrow r_i)t_{i-1},$$

and thus $\sigma(x_i \leftarrow r_i)$ is a unifier of s_{i-1}, t_{i-1}.

Conversely, suppose s_{i-1}, t_{i-1} are unifiable, then for some substitution

$$\sigma = \begin{pmatrix} x_i \leftarrow \sigma x_i \\ \vdots \\ x_n \leftarrow \sigma x_n \end{pmatrix}$$

we have $\sigma s_{i-1} = \sigma t_{i-1}$. From $\sigma x_i = \sigma r_i(x_{i+1}, \dots, x_n)$ we have

$$\sigma x_i = r_i(\sigma x_{i+1}, \dots, \sigma x_n).$$

Thus we can write σ as a composition of two substitutions as follows:

$$\sigma = \begin{pmatrix} x_{i+1} & \leftarrow & \sigma x_{i+1} \\ & \vdots & \\ x_n & \leftarrow & \sigma x_n \end{pmatrix} (x_i \leftarrow r_i),$$

so $\sigma = \sigma'(x_i \leftarrow r_i)$. Then

$$\begin{aligned} \sigma s_{i-1} = \sigma'(x_i \leftarrow r_i)s_{i-1} &= \sigma' s_i \\ \sigma t_{i-1} = \sigma'(x_i \leftarrow r_i)t_{i-1} &= \sigma' t_i. \end{aligned}$$

Because σ unifies s_{i-1}, t_{i-1} it follows that σ' unifies s_i, t_i. □

Lemma 3.12.22 If $s_k = t_k$, then the terms s, t are unifiable. If s_k, t_k fail the CSC, then s, t are not unifiable.

Proof. If $s_k = t_k$, then clearly the terms s_k, t_k are unifiable, namely, we can use the identity substitution. Then by Lemma 3.12.21 it follows that s, t are unifiable.

Let us look at the various possibilities for s_k, t_k to fail the CSC. As we enter the kth iteration of the WHILE loop we picture the terms s_k, t_k and their critical subterms s'_k, t'_k:

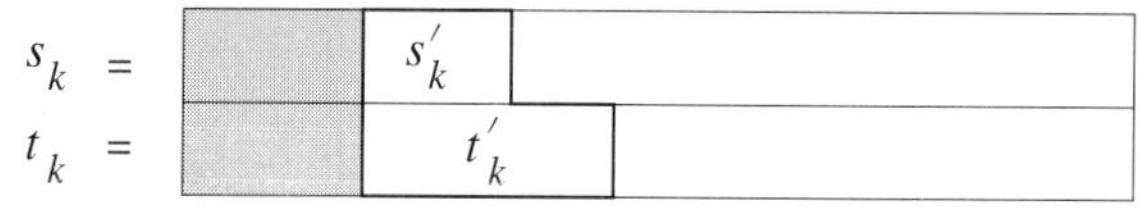

Figure 3.12.23

Both of the critical subterms must exist, as it is impossible to have s_k be a proper initial segment or a proper extension of t_k by Theorem 3.12.17 (p. 197). So we have the two critical subterms s'_k, t'_k of s_k, t_k.

If neither of s'_k, t'_k is a variable, then each is either a constant symbol or starts with a function symbol. We look first at the case in which both start with function symbols. Say we have

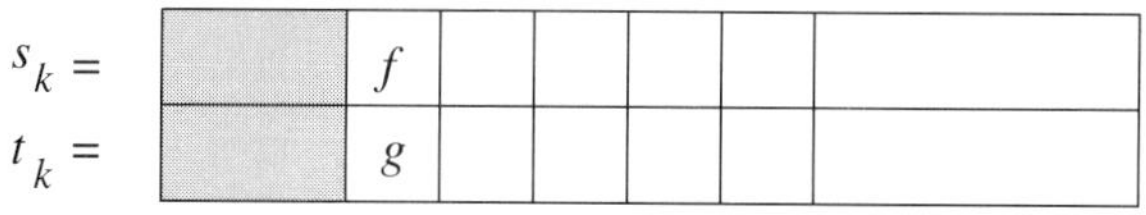

Figure 3.12.24 Clashing function symbols

Note that the function symbols must be different, else their squares would belong to the shaded region. We claim that no unification is possible for s_k, t_k in this case.

Any substitution would send both shaded areas to the same string, but the f and g would not change; they would merely stay in the same place or be displaced to the right:

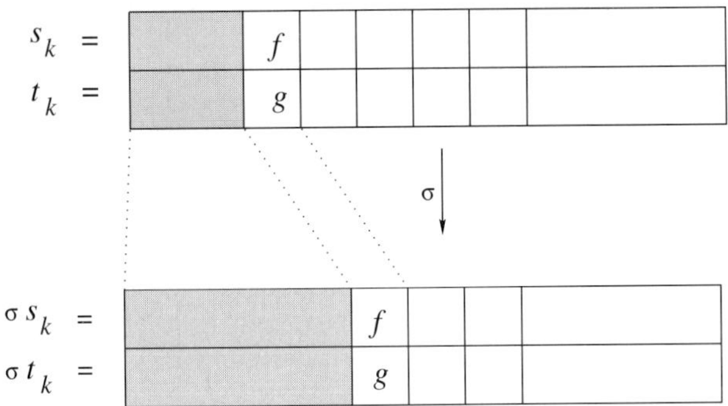

Figure 3.12.25 Clashing function symbols continue to clash after substitution

Thus no substitution σ can be a unifier of s_k, t_k. But then s and t are not unifiable.

The cases that one or both of s'_k, t'_k are a constant symbol are proved in the same manner.

The other possibility for the CSC to fail is if one of the critical subterms is a variable, say x, and that variable occurs in the other critical subterm r. We know $x \neq r$, else there would be another shaded square. So, letting s'_k be x, the situation looks like the following:

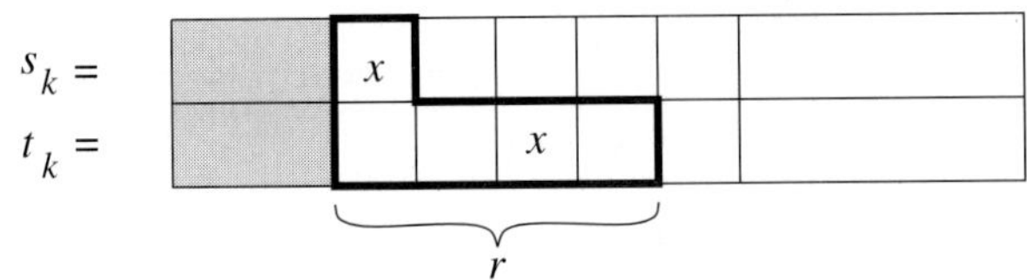

Figure 3.12.26 A variable in a first unshaded square

Clearly, the x cannot be the first symbol in t'_k, for otherwise that square would be shaded. Thus t'_k is more than a variable. Now any substitution σ gives the following:

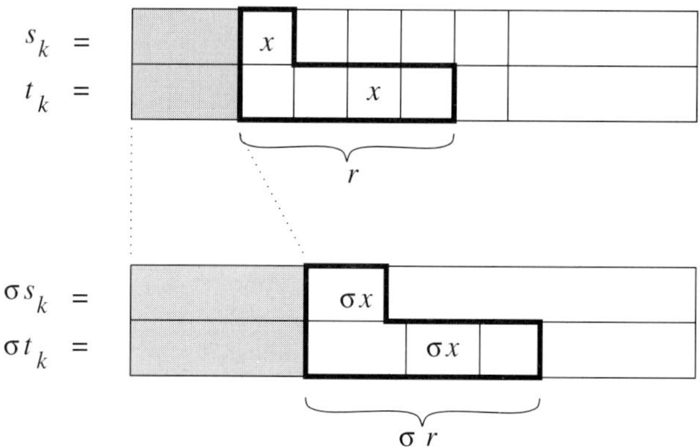

Figure 3.12.27 A variable in a first unshaded square that also appears in t

If $\sigma s_k, \sigma t_k$ were equal, we would have the situation in which the term σx is a proper initial segment of the term σr. This is impossible by Theorem 3.12.17 (p. 197).

This finishes our proof that if the CSC fails then the terms s_k, t_k are not unifiable, and thus the same applies to our original terms s, t.

Now, we know that s, t are unifiable iff $s_k = t_k$, so the algorithm correctly identifies the cases where s, t are unifiable. □

The next part of the proof is to show that if s, t are unifiable, then the output of the algorithm, namely, μ_k, is the most general unifier of s, t.

Lemma 3.12.28 If $s_k = t_k$, then μ_k is the most general unifier of s, t.

Proof. First, we have

$$\mu_k s = s_k = t_k = \mu_k t,$$

so μ_k is a unifier of s and t.

Now, let σ be any unifier of the terms s, t. Then by Lemma 3.12.21 (p. 202) there are σ_i such that

$$\begin{aligned}
\sigma &= \sigma_1(x_1 \leftarrow r_1) \\
\sigma_1 &= \sigma_2(x_2 \leftarrow r_2) \\
&\vdots \\
\sigma_{k-1} &= \sigma_k(x_k \leftarrow r_k).
\end{aligned}$$

Thus we have

$$\begin{aligned}\sigma &= \sigma_k(x_k \leftarrow r_k)\cdots(x_1 \leftarrow r_1)\\ &= \sigma_k\mu_k,\end{aligned}$$

so indeed μ_k is more general than σ. Thus μ_k is the most general unifier of s, t. □

Combining Lemmas 3.12.19–3.12.28 we have proved the following:

Theorem 3.12.29 The unification algorithm is correct.

3.12.6 Unification of Finitely Many Terms

It is straightforward to extend the notion of unifiers to finitely many terms $s_1, \ldots, s_n$.

Definition 3.12.30 We say that σ is a unifier of $s_1, \ldots, s_n$ if

$$\sigma s_1 = \sigma s_2 = \cdots = \sigma s_n.$$

As before, we have the notion of "more general" and "most general" unifiers. A simple iteration of our previous algorithm for a pair of terms handles this more general case and provides a proof of the following.

Theorem 3.12.31 Given terms $s_1, \ldots, s_n$ that are unifiable, there is a most general unifier μ, i.e., μ is more general than any unifier σ.

Proof. Let σ be a unifier of $s_1, \ldots, s_n$. We choose μ_1 to be the most general unifier of the terms s_1, s_2. Then we have $\sigma = \tau_1\mu_1$, for some τ_1.

Now, τ_1 is a unifier of $\mu_1 s_1 = \mu_1 s_2, \mu_1 s_3, \ldots, \mu_1 s_n$. Let μ_2 be a most general unifier of $\mu_1 s_2, \mu_1 s_3$, and let μ_3 be a most general unifier of $\mu_2\mu_1 s_3, \mu_2\mu_1 s_4$. Continuing, we obtain $\mu_1, \cdots, \mu_{n-1}$.

Let $\mu = \mu_{n-1} \cdots \mu_1$. Then μ is a unifier of $s_1, \ldots, s_n$, and μ is more general than σ. As the definition of μ does not depend on σ, it follows that μ is the most general unifier of $s_1, \ldots, s_n$. □

The algorithm to determine if the set of unifiers of $s_1, \ldots, s_n$ is empty or not is to try to construct the sequence of $\mu_1, \ldots, \mu_{n-1}$. If for some i, μ_i does not exist, then there is no unifier. Otherwise the composition of the μ_i (as in the proof) gives the desired most general unifier.

A similar procedure applies to determining if a finite set of pairs of terms $(s_1, s'_1), \ldots, (s_n, s'_n)$ is unifiable.

Definition 3.12.32 A finite set $(s_1, s'_1), \ldots, (s_n, s'_n)$ of pairs of terms is *unifiable* if there is a substitution σ such that

$$\sigma s_1 = \sigma s'_1, \ldots, \sigma s_n = \sigma s'_n.$$

If the set of pairs $(s_1, s'_1), \ldots, (s_n, s'_n)$ is unifiable, then a unifier μ is the most general unifier of the set of pairs if for every unifier σ there is a τ such that $\sigma = \tau\mu$.

The procedure to determine if a set of pairs is unifiable, and if so to find the most general unifier, is as follows. Let μ_1 be a most general unifier of s_1, s'_1. Then let μ_2 be a most general unifier of $\mu_1 s_2, \mu_1 s'_2$. Then let μ_3 be a most general unifier of $\mu_2\mu_1 s_3, \mu_2\mu_1 s'_3$, and so forth.

If this procedure can continue for $n-1$ steps, then $\mu = \mu_{n-1} \cdots \mu_1$ is the most general unifier of the pairs. Otherwise there is no unifier for the set of pairs.

Exercises

In the following problems, determine if the three terms are unifiable, and if so, find the most general unifier.

3.12.9 ffy, $ffffz$, and $fffy$

3.12.10 ffx, ffy, and $ffggz$

3.12.11 $(x + y) \cdot (u + v)$, $u \cdot (z \cdot (x \cdot y))$, and $x \cdot (y + z)$

3.12.12 $(y_1 \vee y_2)' \wedge y_3$, $z'_1 \wedge (z_2 \vee z'_3)$, and $x_1 \wedge (x_2 \vee x_3)$

3.12.13 $z_1 \cdot (z_2^{-1} \cdot z_2)$, $y_1 \cdot (y_2 \cdot y_1)^{-1}$, and $x_1 \cdot (x_2 \cdot x_3)$

3.13 TERM REWRITE SYSTEMS (TRSs)

The main purpose of a set of term rewrite rules $\mathcal{R}$ for a set of equations $\mathcal{E}$ is to provide a simple method to reduce any given term t to a normal form $n(t)$. Then we can determine if $s \approx s'$ is a consequence of $\mathcal{E}$ simply by checking if $n(s) = n(s')$, i.e., if the normal forms $n(s)$ and $n(s')$ are the same. Our goal in this section is to set up the definitions and present the basic theorem that motivates the Knuth–Bendix procedure.

3.13.1 Definition of a TRS

Definition 3.13.1 A *term rewrite (rule)* is an expression $s \longrightarrow t$, sometimes called a *directed equation*, where s and t are terms. s is called the *head* of the rule, and t the *tail* of the rule.

A *term rewrite system*, abbreviated TRS, is a set $\mathcal{R}$ of term rewrite rules.

Example 3.13.2 A simple system $\mathcal{R}$ of three term rewrite rules used for monoids is

$$R = \left\{ \begin{array}{l} (x \cdot y) \cdot z \longrightarrow x \cdot (y \cdot z) \\ 1 \cdot x \longrightarrow x \\ x \cdot 1 \longrightarrow x \end{array} \right\}.$$

The next convention will save considerable subsequent clarification of which language we are using.

Notational Convention 3.13.3 If the language $\mathcal{L}$ is not specified in the discussion of a TRS $\mathcal{R}$, we will assume that the language consists of precisely the function symbols occurring in the rewrite rules of $\mathcal{R}$, with their specified arities.

Definition 3.13.4 An *elementary* rewrite obtained from $s \longrightarrow t$ is a rewrite of the form

$$r(\cdots \sigma s \cdots) \longrightarrow r(\cdots \sigma t \cdots),$$

where σ is a substitution, r a term, and *one* occurrence of σs on the left has been replaced by σt on the right.

Example 3.13.5 The rewrite

$$y \cdot (((x \cdot y) \cdot x) \cdot (y \cdot x)) \longrightarrow y \cdot ((x \cdot y) \cdot (x \cdot (y \cdot x)))$$

is an elementary rewrite obtained from $(x \cdot y) \cdot z \longrightarrow x \cdot (y \cdot z)$ by using the substitution

$$\sigma = \begin{pmatrix} x \leftarrow x \cdot y \\ y \leftarrow x \\ z \leftarrow y \cdot x \end{pmatrix}.$$

Definition 3.13.6 If $\mathcal{R}$ is a set of term rewrite rules, we use the expression $s \longrightarrow_{\mathcal{R}} t$ to indicate that $s \longrightarrow t$ is an elementary rewrite of some term rewrite $p \longrightarrow q$ in $\mathcal{R}$.

We say $s \longrightarrow_{\mathcal{R}}^{+} s'$ if there is a finite sequence of elementary rewrites $t_i \longrightarrow_{\mathcal{R}} t_{i+1}$ such that

$$s = t_1 \longrightarrow_{\mathcal{R}} t_2 \longrightarrow_{\mathcal{R}} \cdots \longrightarrow_{\mathcal{R}} t_n = s'.$$

The notation $s \longrightarrow_{\mathcal{R}}^{*} s'$ means that one of $s = s'$ and $s \longrightarrow_{\mathcal{R}}^{+} s'$ holds.

Example 3.13.7 Using the $\mathcal{R}$ in Example 3.13.2 we have

$$((x \cdot 1) \cdot (1 \cdot y)) \cdot (z \cdot 1) \longrightarrow_{\mathcal{R}}^{+} x \cdot (y \cdot z),$$

since

$$\begin{aligned} ((x \cdot 1) \cdot (1 \cdot y)) \cdot (z \cdot 1) &\longrightarrow_{\mathcal{R}} (x \cdot (1 \cdot y)) \cdot (z \cdot 1) \longrightarrow_{\mathcal{R}} \\ (x \cdot y) \cdot (z \cdot 1) &\longrightarrow_{\mathcal{R}} (x \cdot y) \cdot z \longrightarrow_{\mathcal{R}} x \cdot (y \cdot z). \end{aligned}$$

3.13.2 Terminating TRSs

Definition 3.13.8 Given a set of term rewrite rules $\mathcal{R}$, a sequence of elementary rewrites

$$s_1 \longrightarrow_{\mathcal{R}} \cdots \longrightarrow_{\mathcal{R}} s_n$$

is *terminating* if no $s_n \longrightarrow_{\mathcal{R}} s_{n+1}$ is possible.

$\mathcal{R}$ is *terminating* if every sequence of elementary rewrites

$$s_1 \longrightarrow_{\mathcal{R}} \cdots$$

is finite, i.e., $\longrightarrow_{\mathcal{R}}^{+}$ is a *well founded*[3] relation.

Thus if $\mathcal{R}$ is terminating, then starting from any term and applying rewrites using $\longrightarrow_{\mathcal{R}}$, we must in a finite number of steps come to a terminating term.

Example 3.13.9 There is one degenerate case that we want to note, namely, if $\mathcal{R}$ has a rule of the form $x \longrightarrow t$, x being a variable, then $\mathcal{R}$ is *not* terminating. This follows from the fact that either (a) x does not appear in t, in which case we have the elementary rewrite $t \longrightarrow_{\mathcal{R}} t$ from $x \longrightarrow t$ and thus an infinite sequence of elementary rewrites

$$t \longrightarrow_{\mathcal{R}} t \longrightarrow_{\mathcal{R}} t \longrightarrow_{\mathcal{R}} t \cdots ;$$

or (b) x does appear in t, in which case we can apply the rule $x \longrightarrow t$ to the term t, and this will yield a term with x in it, etc., giving rise to an infinite sequence of elementary rewrites, starting with x.

[3] A binary relation $>$ on a set A is *well founded* if there is no infinite sequence a_n of elements of A such that $a_1 > a_2 > \cdots$.

Example 3.13.10 If a set of rules $\mathcal{R}$ is such that $\longrightarrow_{\mathcal{R}}$ sends terms to terms of shorter length, then it is terminating, e.g., the set

$$R = \left\{ \begin{array}{c} 1 \cdot x \longrightarrow x \\ x \cdot 1 \longrightarrow x \end{array} \right\}$$

is terminating. However, we do not need $\longrightarrow_{\mathcal{R}}$ to shorten the terms to have a terminating system. For example, the single rewrite

$$(x \cdot y) \cdot z \longrightarrow x \cdot (y \cdot z)$$

gives a terminating system, since it is shifting the parentheses "to the right."

Combining these two observations we see that the set of rules $\mathcal{R}$ from Example 3.13.2 is terminating.

The following gives a very simple and useful method for showing that some TRSs are terminating. Recall that $|t|$, the *length* of the term t, for t in prefix form, is defined to be the number of symbols in t. Now, let us define $|t|_x$, the *x–length* of t, for x a variable, to be the number of occurrences of x in t. (Thus, e.g., $|fxfxy|_x = 2$.)

Theorem 3.13.11 Let $\mathcal{R}$ be a term rewrite system with the property that for each $s \longrightarrow t \in \mathcal{R}$ we have

a. $|s| > |t|$, and
b. $|s|_x \geq |t|_x$ for every variable x.

Then $\mathcal{R}$ is a terminating TRS.

Proof. For r a term let $|r|_F$ be the number of occurrences of function symbols and constants in r. Then we have

$$|r| = |r|_F + \sum_x |r|_x.$$

If σ is a substitution, then

$$|\sigma r| = |r|_F + \sum_x |r|_x \cdot |\sigma x|,$$

as each occurrence of a variable x is replaced by an occurrence of σx.

Now, we assume $\mathcal{R}$ satisfies hypotheses (a) and (b). For $s \longrightarrow t \in \mathcal{R}$ we have $|s| > |t|$ by (a), and thus

$$|s|_F + \sum_x |s|_x > |t|_F + \sum_x |t|_x.$$

From (b) we have $|s|_x \geq |t|_x$, for all variables x. Thus, given any substitution σ,

$$|s|_F + \sum_x |s|_x \cdot |\sigma x| > |t|_F + \sum_x |t|_x \cdot |\sigma x|,$$

and consequently,

$$|\sigma s| > |\sigma t|.$$

From this it follows that for any elementary rewrite $s' \longrightarrow t'$ for $\mathcal{R}$,

$$|s'| > |t'|.$$

Thus the rewrites in $\longrightarrow_{\mathcal{R}}$ are all length–reducing. Since the length of a term r is a positive integer, it follows that we can apply at most $|r|$ successive elementary rewrites to r before encountering a terminal term. Thus $\mathcal{R}$ is terminating. □

Example 3.13.12 The system $\mathcal{R} = \{fx + gx \longrightarrow x \cdot x\}$ is terminating, where $+, \cdot$ are binary, and f, g are unary, as well as the system $\mathcal{R} = \{fgfx \longrightarrow ggx,\ gffx \longrightarrow fx\}$.

3.13.3 Normal Form TRSs

Now we start to examine the kind of TRSs that interest us, namely, those that produce normal forms.

Definition 3.13.13 A set $\mathcal{R}$ of term rewrite rules is a *normal form* TRS if $\mathcal{R}$ is a *uniquely* terminating TRS, that is, $\mathcal{R}$ is terminating, and for any given term s, all terminating sequences $s \longrightarrow_{\mathcal{R}} \cdots$ terminate with the same term, $n_{\mathcal{R}}(s)$, called the *normal form* of s (with respect to $\mathcal{R}$).

If for a set of equations $\mathcal{E}$ we have a normal form TRS $\mathcal{R}$ such that

$$\mathcal{E} \models s \approx t \qquad \text{iff} \qquad n_{\mathcal{R}}(s) = n_{\mathcal{R}}(t),$$

then we say $\mathcal{R}$ is a *normal form TRS for $\mathcal{E}$*.

Example 3.13.14 The system $\mathcal{R}$ from Example 3.13.2 is a normal form TRS, since the result of applying the rules in any order to a term s will result in the term 1 if there are no variables present in s; otherwise, the rules will have the effect of deleting all the 1's and moving the parentheses to the right.

Example 3.13.15 In 1951 Trevor Evans showed that the equations $\mathcal{E}$ in the table below defining *loops* $(L, /, \cdot, \backslash, 1)$ have a normal form TRS $\mathcal{R}$ as given on the right:

LOOPS

$\mathcal{E}$			$\mathcal{R}$		
$x \cdot (x\backslash y)$	$\approx$	y	$x\backslash x$	$\longrightarrow$	1
$x\backslash(x \cdot y)$	$\approx$	y	x/x	$\longrightarrow$	1
$(y/x) \cdot x$	$\approx$	y	$1 \cdot x$	$\longrightarrow$	x
$(y \cdot x)/x$	$\approx$	y	$x \cdot 1$	$\longrightarrow$	x
$x\backslash x$	$\approx$	y/y	$1\backslash x$	$\longrightarrow$	x
$x \cdot 1$	$\approx$	x	$x/1$	$\longrightarrow$	x
$1 \cdot x$	$\approx$	x	$x \cdot (x\backslash y)$	$\longrightarrow$	y
			$(x/y) \cdot y$	$\longrightarrow$	x
			$x\backslash(x \cdot y)$	$\longrightarrow$	y
			$(y \cdot x)/x$	$\longrightarrow$	y
			$x/(y\backslash x)$	$\longrightarrow$	y
			$(x/y)\backslash x$	$\longrightarrow$	$y.$

We can easily see that the TRS $\mathcal{R}$ on the right is terminating by Theorem 3.13.11. However, the proof that it is a normal form TRS for $\mathcal{E}$ is nontrivial.

To see how the rules can be used to study loops, note that the terms $(x \cdot y) \cdot z$ and $x \cdot (y \cdot z)$ are both in normal form. Thus the associative law does *not* follow from the loop axioms. For a positive result, consider the following. Let $t(x, y, z) = (x/(y\backslash y)) \cdot (y\backslash z)$. To verify that $\mathcal{E} \models t(x, x, y) \approx t(y, x, x) \approx y$ we calculate the normal forms of the two sides:

$$\begin{aligned} t(x, x, y) &= (x/(x\backslash x)) \cdot (x\backslash y) \longrightarrow_{\mathcal{R}} (x/1) \cdot (x\backslash y) \\ &\qquad \longrightarrow_{\mathcal{R}} x \cdot (x\backslash y) \longrightarrow_{\mathcal{R}} y \\ t(y, x, x) &= (y/(x\backslash x)) \cdot (x\backslash x) \longrightarrow_{\mathcal{R}} (y/(x\backslash x)) \cdot 1 \\ &\qquad \longrightarrow_{\mathcal{R}} (y/(x\backslash x)) \longrightarrow_{\mathcal{R}} y/1 \longrightarrow_{\mathcal{R}} y. \end{aligned}$$

Example 3.13.16 Another example of a normal form TRS for a set of equations, due to Knuth and Bendix in 1967, takes $\mathcal{E}$ to be equations that define groups using *left identity* and *left inverse* axioms, rather than the ones used in our axioms $\mathcal{G}$ for groups.

GROUPS

$\mathcal{E}$	$\mathcal{R}$
$1 \cdot x \approx x$	$1^{-1} \longrightarrow 1$
$x^{-1} \cdot x \approx 1$	$x \cdot 1 \longrightarrow x$
$(x \cdot y) \cdot z \approx x \cdot (y \cdot z)$	$1 \cdot x \longrightarrow x$
	$(x^{-1})^{-1} \longrightarrow x$
	$x^{-1} \cdot x \longrightarrow 1$
	$x \cdot x^{-1} \longrightarrow 1$
	$x^{-1} \cdot (x \cdot y) \longrightarrow y$
	$x \cdot (x^{-1} \cdot y) \longrightarrow y$
	$(x \cdot y)^{-1} \longrightarrow y^{-1} \cdot x^{-1}$
	$(x \cdot y) \cdot z \longrightarrow x \cdot (y \cdot z).$

In this case it is not immediate that the system $\mathcal{R}$ is terminating, much less a normal form TRS. We will look at this example in Secs. 3.14 and 3.15.

Definition 3.13.17 For a TRS $\mathcal{R}$ define the associated set of equations $\mathcal{E}(\mathcal{R})$ by

$$\mathcal{E}(\mathcal{R}) = \{s \approx t : s \longrightarrow t \in \mathcal{R}\}.$$

Lemma 3.13.18 For $\mathcal{R}$ a TRS we have

$$\mathcal{E}(\mathcal{R}) \models s \approx t \qquad \text{iff} \qquad s(\longrightarrow_{\mathcal{R}} \cup {}_{\mathcal{R}}\!\longleftarrow)^* t,$$

where ${}_{\mathcal{R}}\!\longleftarrow$ is the converse of $\longrightarrow_{\mathcal{R}}$.

Proof. The direction ($\Longleftarrow$) follows from the definition of $\mathcal{E}(\mathcal{R})$. For the converse we note that $\mathcal{E}(\mathcal{R}) \models s \approx t$ means there is a chained derivation of $s \approx t$. Replacing each step of the chained derivation with the corresponding elementary rewrite obtained from $\mathcal{R}$ finishes the proof. □

Thus the equations $s \approx t$ that are consequences of $\mathcal{E}(\mathcal{R})$ are precisely those that can be obtained by elementary rewrites $s_i \longrightarrow_{\mathcal{R}} t_i$, or their converses $t_i\,{}_{\mathcal{R}}\!\longleftarrow s_i$, e.g., $s \longrightarrow_R s_1\,{}_{\mathcal{R}}\!\longleftarrow s_2\,{}_{\mathcal{R}}\!\longleftarrow t$ would be permitted.

The following simple fact can be quite useful in studying normal form TRSs.

Proposition 3.13.19 If $\mathcal{R}$ is a normal form TRS, then it is a normal form TRS for $\mathcal{E}(\mathcal{R})$.

Proof. Certainly, if two terms s and t have the same normal form with respect to $\mathcal{R}$ then $\mathcal{E}(\mathcal{R}) \models s \approx t$. Conversely, suppose $s \approx t$ is a consequence of $\mathcal{E}(\mathcal{R})$. Then $s(\longrightarrow_{\mathcal{R}} \cup_{\mathcal{R}} \longleftarrow)^* t$ holds by Lemma 3.13.18. Now it is easy to see that $s_i \longrightarrow_{\mathcal{R}} s_{i+1}$ implies $n_{\mathcal{R}}(s_i) = n_{\mathcal{R}}(s_{i+1})$. Consequently, $n_{\mathcal{R}}(s) = n_{\mathcal{R}}(t)$. □

Notational Convention 3.13.20 It will be convenient in the following exercises, and later in the text, to abbreviate a string $f \cdots f$ of n unary function symbols as f^n. This will allow us to write f^4x rather than $ffffx$, etc. The notation f^0x will mean simply x.

Exercises

3.13.1 Explain why each of the following $\mathcal{R}$'s is a normal form TRS and describe the possible normal forms. Also give the normal form of the term t following the description of $\mathcal{R}$. We assume that the language of $\mathcal{R}$ consists precisely of the function symbols that appear in $\mathcal{R}$. (In the following f, g are unary; $+$ is binary, 0 is a constant.)

a. $\mathcal{R} = \{ffx \longrightarrow x\}$, $t = f^8x$
b. $\mathcal{R} = \{ffx \longrightarrow fx\}$, $t = f^8x$
c. $\mathcal{R} = \{fgx \longrightarrow x\}$, $t = fffggx$
d. $\mathcal{R} = \{fgx \longrightarrow fx\}$, $t = fffggx$
e. $\mathcal{R} = \{fgx \longrightarrow gx\}$, $t = fffggx$
f. $\mathcal{R} = \{ffx \longrightarrow x,\ ggx \longrightarrow x,\ fgx \longrightarrow gfx\}$, $t = fffggx$
g. $\mathcal{R} = \{fffx \longrightarrow x,\ ggx \longrightarrow x,\ fgx \longrightarrow gffx\}$, $t = ffgx$
h. $\mathcal{R} = \{x + y \longrightarrow x\}$, $t = (((x + y) + z) + u) + v$
i. $\mathcal{R} = \{x + y \longrightarrow 0\}$, $t = (((x + y) + z) + u) + v$

3.13.4 Critical Pairs

Definition 3.13.21 A *critical pair* of a TRS $\mathcal{R}$ is a pair (s, t) of terms obtained as follows:
Take two (not necessarily distinct) rules of $\mathcal{R}$,

$$\begin{aligned} s_1 &\longrightarrow t_1 \\ s_2 &\longrightarrow t_2, \end{aligned}$$

and rename the variables in one of them if necessary so that s_1 and s_2 have *no variables in common*. Choose an occurrence of a *nonvariable* subterm s_1' of s_1 such that s_1', s_2 are unifiable. Find a most general unifier μ for s_1' and s_2.

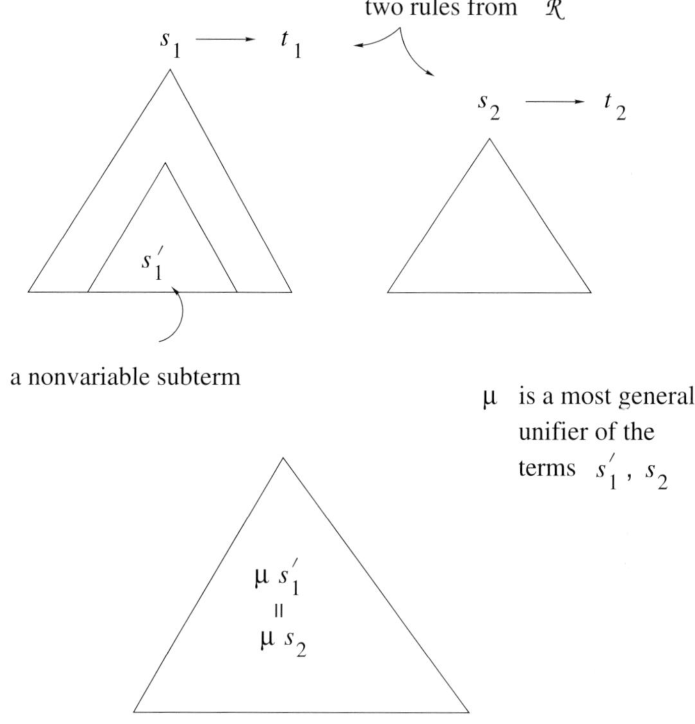

Figure 3.13.22 Preliminaries to finding a critical pair

Then, with $s_1 = r_1(\cdots s_1' \cdots)$ we have

$$\begin{aligned} \mu s_1 &= \mu r_1(\cdots s_1' \cdots) \\ &= \widehat{r}_1(\cdots \mu s_1' \cdots) \\ &= \widehat{r}_1(\cdots \mu s_2 \cdots). \end{aligned}$$

This leads to two results s and t (possibly the same) that form a *critical pair* (s, t), obtained by applying rewrite rules to μs_1, namely,

$$s = \mu t_1 \text{ and } t = \widehat{r}_1(\cdots \mu s_2 \cdots).$$

The critical pair (s, t) is *trivial* if $s = t$.

We can visualize the formation of a critical pair as follows:

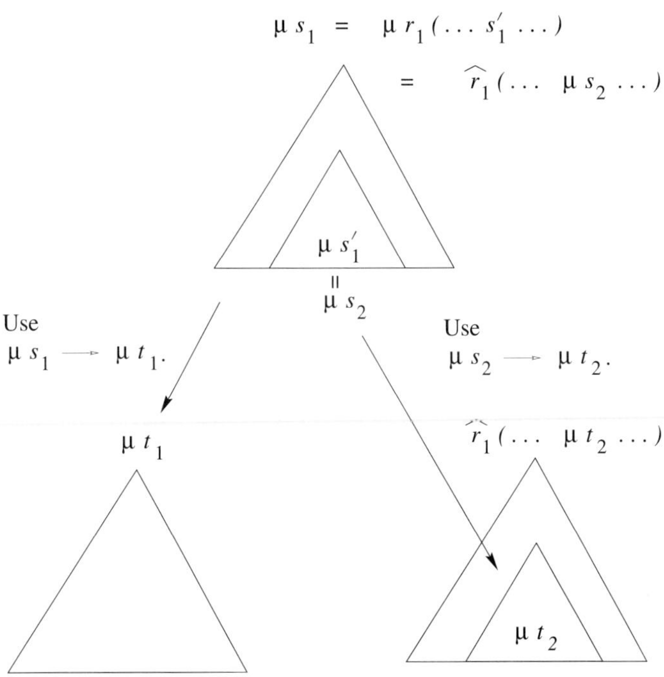

Figure 3.13.23 Forming the critical pair $(\mu t_1, \widehat{r}_1(\cdots \mu t_2 \cdots))$

Thus we see that critical pairs of $\mathcal{R}$ are special cases of the pairs of terms s, t obtained from elementary rewrites with a common starting term:

$$\begin{array}{lll} \mu s_1 & \longrightarrow_{\mathcal{R}} & s \\ \mu s_1 & \longrightarrow_{\mathcal{R}} & t. \end{array}$$

The rule $\mu s_1 \longrightarrow \mu t_1$ rewrites all the starting term μs_1, and the rule $\mu s_2 \longrightarrow \mu t_2$ rewrites part (or all) of μs_1.

Next, we look at two examples of deriving critical pairs from a *single* rule.

Example 3.13.24 Suppose $s_1 \longrightarrow t_1$ and $s_2 \longrightarrow t_2$ are both $ffx \longrightarrow x$. To find the critical pairs we first rename the variables in the second rule to be disjoint from the first, e.g., we use

$$ffx \longrightarrow x \qquad \text{for } s_1 \longrightarrow t_1 \tag{28}$$

$$ffu \longrightarrow u \qquad \text{for } s_2 \longrightarrow t_2. \tag{29}$$

The nonvariable subterms s_1' of s_1 are ffx and fx.

For the first choice, $s_1' = ffx$, we need to find the most general unifier μ of s_1' and s_2, i.e., of ffx and ffu. This is easily seen to be given by $\mu = (u \leftarrow x)$. Applying μ to (28), (29) gives

$$\begin{array}{ll} \underline{ffx} \longrightarrow x & \text{for } \mu s_1 \longrightarrow \mu t_1 \\ ffx \longrightarrow x & \text{for } \mu s_2 \longrightarrow \mu t_2. \end{array}$$

The relevant (and only) occurrence of $\mu s_1'$ in μs_1 has been underlined. The two rules are to be applied to $\mu s_1 = \underline{ffx}$, with the second rule applied to the underlined portion. This gives the pair of term rewrites

$$\begin{array}{l} ffx \longrightarrow x \\ ffx \longrightarrow x, \end{array}$$

which yields the (trivial) critical pair (x, x).

For the second choice, $s_1' = fx$, we need to find the most general unifier μ of s_1' and s_2, i.e., of fx and ffu. This is easily found to be given by $\mu = (x \leftarrow fu)$. Applying μ to (28), (29) gives

$$\begin{array}{ll} f\underline{ffu} \longrightarrow fu & \text{for } \mu s_1 \longrightarrow \mu t_1 \\ ffu \longrightarrow u & \text{for } \mu s_2 \longrightarrow \mu t_2. \end{array}$$

The relevant (and only) occurrence of $\mu s_1'$ in μs_1 has been underlined. The two rules are to be applied to $\mu s_1 = f\underline{ffu}$, with the second rule applied to the underlined portion. This gives the pair of term rewrites

$$\begin{array}{l} fffu \longrightarrow fu \\ fffu \longrightarrow fu, \end{array}$$

which yields another (trivial) critical pair (fu, fu), or changing variables, the critical pair (fx, fx).

Thus we obtain in total two critical pairs in this example, namely, (x, x) and (fx, fx).

Example 3.13.25 Suppose $s_1 \longrightarrow t_1$ and $s_2 \longrightarrow t_2$ are both $(x \cdot y) \cdot z \longrightarrow x \cdot (y \cdot z)$. To find the critical pairs we first rename the variables in the second rule to be disjoint from the first, e.g., we use

$$\begin{array}{lll} (x \cdot y) \cdot z \longrightarrow x \cdot (y \cdot z) & \text{for } s_1 \longrightarrow t_1 & (30) \\ (u \cdot v) \cdot w \longrightarrow u \cdot (v \cdot w) & \text{for } s_2 \longrightarrow t_2. & (31) \end{array}$$

The nonvariable subterms s_1' of s_1 are $(x \cdot y) \cdot z$ and $x \cdot y$.

For the first choice, $s_1' = (x \cdot y) \cdot z$, we need to find the most general unifier μ of s_1' and s_2, i.e., of $(x \cdot y) \cdot z$ and $(u \cdot v) \cdot w$. This is easily seen to be given by $\mu = \begin{pmatrix} u \leftarrow x \\ v \leftarrow y \\ w \leftarrow z \end{pmatrix}$. Applying μ to (30), (31) gives

$$\begin{array}{ll} \underline{(x \cdot y) \cdot z} \longrightarrow x \cdot (y \cdot z) & \text{for } \mu s_1 \longrightarrow \mu t_1 \\ (x \cdot y) \cdot z \longrightarrow x \cdot (y \cdot z) & \text{for } \mu s_2 \longrightarrow \mu t_2. \end{array}$$

The relevant (and only) occurrence of $\mu s_1'$ in μs_1 has been underlined. The two rules are to be applied to $\mu s_1 = \underline{(x \cdot y) \cdot z}$, with the second rule applied to the underlined portion. This gives the pair of term rewrites

$$\begin{array}{l} (x \cdot y) \cdot z \longrightarrow x \cdot (y \cdot z) \\ (x \cdot y) \cdot z \longrightarrow x \cdot (y \cdot z), \end{array}$$

which yields the (trivial) critical pair $\big(x \cdot (y \cdot z), x \cdot (y \cdot z)\big)$.

For the second choice, $s_1' = x \cdot y$, we need to find the most general unifier μ of s_1' and s_2, i.e., of $x \cdot y$ and $(u \cdot v) \cdot w$. Applying our unification algorithm gives

$\cdot$	x	y		
$\cdot$	$\cdot$	u	v	w

$\downarrow \quad (x \leftarrow u \cdot v)$

$\cdot$	$\cdot$	u	v	y
$\cdot$	$\cdot$	u	v	w

$\downarrow \quad (y \leftarrow w)$

$\cdot$	$\cdot$	u	v	w
$\cdot$	$\cdot$	u	v	w

.

Thus the two terms are unifiable, and the most general unifier is given by

$$\mu = \begin{pmatrix} x \leftarrow u \cdot v \\ y \leftarrow w \end{pmatrix}.$$

Applying μ to (30), (31) gives

$$\begin{array}{ll} (\underline{(u \cdot v) \cdot w}) \cdot z \longrightarrow (u \cdot v) \cdot (w \cdot z) & \text{for } \mu s_1 \longrightarrow \mu t_1 \\ (u \cdot v) \cdot w \longrightarrow u \cdot (v \cdot w) & \text{for } \mu s_2 \longrightarrow \mu t_2. \end{array}$$

The relevant (and only) occurrence of $\mu s_1'$ in μs_1 has been underlined. The two rules are to be applied to $\mu s_1 = (\underline{(u \cdot v) \cdot w}) \cdot z$, with the second rule applied to the underlined portion. This gives the pair of term rewrites

$$\begin{aligned}((u \cdot v) \cdot w) \cdot z &\longrightarrow (u \cdot v) \cdot (w \cdot z) \\ ((u \cdot v) \cdot w) \cdot z &\longrightarrow (u \cdot (v \cdot w)) \cdot z,\end{aligned}$$

which yields another critical pair: $\big((u \cdot v) \cdot (w \cdot z),\, (u \cdot (v \cdot w)) \cdot z\big)$.

The next example looks at finding the critical pairs for two distinct rewrite rules.

Example 3.13.26 Let the rewrite rules $s_1 \longrightarrow t_1$ and $s_2 \longrightarrow t_2$ be given by

$$\begin{aligned}(x \cdot y)^{-1} \cdot (x \cdot z) &\longrightarrow y^{-1} \cdot z \\ (x \cdot y) \cdot z &\longrightarrow x \cdot (y \cdot z).\end{aligned}$$

Let us look at all the critical pairs we can form from this pair. First, we rename the variables in the second term so that s_1 and s_2 have no variables in common, e.g., we use

$$(x \cdot y)^{-1} \cdot (x \cdot z) \longrightarrow y^{-1} \cdot z \qquad \text{for } s_1 \longrightarrow t_1 \tag{32}$$

$$(u \cdot v) \cdot w \longrightarrow u \cdot (v \cdot w) \qquad \text{for } s_2 \longrightarrow t_2. \tag{33}$$

Next, we make a column list for each of the terms s_1 and s_2 and their nonvariable subterms and then match the head of each column with the members of the other that are unifiable with it as follows:

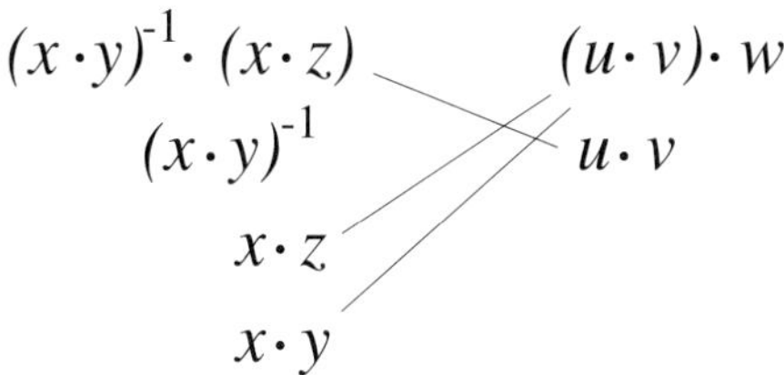

Figure 3.13.27 Unifiable subterms

Thus we have three possibilities to consider. The first case we examine is the pair $s_1 = (x \cdot y)^{-1} \cdot (x \cdot z)$ and $s_2' = u \cdot v$. (In this case the roles of s_1 and s_2 are reversed from our earlier general discussion, namely, μs_2 will be the term to be rewritten in two different ways to obtain the critical pair.)

The most general unifier of s_1 and s_2' is given by

$$\mu = \begin{pmatrix} u \leftarrow (x \cdot y)^{-1} \\ v \leftarrow x \cdot z \end{pmatrix}.$$

Applying μ to (32), (33) gives

$$\begin{array}{rll} (x \cdot y)^{-1} \cdot (x \cdot z) \longrightarrow y^{-1} \cdot z & & \text{for } \ \mu s_1 \longrightarrow \mu t_1 \\ (\underline{(x \cdot y)^{-1} \cdot (x \cdot z)}) \cdot w \longrightarrow (x \cdot y)^{-1} \cdot ((x \cdot z) \cdot w) & & \text{for } \ \mu s_2 \longrightarrow \mu t_2. \end{array}$$

The relevant (and only) occurrence of $\mu s_2'$ in μs_2 is underlined. The two rules are to be applied to $\mu s_2 = (\underline{(x \cdot y)^{-1} \cdot (x \cdot z)}) \cdot w$, with the first rule applied to the underlined portion. This gives the pair of term rewrites

$$\begin{array}{l} ((x \cdot y)^{-1} \cdot (x \cdot z)) \cdot w \longrightarrow (y^{-1} \cdot z) \cdot w \\ ((x \cdot y)^{-1} \cdot (x \cdot z)) \cdot w \longrightarrow (x \cdot y)^{-1} \cdot ((x \cdot z) \cdot w), \end{array}$$

which gives the critical pair $\big((y^{-1} \cdot z) \cdot w,\ (x \cdot y)^{-1} \cdot ((x \cdot z) \cdot w)\big)$.

The second case we examine is the pair $s_1' = x \cdot z$ and $s_2 = (u \cdot v) \cdot w$. The most general unifier of s_1' and s_2 is given by

$$\mu = \begin{pmatrix} x \leftarrow u \cdot v \\ z \leftarrow w \end{pmatrix}.$$

Applying μ to (32), (33) gives

$$\begin{array}{rll} ((u \cdot v) \cdot y)^{-1} \cdot (\underline{(u \cdot v) \cdot w}) \longrightarrow y^{-1} \cdot w & & \text{for } \ \mu s_1 \longrightarrow \mu t_1 \\ (u \cdot v) \cdot w \longrightarrow u \cdot (v \cdot w) & & \text{for } \ \mu s_2 \longrightarrow \mu t_2. \end{array}$$

The relevant (and only) occurrence of $\mu s_1'$ in μs_1 is underlined. These two rules are to be applied to $\mu s_1 = ((u \cdot v) \cdot y)^{-1} \cdot (\underline{(u \cdot v) \cdot w})$, with the second rule applied to the underlined portion. This gives the pair of term rewrites

$$\begin{array}{l} ((u \cdot v) \cdot y)^{-1} \cdot ((u \cdot v) \cdot w) \longrightarrow y^{-1} \cdot w \\ ((u \cdot v) \cdot y)^{-1} \cdot ((u \cdot v) \cdot w) \longrightarrow ((u \cdot v) \cdot y)^{-1} \cdot (u \cdot (v \cdot w)), \end{array}$$

which gives the critical pair $\big(y^{-1} \cdot w,\ ((u \cdot v) \cdot y)^{-1} \cdot (u \cdot (v \cdot w))\big)$.

The third case to examine is the pair $s_1' = x \cdot y$ and $s_2 = (u \cdot v) \cdot w$. The most general unifier of s_1' and s_2 is given by

$$\mu = \begin{pmatrix} x \leftarrow u \cdot v \\ y \leftarrow w \end{pmatrix}.$$

Applying μ to (32), (33) gives

$$\begin{array}{rll} (\underline{(u \cdot v) \cdot w})^{-1} \cdot ((u \cdot v) \cdot z) \longrightarrow w^{-1} \cdot z & & \text{for } \ \mu s_1 \longrightarrow \mu t_1 \\ (u \cdot v) \cdot w \longrightarrow u \cdot (v \cdot w) & & \text{for } \ \mu s_2 \longrightarrow \mu t_2. \end{array}$$

The relevant (and only) occurrence of $\mu s_1'$ in μs_1 is underlined. These two rules are to be applied to $\mu s_1 = \underline{((u \cdot v) \cdot w)}^{-1} \cdot ((u \cdot v) \cdot z)$, with the second rule applied to the underlined portion. This gives the pair of term rewrites

$$((u \cdot v) \cdot w)^{-1} \cdot ((u \cdot v) \cdot z) \longrightarrow w^{-1} \cdot z$$
$$((u \cdot v) \cdot w)^{-1} \cdot ((u \cdot v) \cdot z) \longrightarrow (u \cdot (v \cdot w))^{-1} \cdot ((u \cdot v) \cdot z),$$

which gives the critical pair $\left(w^{-1} \cdot z,\, (u \cdot (v \cdot w))^{-1} \cdot ((u \cdot v) \cdot z)\right)$.

The final example considers repeat occurrences of a subterm.

Example 3.13.28 Let the rewrite rules $s_1 \longrightarrow t_1$ and $s_2 \longrightarrow t_2$ be given by

$$(x + y) + (x + y) \longrightarrow 0$$
$$x + y \longrightarrow x.$$

Let us look at the critical pairs we can form from this pair. First, we rename the variables in the second term so that s_1 and s_2 have no variables in common, e.g., we use:

$$(x + y) + (x + y) \longrightarrow 0 \qquad \text{for } s_1 \longrightarrow t_1 \qquad (34)$$
$$u + v \longrightarrow u \qquad \text{for } s_2 \longrightarrow t_2. \qquad (35)$$

Since the only nonvariable subterm of $u + v$ is itself, it follows that we will have to use some s_1' and s_2 to form critical pairs.

The first choice to consider is $s_1' = (x+y)+(x+y)$ and $s_2 = u+v$. The most general unifier of s_1' and s_2 is given by

$$\mu = \begin{pmatrix} u \leftarrow x + y \\ v \leftarrow x + y \end{pmatrix}.$$

Applying μ to (34), (35) gives

$$\underline{(x + y) + (x + y)} \longrightarrow 0 \qquad \text{for } \mu s_1 \longrightarrow \mu t_1$$
$$(x + y) + (x + y) \longrightarrow x + y \qquad \text{for } \mu s_2 \longrightarrow \mu t_2.$$

The relevant (and only) occurrence of $\mu s_1'$ in μs_1 is underlined. The two rules are to be applied to $\mu s_1 = \underline{(x + y) + (x + y)}$, with the first rule applied to the underlined portion. This gives the pair of term rewrites

$$(x + y) + (x + y) \longrightarrow 0$$
$$(x + y) + (x + y) \longrightarrow x + y,$$

which gives the critical pair $(0, x + y)$.

The second case we examine is the pair $s_1' = x+y$ and $s_2 = u+v$, where we choose the *first* occurrence of s_1' in s_1. The most general unifier of s_1' and s_2 is given by

$$\mu = \begin{pmatrix} u \leftarrow x \\ v \leftarrow y \end{pmatrix}.$$

Applying μ to (34), (35) gives

$$\begin{aligned} \underline{(x+y)} + (x+y) &\longrightarrow 0 && \text{for } \mu s_1 \longrightarrow \mu t_1 \\ x+y &\longrightarrow x && \text{for } \mu s_2 \longrightarrow \mu t_2. \end{aligned}$$

The relevant (but *not* the only) occurrence of $\mu s_1'$ in μs_1 is underlined. These two rules are to be applied to $\mu s_1 = \underline{(x+y)} + (x+y)$, with the second rule applied to the underlined portion. This gives the pair of term rewrites

$$\begin{aligned} (x+y) + (x+y) &\longrightarrow 0 \\ (x+y) + (x+y) &\longrightarrow x + (x+y), \end{aligned}$$

which gives the critical pair $\big(0, x + (x+y)\big)$.

The third case we examine is the pair $s_1' = x+y$ and $s_2 = u+v$, where we choose the *second* occurrence of s_1' in s_1. The most general unifier of s_1' and s_2 is given by

$$\mu = \begin{pmatrix} u \leftarrow x \\ v \leftarrow y \end{pmatrix}.$$

Applying μ to (34), (35) gives

$$\begin{aligned} (x+y) + \underline{(x+y)} &\longrightarrow 0 && \text{for } \mu s_1 \longrightarrow \mu t_1 \\ x+y &\longrightarrow x && \text{for } \mu s_2 \longrightarrow \mu t_2. \end{aligned}$$

The relevant (but *not* the only) occurrence of $\mu s_1'$ in μs_1 is underlined. These two rules are to be applied to $\mu s_1 = (x+y) + \underline{(x+y)}$, with the second rule applied to the underlined portion. This gives the pair of term rewrites

$$\begin{aligned} (x+y) + (x+y) &\longrightarrow 0 \\ (x+y) + (x+y) &\longrightarrow (x+y) + x, \end{aligned}$$

which gives the critical pair $\big(0, (x+y) + x\big)$.

Exercises

3.13.2 Find the critical pairs for the following single–rule term rewrite systems. (f, g are unary; $+, \cdot$ are binary.)

a. $\mathcal{R} = \{fffx \longrightarrow x\}$
b. $\mathcal{R} = \{fffx \longrightarrow fx\}$
c. $\mathcal{R} = \{fffx \longrightarrow ffx\}$
d. $\mathcal{R} = \{fgx \longrightarrow fx\}$
e. $\mathcal{R} = \{fgx \longrightarrow gfx\}$
f. $\mathcal{R} = \{fgfx \longrightarrow gfgx\}$
g. $\mathcal{R} = \{x \cdot (y + z) \longrightarrow (x \cdot y) + (x \cdot z)\}$
h. $\mathcal{R} = \{(x \cdot y) + (x \cdot z) \longrightarrow x \cdot (y + z)\}$

3.13.3 Find the critical pairs for the following two–rule term rewrite systems. ($f, g, {}^{-1}$ are unary; $+, \cdot$ are binary, 1 is a constant.)

a. $\mathcal{R} = \{fx \longrightarrow x,\ fgx \longrightarrow gfx\}$
b. $\mathcal{R} = \{ffffx \longrightarrow fx,\ fffx \longrightarrow ffx\}$
c. $\mathcal{R} = \{fgfx \longrightarrow gfgx,\ ffx \longrightarrow fx\}$
d. $\mathcal{R} = \{x \cdot 1 \longrightarrow x,\ (x \cdot y) \cdot z \longrightarrow x \cdot (y \cdot z)\}$
e. $\mathcal{R} = \{x \cdot 1 \longrightarrow x,\ x \cdot x^{-1} \longrightarrow 1\}$
f. $\mathcal{R} = \{x \cdot x^{-1} \longrightarrow 1,\ (x \cdot y) \cdot z \longrightarrow x \cdot (y \cdot z)\}$
g. $\mathcal{R} = \{(x \cdot y) \cdot z \longrightarrow x \cdot (y \cdot z),\ x \cdot (y + z) \longrightarrow (x \cdot y) + (x \cdot z)\}$

3.13.5 Critical Pairs Lemma

Definition 3.13.29 A TRS $\mathcal{R}$ is *confluent on critical pairs* if, given any critical pair (s, t), there is a term r such that

$$\begin{array}{lll} s & \longrightarrow^*_{\mathcal{R}} & r \\ t & \longrightarrow^*_{\mathcal{R}} & r. \end{array}$$

Remark 3.13.30 Of course we need to check only *nontrivial* critical pairs (s, t) in the definition to show that $\mathcal{R}$ is confluent on critical pairs.

The following fundamental result of Knuth and Bendix lies at the center of methods searching for a normal form TRS for a set of equations $\mathcal{E}$.

Theorem 3.13.31 [CRITICAL PAIRS LEMMA] Suppose $\mathcal{R}$ is a terminating TRS. Then $\mathcal{R}$ is a normal form TRS iff $\mathcal{R}$ is confluent on critical pairs.

Proof. The proof is deferred to Sec. 3.13.7 □

Remark 3.13.32 Thus if $\mathcal{R}$ is a terminating TRS such that all critical pairs are trivial, then it must be a normal form TRS.

One simple way to find normal form TRSs is to use the Critical Pairs Lemma in conjunction with Theorem 3.13.11 (p. 210).

Corollary 3.13.33 Let $\mathcal{R}$ be a term rewrite system with the property that for each $s \longrightarrow t \in \mathcal{R}$ we have

a. $|s| > |t|$,
b. $|s|_x \geq |t|_x$ for every variable x, and
c. $\mathcal{R}$ is confluent on critical pairs.

Then $\mathcal{R}$ is a normal form TRS.

Example 3.13.34 The term rewrite system $\mathcal{R} = \{fffx \longrightarrow x\}$ is a normal form TRS as it is easily seen to satisfy (a) and (b), and the only critical pairs are trivial, so it satisfies (c).

Corollary 3.13.35 There is an algorithm to determine if a finite and terminating TRS $\mathcal{R}$ is a normal form TRS.

Proof. By Theorem 3.13.31 it suffices to check for confluence on critical pairs. Since there are only finitely many critical pairs (s, t) that we can obtain from $\mathcal{R}$, we apply the elementary rewrite rules (in any manner desired) to s and t until we come to terminal terms, say $s \longrightarrow s'$ and $t \longrightarrow t'$. Such must happen as $\mathcal{R}$ terminates. Now we can determine if a rewrite rule can be applied to a term, so we can determine if we have reached terminal terms. If for each critical pair $s' = t'$, then $\mathcal{R}$ is a normal form TRS. Otherwise it is not. □

How do we find a normal form TRS $\mathcal{R}$ for a given $\mathcal{E}$? Such is the goal of various versions of the Knuth–Bendix procedure, one of which we will examine in Sec. 3.15.

Exercises

3.13.4 Assume we have already proved that the following single–rule TRSs are terminating. (Indeed this is the case.) Use the Critical Pairs Lemma to determine which of the following are normal form term rewrite systems. (f, g are unary; $+, \cdot$ are binary.)

a. $\mathcal{R} = \{ffx \longrightarrow fx\}$
b. $\mathcal{R} = \{fffx \longrightarrow fx\}$
c. $\mathcal{R} = \{fffx \longrightarrow ffx\}$
d. $\mathcal{R} = \{fgx \longrightarrow fx\}$
e. $\mathcal{R} = \{fgx \longrightarrow gfx\}$
f. $\mathcal{R} = \{fgfx \longrightarrow gfgx\}$
g. $\mathcal{R} = \{x \cdot (y + z) \longrightarrow (x \cdot y) + (x \cdot z)\}$

3.13.5 Assume we have already proved that the following two–rule TRSs are terminating. (Indeed this is the case.) Use the Critical Pairs Lemma to determine which of the following are normal form term rewrite systems. ($f, g, {}^{-1}$ are unary; $+, \cdot$ are binary, 1 is a constant.)

a. $\mathcal{R} = \{fx \longrightarrow x,\ fgx \longrightarrow gfx\}$
b. $\mathcal{R} = \{ffffx \longrightarrow fx,\ fffx \longrightarrow ffx\}$
c. $\mathcal{R} = \{fgfx \longrightarrow gfgx,\ ffx \longrightarrow fx\}$
d. $\mathcal{R} = \{x \cdot 1 \longrightarrow x,\ (x \cdot y) \cdot z \longrightarrow x \cdot (y \cdot z)\}$
e. $\mathcal{R} = \{x \cdot 1 \longrightarrow x,\ x \cdot x^{-1} \longrightarrow 1\}$
f. $\mathcal{R} = \{x \cdot x^{-1} \longrightarrow 1,\ (x \cdot y) \cdot z \longrightarrow x \cdot (y \cdot z)\}$
g. $\mathcal{R} = \{(x \cdot y) \cdot z \longrightarrow x \cdot (y \cdot z),\ x \cdot (y + z) \longrightarrow (x \cdot y) + (x \cdot z)\}$

3.13.6 Terms as Strings

Before tackling the proof of the Critical Pairs Lemma we need to give a more precise formulation of elementary term rewrites by considering the effect of substitution and replacement on strings.

If we have fixed our language $\mathcal{L}$ of algebras and set of variables X, then we first observe what has been noted before, that terms are certain strings on the alphabet $X \cup \mathcal{L}$, i.e., we have $T(X) \subset (X \cup \mathcal{L})^*$.

Recall (from Example 3.4.12 on p. 156) that we have the natural binary operation of concatenation on $(X \cup \mathcal{L})^*$.

This makes it possible to have a more precise notation for an *occurrence* of a subterm, namely, if s' is a subterm of s, then the notation

$$s = As'B$$

subdivides string s into part A before a particular occurrence of s', the occurrence of s', and part B after the occurrence of s'. We see that this is more precise than our previous (and convenient) notation for subterms, namely, $s = r(\cdots s' \cdots)$.

Now we have a precise notation for a replacement of an occurrence, namely, replacing the indicated occurrence of s' in $As'B$ with t' gives the term $At'B$. Actually this is one of our basic facts about working with terms, that replacement in a term gives a term.

Fact 3.13.36 If AsB is a term and s is a term, then for any term t, AtB is also a term.

Next, let us take a brief look at substitutions. Actually, we can extend the notion of substitutions to the set of strings $(X \cup \mathcal{L})^*$, namely, if we have

$$\sigma = \begin{pmatrix} x_1 \leftarrow A_1 \\ \vdots \\ x_n \leftarrow A_n \end{pmatrix},$$

where the A_i are any strings in $(X \cup \mathcal{L})^*$, then given any string $S(x_1, \ldots, x_n)$, we have the result of simultaneous substitution of the A_i for the x_i, namely, $\sigma S = S(A_1, \cdots, A_n)$. Such *string* substitutions σ are characterized by

$$\begin{array}{ll} \sigma\Lambda = \Lambda & \text{for } \Lambda \text{ the empty string} \\ \sigma f = f & \text{for } f \in \mathcal{F} \\ \sigma c = c & \text{for } c \in \mathcal{C} \\ \sigma x \in (X \cup \mathcal{L})^* & \text{for } x \in X \\ \sigma(AB) = \sigma(A)\sigma(B). & \end{array}$$

We will continue to work only with *term* substitutions, namely, those substitutions σ that map terms to terms. We need only to change the fourth condition above to $\sigma x \in T(X)$. But now we view term substitutions as applying to any string from $(X \cup \mathcal{L})^*$

From this point of view, if we are given a term substitution σ and a term $s = AtB$, where t is also a term, then we have $\sigma s = \sigma A \sigma t \sigma B$.

To avoid parentheses, we will understand that σAB means $\sigma(A)B$, i.e., σ applies only to the item immediately following it.

Now we can detail the fact that $s \longrightarrow_{\mathcal{R}} t$ by saying that for some $A, B \in (X \cup \mathcal{L})^*$, for some term substitution σ, and some $s' \longrightarrow t' \in \mathcal{R}$ we have

$$\begin{aligned} s &= A\sigma s'B \\ t &= A\sigma t'B. \end{aligned}$$

Thus the typical elementary rewrite looks like $A\sigma s'B \longrightarrow A\sigma t'B$. Next, we review two of our most fundamental facts about elementary rewrites.

Fact 3.13.37 $\longrightarrow_{\mathcal{R}}$ is closed under replacement.

Suppose CsD is a term, with the indicated occurrence of the term s, and assume $s \longrightarrow_{\mathcal{R}} t$ holds. Then choosing A, B, σ, s', t' as before, we see that

$$CsD = CA\sigma s'BD \longrightarrow_{\mathcal{R}} CA\sigma t'BD = CtD.$$

Fact 3.13.38 $\longrightarrow_{\mathcal{R}}$ is closed under substitution.

Suppose $s \longrightarrow_{\mathcal{R}} t$ holds. Let τ be a term substitution. Then choosing A, B, σ, s', t' as before, we see that

$$\begin{array}{rcl} \tau s & = & \tau(A\sigma s' B) \\ & = & \tau A \tau\sigma s' \tau B \\ & \longrightarrow_{\mathcal{R}} & \tau A \tau\sigma t' \tau B \\ & = & \tau(A\sigma t' B) \\ & = & \tau t. \end{array}$$

The following two facts are easy extensions of the preceding two facts.

Fact 3.13.39 $\longrightarrow_{\mathcal{R}}^{*}$ is closed under replacement.

Fact 3.13.40 $\longrightarrow_{\mathcal{R}}^{*}$ is closed under substitution.

Now we come to a basic fact that we will need.

Fact 3.13.41 If s, t are terms and σ is a term substitution such that

$$\sigma s = AtB$$

for some strings A, B, then there are strings C, D and a term s' such that

$$s = Cs'D$$

and either

a. s' is a variable, say x, and $\sigma s = \sigma C \sigma x \sigma D = \sigma C A' t B' \sigma D$, or
b. s' is not a variable and $\sigma C = A$, $\sigma s' = t$, and $\sigma D = B$.

This fact is easiest to see with the following diagrams:

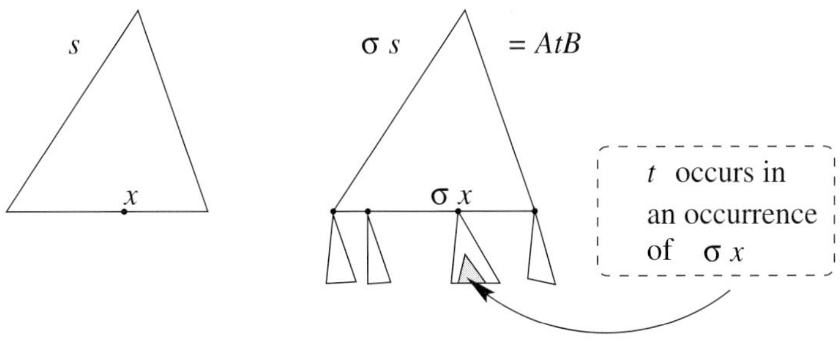

Figure 3.13.42 t occurs in, or equals, σx

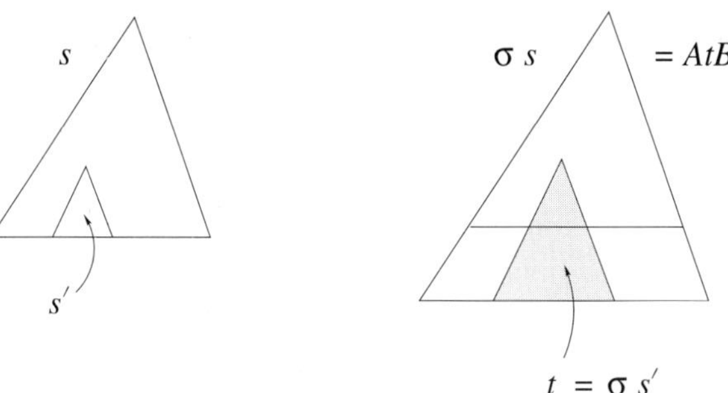

Figure 3.13.43 $t = \sigma s'$, and s' is not a variable

3.13.7 Confluence

When checking $\mathcal{R}$ for being a normal form TRS our objective is to avoid examining infinitely many sequences of elementary rewrites. A first step in this direction is to examine the notion of *confluence*.

Definition 3.13.44 Let $\mathcal{R}$ be a TRS. Then

a. $\mathcal{R}$ is *confluent* if for any terms s, s', s'' with

$$\begin{array}{lll} s & \longrightarrow^*_{\mathcal{R}} & s' \\ s & \longrightarrow^*_{\mathcal{R}} & s'' \end{array}$$

there exists a term t such that

$$\begin{array}{lll} s' & \longrightarrow^*_{\mathcal{R}} & t \\ s'' & \longrightarrow^*_{\mathcal{R}} & t. \end{array}$$

This condition is often depicted as follows, where the *dashed lines with an arrow* mean $\longrightarrow^*_{\mathcal{R}}$:

Whenever

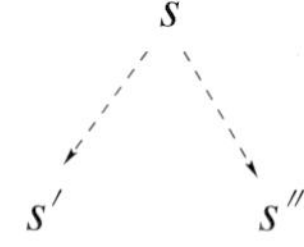

there is a term t such that

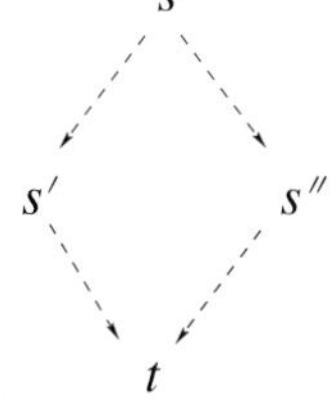

Figure 3.13.45 Confluence

b. $\mathcal{R}$ is *locally confluent* if

$$\left.\begin{array}{lcl} s & \longrightarrow_{\mathcal{R}} & s' \\ s & \longrightarrow_{\mathcal{R}} & s'' \end{array}\right\} \Longrightarrow \exists t \left\{\begin{array}{lcl} s' & \longrightarrow^{*}_{\mathcal{R}} & t \\ s'' & \longrightarrow^{*}_{\mathcal{R}} & t. \end{array}\right.$$

This condition is depicted as follows, where the *solid lines with an arrow* mean $\longrightarrow_{\mathcal{R}}$:

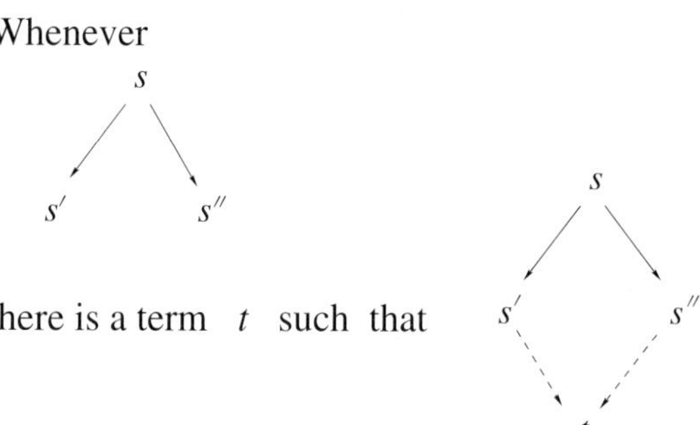

Figure 3.13.46 Local confluence

An obvious reason for studying confluence is the following result:

Lemma 3.13.47 A normal form TRS is confluent, and a confluent TRS is locally confluent.

Proof. Let $\mathcal{R}$ be a normal form TRS. Suppose we have terms s, s', s'' such that

$$\begin{array}{lcl} s & \longrightarrow^{*}_{\mathcal{R}} & s' \\ s & \longrightarrow^{*}_{\mathcal{R}} & s''. \end{array}$$

Then we must have

$$\begin{array}{lcl} s' & \longrightarrow^{*}_{\mathcal{R}} & n(s) \\ s'' & \longrightarrow^{*}_{\mathcal{R}} & n(s), \end{array}$$

where $n(s)$ is the normal form of s. Thus $\mathcal{R}$ is confluent.

Clearly, confluence implies local confluence. □

Example 3.13.48 Let $\mathcal{R} = \{fgfx \longrightarrow gx,\ gfgx \longrightarrow fx\}$, where f, g are unary. Then we have $fgfgx \longrightarrow ffx$ (by applying the second rule), and $fgfgx \longrightarrow ggx$ (by applying the first rule). Now, no rule applies to either ffx or ggx, so $\mathcal{R}$ is not locally confluent. However, it is easy to see that it is terminating.

Example 3.13.49 The system $\mathcal{R} = \{x \longrightarrow y\}$ is clearly confluent, since any term can elementarily rewrite to any other term. But $\mathcal{R}$ is not terminating.

Lemma 3.13.50 Suppose $\mathcal{R}$ is a terminating TRS, i.e., such that $\longrightarrow^{+}_{\mathcal{R}}$ is well founded. Then $\mathcal{R}$ is confluent iff $\mathcal{R}$ is locally confluent.

Proof. The direction ($\Rightarrow$) is clear, since local confluence is a special case of confluence. For the converse suppose $\mathcal{R}$ is locally confluent.

Let us say that a term s is $\mathcal{R}$*–confluent* if whenever $s \longrightarrow^{*}_{\mathcal{R}} s'$ and $s \longrightarrow^{*}_{\mathcal{R}} s''$ there is a term t such that $s' \longrightarrow^{*}_{\mathcal{R}} t$ and $s'' \longrightarrow^{*}_{\mathcal{R}} t$.

If $\mathcal{R}$ is not confluent then there must be a non–$\mathcal{R}$–confluent term s_1. Let $s_1 \longrightarrow_{\mathcal{R}} s_2 \longrightarrow_{\mathcal{R}} \cdots$ be a sequence of elementary rewrites between non–$\mathcal{R}$–confluent terms s_i. By well foundedness this sequence must be finite.

So we choose a non–$\mathcal{R}$–confluent term s such that any term t with $s \longrightarrow^{+}_{\mathcal{R}} t$ is $\mathcal{R}$–confluent, and we choose rewrites

$$\begin{array}{lll} s & \longrightarrow^{*}_{\mathcal{R}} & s' \\ s & \longrightarrow^{*}_{\mathcal{R}} & s'' \end{array} \tag{36}$$

such that there is no term t satisfying

$$\begin{array}{lll} s' & \longrightarrow^{*}_{\mathcal{R}} & t \\ s'' & \longrightarrow^{*}_{\mathcal{R}} & t\,. \end{array} \tag{37}$$

Certainly, $s' \neq s''$, and both are different from s, otherwise there would exist a term t to make (37) true.

Next, we choose terms s_1 and s_2 such that

$$\begin{array}{lllll} s & \longrightarrow_{\mathcal{R}} & s_1 & \longrightarrow^{*}_{\mathcal{R}} & s' \\ s & \longrightarrow_{\mathcal{R}} & s_2 & \longrightarrow^{*}_{\mathcal{R}} & s'' \end{array}$$

hold, i.e., we have

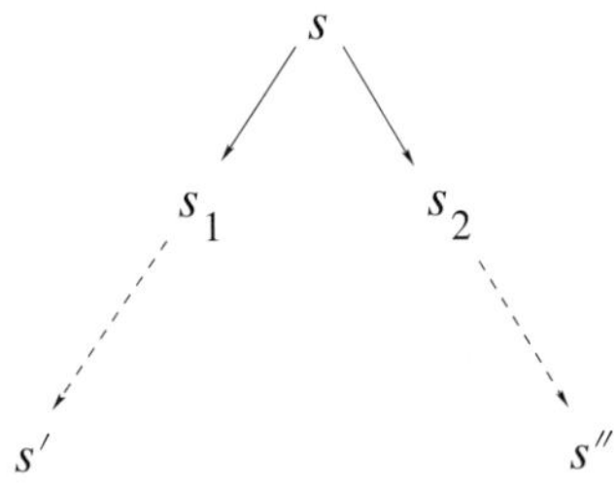

Figure 3.13.51 Assuming a failure of confluence

By local confluence there must be a term s_0 such that

$$\begin{array}{lll} s_1 & \longrightarrow^*_{\mathcal{R}} & s_0 \\ s_2 & \longrightarrow^*_{\mathcal{R}} & s_0 \end{array}$$

holds:

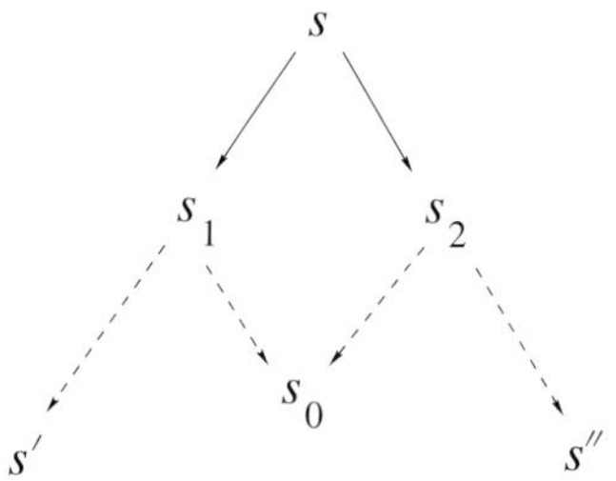

Figure 3.13.52 Applying local confluence

Since $s \longrightarrow_{\mathcal{R}} s_1$ and $s \longrightarrow_{\mathcal{R}} s_2$, it follows by our assumption on s that both s_1 and s_2 are $\mathcal{R}$–confluent terms. Thus there must be terms t_1 and t_2 such that

$$\begin{array}{lll} s' & \longrightarrow^*_{\mathcal{R}} & t_1 \\ s_0 & \longrightarrow^*_{\mathcal{R}} & t_1 \end{array}$$

and

$$\begin{array}{lll} s'' & \longrightarrow^*_{\mathcal{R}} & t_2 \\ s_0 & \longrightarrow^*_{\mathcal{R}} & t_2\,. \end{array}$$

This condition is depicted as follows:

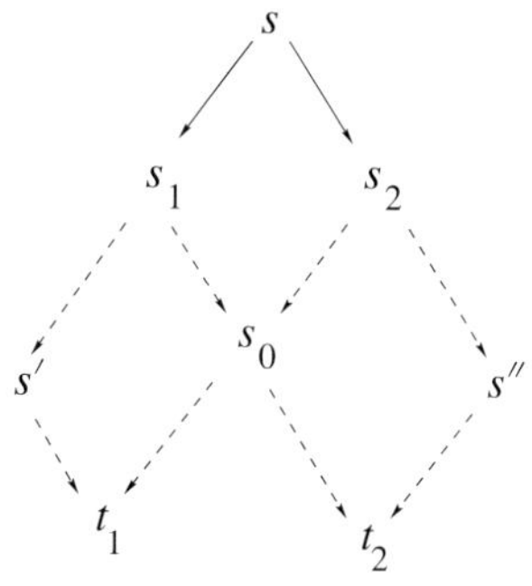

Figure 3.13.53 s_1 and s_2 are $\mathcal{R}$–confluent

We also have $s \longrightarrow^{+}_{\mathcal{R}} s_0$, so s_0 is $\mathcal{R}$–confluent. This guarantees the existence of a term t such that

$$\begin{array}{lcl} t_1 & \longrightarrow^{*}_{\mathcal{R}} & t \\ t_2 & \longrightarrow^{*}_{\mathcal{R}} & t. \end{array}$$

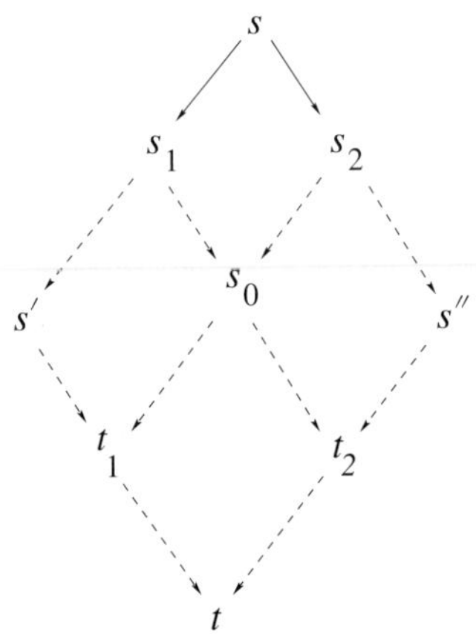

Figure 3.13.54 s_0 is also $\mathcal{R}$–confluent

Now this implies

$$\begin{array}{lcl} s' & \longrightarrow^{*}_{\mathcal{R}} & t \\ s'' & \longrightarrow^{*}_{\mathcal{R}} & t, \end{array}$$

which gives a contradiction. □

The following important lemma was used in a special case (loops) by Evans in 1951. The general proof is no more difficult than Evans's special case.

Lemma 3.13.55 Given a terminating TRS $\mathcal{R}$,

$\mathcal{R}$ is a normal form TRS iff $\mathcal{R}$ is confluent
iff $\mathcal{R}$ is locally confluent.

Proof. If $\mathcal{R}$ is a normal form TRS, then Lemma 3.13.47 (p. 229) says that $\mathcal{R}$ is confluent. If $\mathcal{R}$ is terminating and confluent, then it is easily seen to be a normal form TRS, for if we have two terminating sequences of rewrites from s, say

$$\begin{array}{lcl} s & \longrightarrow^{*}_{\mathcal{R}} & s' \\ s & \longrightarrow^{*}_{\mathcal{R}} & s'', \end{array}$$

then $s' = s''$, for otherwise confluence would guarantee that s' and s'' were not both terminal terms.

So we need only to show that local confluence implies confluence. But this is precisely the content of Lemma 3.13.50. □

Now we proceed to show, for a *finite* and *terminating* TRS, that we can effectively test for local confluence. This is the fundamental result of Knuth and Bendix.

It will be convenient now to call a pair of rewrites $s \longrightarrow_{\mathcal{R}} s'$, $s \longrightarrow_{\mathcal{R}} s''$ with the same head term an $\mathcal{R}$–*split*. We will say an $\mathcal{R}$–split $s \longrightarrow_{\mathcal{R}} s'$, $s \longrightarrow_{\mathcal{R}} s''$ is $\mathcal{R}$–confluent if there is a term t such that $s' \longrightarrow_{\mathcal{R}}^{*} t$, $s'' \longrightarrow_{\mathcal{R}}^{*} t$.

In this notation the definition of the local confluence of $\mathcal{R}$ can be rephrased as follows:

$\mathcal{R}$ is locally confluent iff $\mathcal{R}$–splits are $\mathcal{R}$–confluent.

Being confluent on critical pairs is clearly part of the requirement for being locally confluent. We want to show that this is also sufficient. We will reduce the local confluence property to the confluence on critical pairs property in a sequence of stages. This is where we will need our work in the section on strings.

When we have $\mathcal{R}$–splits

$$\begin{array}{lll} s & \longrightarrow_{\mathcal{R}} & s' \\ s & \longrightarrow_{\mathcal{R}} & s'' \end{array}$$

that we want to analyze for $\mathcal{R}$–confluence, we will express the rewrites in more detail, to exhibit the rewriting of occurrences of subterms, as follows:

$$\begin{aligned} s &= A_1\sigma_1 s_1 B_1 = A_2\sigma_2 s_2 B_2 \\ s' &= A_1\sigma_1 t_1 B_1 \\ s'' &= A_2\sigma_2 t_2 B_2, \end{aligned}$$

where $s_1 \longrightarrow t_1$ and $s_2 \longrightarrow t_2$ are from $\mathcal{R}$.

Then the $\mathcal{R}$–splits to be tested for $\mathcal{R}$–confluence look like

$$\begin{array}{lll} s = A_1\sigma_1 s_1 B_1 & \longrightarrow_{\mathcal{R}} & A_1\sigma_1 t_1 B_1 \\ s = A_2\sigma_2 s_2 B_2 & \longrightarrow_{\mathcal{R}} & A_2\sigma_2 t_2 B_2. \end{array} \tag{38}$$

At this point we want to introduce a notational simplification by requiring that different rules in $\mathcal{R}$ have *no variables in common* and that for each rule $s \longrightarrow t$ in $\mathcal{R}$ there is also a rule $\hat{s} \longrightarrow \hat{t}$ in $\mathcal{R}$ that is the same as $s \longrightarrow t$ except different variables appear.

If we take a given TRS $\mathcal{R}'$ and transform it into such a TRS $\mathcal{R}$ by simply renaming variables (and taking two copies of each rule, with different variables) then we can readily check that $s \longrightarrow_{\mathcal{R}} t$ holds iff $s \longrightarrow_{\mathcal{R}'} t$, for any terms s, t. Thus the properties of being terminating, confluent, locally confluent, or a normal form TRS are the same for both $\mathcal{R}$ and $\mathcal{R}'$. They also have the same critical pairs, up to naming of the variables. So the property of being confluent on critical pairs is the same for both. (Having two copies of each rule in $\mathcal{R}$ allows us to find critical pairs whose construction requires that we use the same rule twice, i.e., for both $s_1 \longrightarrow t_1$ and $s_2 \longrightarrow t_2$, without having to rename the variables.)

Having made this simplifying assumption on $\mathcal{R}$ we see that the $\mathcal{R}$–splits that need to be proved $\mathcal{R}$–confluent in order to prove $\mathcal{R}$ is locally confluent can be expressed in the form

$$\begin{aligned} s = A_1\sigma s_1 B_1 &\longrightarrow_{\mathcal{R}} A_1\sigma t_1 B_1 \\ s = A_2\sigma s_2 B_2 &\longrightarrow_{\mathcal{R}} A_2\sigma t_2 B_2, \end{aligned} \tag{39}$$

since the pair of substitutions σ_1, σ_2 in (38) can be replaced by a single substitution σ that agrees with σ_1 on the variables of s_1, and with σ_2 on the variables of s_2.

Lemma 3.13.56 $\mathcal{R}$ is locally confluent iff $\mathcal{R}$–splits of the form

$$\begin{aligned} s = A\sigma s_1 B &\longrightarrow_{\mathcal{R}} A\sigma t_1 B \\ s = AC\sigma s_2 DB &\longrightarrow_{\mathcal{R}} AC\sigma t_2 DB \end{aligned} \tag{40}$$

are $\mathcal{R}$–confluent, i.e., we need to check $\mathcal{R}$–confluence only for the following cases:

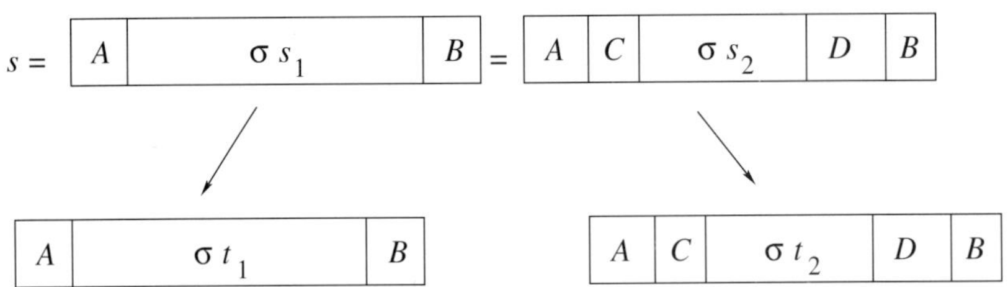

Figure 3.13.57 $\mathcal{R}$–splits to be checked for $\mathcal{R}$–confluence

Proof. The only other possibility for the indicated occurrences of the subterms σs_1 and σs_2 in the string s is that they are *disjoint*. Proper overlapping cannot happen! In the disjoint case we claim $\mathcal{R}$–confluence, namely, we then have the situation

$$s = A\sigma s_1 B\sigma s_2 C \qquad \text{or} \qquad s = A\sigma s_2 B\sigma s_1 C.$$

By symmetry it suffices to consider the first of these. Then the $\mathcal{R}$–split looks like

$$\begin{array}{lll} s = A\sigma s_1 B\sigma s_2 C & \longrightarrow_{\mathcal{R}} & A\sigma t_1 B\sigma s_2 C \\ s = A\sigma s_1 B\sigma s_2 C & \longrightarrow_{\mathcal{R}} & A\sigma s_1 B\sigma t_2 C. \end{array}$$

In this case we can choose $t = A\sigma t_1 B\sigma t_2 C$, as we have

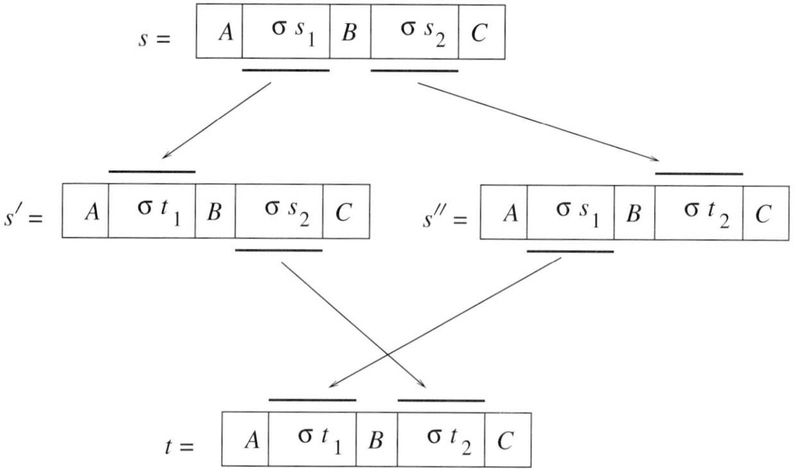

Figure 3.13.58 The disjoint case leads to local confluence

□

Lemma 3.13.59 $\mathcal{R}$ is locally confluent iff the $\mathcal{R}$–splits of the form

$$\begin{array}{lll} s = \sigma s_1 & \longrightarrow_{\mathcal{R}} & \sigma t_1 \\ s = C\sigma s_2 D & \longrightarrow_{\mathcal{R}} & C\sigma t_2 D \end{array} \qquad (41)$$

are $\mathcal{R}$–confluent, i.e.,

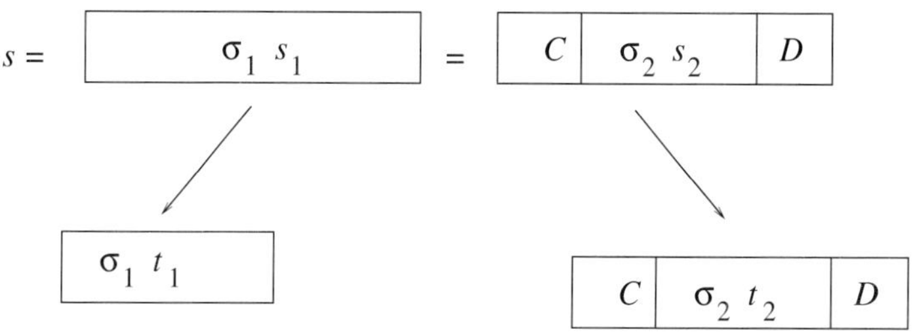

Figure 3.13.60 $\mathcal{R}$–splits to check for $\mathcal{R}$–confluence

Proof. Let us assume all $\mathcal{R}$–splits of the form (41) are $\mathcal{R}$–confluent. To show that $\mathcal{R}$ is locally confluent it suffices, by Lemma 3.13.56, to show that every $\mathcal{R}$–split of the form

$$\begin{array}{rcl} s = A\sigma s_1 B & \longrightarrow_{\mathcal{R}} & A\sigma t_1 B \\ s = AC\sigma s_2 DB & \longrightarrow_{\mathcal{R}} & AC\sigma t_2 DB \end{array} \tag{42}$$

is $\mathcal{R}$–confluent. Now, the $\mathcal{R}$–split

$$\begin{array}{lcl} s_0 = \sigma s_1 & \longrightarrow_{\mathcal{R}} & \sigma t_1 \\ s_0 = C\sigma s_2 D & \longrightarrow_{\mathcal{R}} & C\sigma t_2 D \end{array}$$

is of the form (41), so by our assumption it is $\mathcal{R}$–confluent, and there is a term t such that

$$\begin{array}{lclcl} s_0 = \sigma s_1 & \longrightarrow_{\mathcal{R}} & \sigma t_1 & \overset{*}{\longrightarrow}_{\mathcal{R}} & t \\ s_0 = C\sigma s_2 D & \longrightarrow_{\mathcal{R}} & C\sigma t_2 D & \overset{*}{\longrightarrow}_{\mathcal{R}} & t. \end{array}$$

Then by Fact 3.13.39 (p. 227), we have

$$\begin{array}{rclcl} s = As_0B = A\sigma s_1 B & \longrightarrow_{\mathcal{R}} & A\sigma t_1 B & \overset{*}{\longrightarrow}_{\mathcal{R}} & AtB \\ s = As_0B = AC\sigma s_2 DB & \longrightarrow_{\mathcal{R}} & AC\sigma t_2 DB & \overset{*}{\longrightarrow}_{\mathcal{R}} & AtB. \end{array}$$

Thus we have $\mathcal{R}$–confluence for all $\mathcal{R}$–splits of the form (40), and by Lemma 3.13.56 it follows that $\mathcal{R}$ is locally confluent. □

Lemma 3.13.61 For local confluence, it suffices to show that $\mathcal{R}$–splits of the form

$$\begin{array}{l} s = \sigma s_1 \longrightarrow \sigma t_1 \\ s = A\sigma s_2 B \longrightarrow_{\mathcal{R}} A\sigma t_2 B, \text{ where} \\ \quad s_1 = Cs_1' D, \text{ and } s_1' \text{ is not a variable} \\ \quad \sigma C = A \\ \quad \sigma D = B \\ \quad \sigma s_1' = \sigma s_2 \end{array} \tag{43}$$

are $\mathcal{R}$–confluent.

Proof. From Lemma 3.13.59 and Fact 3.13.41 (p. 227) we see that we need to consider only the $\mathcal{R}$–splits in (43) and the $\mathcal{R}$–splits where $s = \sigma s_1 = A\sigma s_2 B$ and σs_2 occurs as a subterm of an occurrence of σx, where x is a variable in s_1. So let us check that we always have $\mathcal{R}$–confluence in the second situation.

We write s_1 as $s_1 = C_1 x C_2 x \cdots x C_n$, where no string C_i has an x in it. Then the situation we are dealing with is $\sigma x = D_1 \sigma s_2 D_2$. It will be convenient to write $s_1 = s_1(x, y, \ldots)$ and $t_1 = t_1(x, y, \ldots)$.

For then $\sigma s_1 = s_1(\sigma x, \sigma y, \ldots)$ and $\sigma t_1 = t_1(\sigma x, \sigma y, \ldots)$. Now we have

$$\begin{aligned} s = \sigma s_1 &= s_1(\sigma x, \sigma y, \ldots) \\ &= \sigma C_1 \sigma x \sigma C_2 \sigma x \cdots \sigma x \sigma C_n \\ &= \sigma C_1 \sigma x \cdots \sigma C_{i-1} D_1 \sigma s_2 D_2 \sigma C_i \cdots \sigma x \sigma C_n, \end{aligned}$$

and then we have the $\mathcal{R}$–split

$$\begin{aligned} s = \sigma s_1 &\longrightarrow_{\mathcal{R}} t_1(\sigma x, \sigma y, \ldots) \\ s = \sigma s_1 &\longrightarrow_{\mathcal{R}} \sigma C_1 \sigma x \cdots \sigma C_{i-1} D_1 \sigma t_2 D_2 \sigma C_i \cdots \sigma x \sigma C_n. \end{aligned}$$

Since $\sigma x = D_1 \sigma s_2 D_2$, we can apply the rule $\sigma x \longrightarrow_{\mathcal{R}} D_1 \sigma t_2 D_2$ to each occurrence of σx that corresponds to the first argument of $t_1(\sigma x, \sigma y, \ldots)$ to rewrite σt_1 to $t_1(D_1 \sigma t_2 D_2, \sigma y, \ldots)$. This will be our term t. So we have

$$\begin{aligned} s = \sigma s_1 &\longrightarrow_{\mathcal{R}} t_1(\sigma x, \sigma y, \ldots) \\ &\longrightarrow^*_{\mathcal{R}} t_1(D_1 \sigma t_2 D_2, \sigma y, \ldots) = t. \end{aligned}$$

Now we need to show that we can rewrite the term

$$\sigma C_1 \sigma x \cdots \sigma C_{i-1} D_1 \sigma t_2 D_2 \sigma C_i \cdots \sigma x \sigma C_n \tag{44}$$

to this t. Since we have $\sigma x \longrightarrow_{\mathcal{R}} D_1 \sigma t_2 D_2$, we apply this repeatedly to (44) to see that

$$\begin{aligned} &\sigma C_1 \sigma x \cdots \sigma C_{i-1} D_1 \sigma t_2 D_2 \sigma C_i \cdots \sigma x \sigma C_n \\ \longrightarrow^*_{\mathcal{R}} &\sigma C_1 D_1 \sigma t_2 D_2 \cdots \sigma C_{i-1} D_1 \sigma t_2 D_2 \sigma C_i \cdots D_1 \sigma t_2 D_2 \sigma C_n, \end{aligned}$$

i.e.,

$$\sigma C_1 \sigma x \cdots \sigma C_{i-1} D_1 \sigma t_2 D_2 \sigma C_i \cdots \sigma x \sigma C_n \longrightarrow^*_{\mathcal{R}} s_1(D_1 \sigma t_2 D_2, \sigma y, \ldots). \tag{45}$$

Let the substitution τ be defined by

$$\tau = \begin{pmatrix} x & \leftarrow & D_1 \sigma t_2 D_2 \\ y & \leftarrow & \sigma y \\ z & \leftarrow & \sigma z \\ & \vdots & \end{pmatrix}.$$

Applying τ to

$$s_1(x, y, \ldots) \longrightarrow_{\mathcal{R}} t_1(x, y, \ldots)$$

gives

$$\tau s_1(x, y, \ldots) \longrightarrow_{\mathcal{R}} \tau t_1(x, y, \ldots),$$

and thus

$$s_1(D_1\sigma t_2 D_2, \sigma y, \ldots) \longrightarrow_{\mathcal{R}} t_1(D_1\sigma t_2 D_2, \sigma y, \ldots) = t.$$

Combining this result with (45) we have

$$\sigma C_1 \sigma x \cdots \sigma C_{i-1} D_1 \sigma t_2 D_2 \sigma C_i \cdots \sigma x \sigma C_n \longrightarrow^*_{\mathcal{R}} t.$$

This finishes the proof. □

Lemma 3.13.62 To prove that $\mathcal{R}$ is locally confluent it suffices to show that each $\mathcal{R}$–split of the form (43), with σ the most general unifier of s_1' and s_2, is $\mathcal{R}$–confluent.

Proof. Suppose we are given the following $\mathcal{R}$–split:

$$\begin{aligned} s &= \sigma s_1 \longrightarrow_{\mathcal{R}} \sigma t_1 \\ s &= A\sigma s_2 B \longrightarrow_{\mathcal{R}} A\sigma t_2 B, \text{ where} \\ &\quad s_1 = C s_1' D, \text{ and } s_1' \text{ is not a variable} \\ &\quad \sigma C = A \\ &\quad \sigma D = B \\ &\quad \sigma s_1' = \sigma s_2. \end{aligned} \tag{46}$$

Let μ be a most general unifier of s_1', s_2. Then $\mu s_1' = \mu s_2$, and $\sigma = \tau\mu$ for some τ. From $s_1 = C s_1' D$ we have

$$\mu s_1 = \mu C \mu s_1' \mu D = \mu C \mu s_2 \mu D.$$

This leads to the $\mathcal{R}$–split

$$\begin{aligned} s_0 &= \mu s_1 \longrightarrow \mu t_1 \\ s_0 &= A'\mu s_2 B' \longrightarrow_{\mathcal{R}} A'\mu t_2 B', \text{ where} \\ &\quad s_1 = C s_1' D, \text{ and } s_1' \text{ is not a variable} \\ &\quad \mu C = A' \\ &\quad \mu D = B' \\ &\quad \mu s_1' = \mu s_2. \end{aligned} \tag{47}$$

By our assumptions this $\mathcal{R}$–split is $\mathcal{R}$–confluent, i.e., there is a term t such that

$$\begin{aligned} s_0 &= \mu s_1 \longrightarrow_{\mathcal{R}} \mu t_1 \longrightarrow^*_{\mathcal{R}} t \\ s_0 &= A'\mu s_2 B' \longrightarrow_{\mathcal{R}} A'\mu t_2 B' \longrightarrow^*_{\mathcal{R}} t. \end{aligned}$$

But then by Fact 3.13.40 we can apply τ to obtain

$$\begin{aligned} \tau s_0 &= \tau\mu s_1 \longrightarrow_{\mathcal{R}} \tau\mu t_1 \longrightarrow^*_{\mathcal{R}} \tau t \\ \tau s_0 &= \tau A' \tau\mu s_2 \tau B' \longrightarrow_{\mathcal{R}} \tau A' \tau\mu t_2 \tau B' \longrightarrow^*_{\mathcal{R}} \tau t, \end{aligned}$$

i.e.,

$$\begin{aligned} s &= \sigma s_1 \longrightarrow_{\mathcal{R}} \sigma t_1 \longrightarrow^*_{\mathcal{R}} \tau t \\ s &= A\sigma s_2 B \longrightarrow_{\mathcal{R}} A\sigma t_2 B \longrightarrow^*_{\mathcal{R}} \tau t, \end{aligned}$$

and thus we have satisfied the requirements for local confluence from Lemma 3.13.59 (p. 235). □

The last lemma says that to test for local confluence it suffices to test for confluence on critical pairs.

Lemma 3.13.63 Suppose $\mathcal{R}$ is a TRS. Then $\mathcal{R}$ is locally confluent iff $\mathcal{R}$ is confluent on critical pairs.

Proof. The direction ($\Longrightarrow$) is clear, since being confluent on critical pairs is part of what it means to be locally confluent.

For the converse direction ($\Longleftarrow$), suppose we have an $\mathcal{R}$–split $s \longrightarrow s'$, $s \longrightarrow s''$ of the form (43), with σ being the most general unifier of s_1' and s_2. Then s' and s'' is a critical pair. Because we assumed confluence on critical pairs, there must be a term t such that $s' \longrightarrow_{\mathcal{R}}^{*} t$ and $s'' \longrightarrow_{\mathcal{R}}^{*} t$. But then the $\mathcal{R}$–split is $\mathcal{R}$–confluent. By Lemma 3.13.62 we see that $\mathcal{R}$ is locally confluent. □

This suffices to prove the Critical Pairs Lemma, since Lemma 3.13.50 (p. 230) says that for $\mathcal{R}$ a terminating TRS, $\mathcal{R}$ is a normal form TRS iff $\mathcal{R}$ is locally confluent, and the latter happens iff $\mathcal{R}$ is confluent on critical pairs by Lemma 3.13.62.

Exercises

If $\mathcal{R}$ is a normal form TRS, let NF be the set of normal forms of $\mathcal{R}$, i.e., the set of $n_{\mathcal{R}}(t)$.

3.13.6 For $\mathcal{R}$ a normal form TRS show the following hold.

- **a.** Variables are in normal form.
- **b.** NF is closed under subterms, i.e., if t is a normal form and s is a subterm of t, then s is also a normal form.
- **c.** If $s \longrightarrow t$ is in $\mathcal{R}$, then the variables of t form a subset of of the variables of s.
- **d.** The variables of $n_{\mathcal{R}}(s)$ form a subset of the variables of s.

3.13.7 For $\mathcal{R}$ a normal form TRS, with $n = n_{\mathcal{R}}$, show that

- **a.** $n(n(s)) = n(s)$
- **b.** $n(f(n(s_1), \cdots, n(s_n))) = n(f(s_1, \cdots, s_n))$
- **c.** $n(\sigma n(s)) = n(\sigma s)$ for σ a substitution.

We would like to put a normal form TRS into a canonical form, if possible. This is where the notion of reduced comes in. A TRS $\mathcal{R}$ is *reduced* if for $s \longrightarrow t \in R$, no rule in $\mathcal{R} \setminus \{s \longrightarrow t\}$ can be applied to either s or t. (By this definition any single–rule TRS is reduced.)

3.13.8 Determine which of the following TRSs are reduced.

a. $\mathcal{R} = \{fx \longrightarrow x,\ fgx \longrightarrow gfx\}$
b. $\mathcal{R} = \{ffffx \longrightarrow fx,\ fffx \longrightarrow ffx\}$
c. $\mathcal{R} = \{fgfx \longrightarrow gfgx,\ ffx \longrightarrow fx\}$
d. $\mathcal{R} = \{x \cdot 1 \longrightarrow x,\ (x \cdot y) \cdot z \longrightarrow x \cdot (y \cdot z)\}$
e. $\mathcal{R} = \{x \cdot 1 \longrightarrow x,\ x \cdot x^{-1} \longrightarrow e\}$
f. $\mathcal{R} = \{x \cdot x^{-1} \longrightarrow 1,\ (x \cdot y) \cdot z \longrightarrow x \cdot (y \cdot z)\}$
g. $\mathcal{R} = \{(x \cdot y) \cdot z \longrightarrow x \cdot (y \cdot z),\ x \cdot (y + z) \longrightarrow (x \cdot y) + (x \cdot z)\}$

3.13.9 Determine if the 10 rewrite rules for groups (Example 3.13.16, p. 212) are reduced.

3.13.10 Determine if the 12 rewrite rules for loops (Example 3.13.15, p. 212) are reduced.

Two normal form TRSs $\mathcal{R}_1$ and $\mathcal{R}_2$ are *equivalent* if terms have the same normal forms, i.e., $n_{\mathcal{R}_1} = n_{\mathcal{R}_2}$.

3.13.11 Determine if the following normal form TRSs are equivalent.

a. $\mathcal{R}_1 = \{fffx \longrightarrow fx\}$, and $\mathcal{R}_2 = \{ffx \longrightarrow x\}$
b. $\mathcal{R}_1 = \{fgx \longrightarrow gfx\}$, and $\mathcal{R}_2 = \{fgfx \longrightarrow gfgx\}$
c. $\mathcal{R}_1 = \{ffffx \longrightarrow fffx, ffx \longrightarrow fx\}$, and $\mathcal{R}_2 = \{ffx \longrightarrow fx\}$
d. $\mathcal{R}_1 = \{fx \longrightarrow ggx,\ gx \longrightarrow x\}$, and $\mathcal{R}_2 = \{fx \longrightarrow x,\ gx \longrightarrow x\}$

3.13.12 Show that every finite normal form TRS is equivalent to a finite reduced normal form TRS.

3.13.13 Show that two equivalent reduced normal form TRSs $\mathcal{R}_1$ and $\mathcal{R}_2$ are the same up to the naming of the variables in the rewrite rules, i.e., if $s(x_1, \ldots, x_n) \longrightarrow t(x_1, \ldots, x_n) \in \mathcal{R}_1$ then we have $s(y_1, \ldots, y_n) \longrightarrow t(y_1, \ldots, y_n) \in \mathcal{R}_2$ for some renaming y_i of the variables x_i, and conversely.

3.14 REDUCTION ORDERINGS

Suppose we are given a finite set $\mathcal{E}$ of equations $s \approx t$, and we want to start looking for a normal form TRS $\mathcal{R}$ for $\mathcal{E}$. Knuth and Bendix came up with the idea that one should *preselect* a rather large terminating (but not usually normal form) rewrite system $\mathcal{R}'$ that looks compatible with the equations and then try to find an $\mathcal{R}$ inside of

$\longrightarrow^+_{\mathcal{R}'}$. The relation $\longrightarrow^+_{\mathcal{R}'}$ is called a *reduction ordering* and is simply a terminating system in which we will search for $\mathcal{R}$.

Actually, to find the reduction orderings, we do not usually start with a terminating set of rules $\mathcal{R}'$ but rather we go directly to finding the relation $\longrightarrow^+_{\mathcal{R}'}$. Now let us call this relation $\succ$ and describe $\succ$ abstractly by looking at the properties it has:

a. $\succ$ is *irreflexive*, i.e., we always have $s \not\succ s$, since $\mathcal{R}'$ is terminating.

b. $\succ$ is *transitive*, since the $^+$ operation applied to $\longrightarrow_{\mathcal{R}'}$ gives a transitive relation.

c. $\succ$ is *compatible* with the fundamental operations, namely, $s_i \succ t$ implies $fs_1 \cdots s_i \cdots s_k \succ fs_1 \cdots t \cdots s_k$ for $f \in \mathcal{L}$, since we are allowed to apply rewrites to subterms.

d. $\succ$ is *fully invariant*, i.e., $s \succ t$ implies $\sigma s \succ \sigma t$ for σ a substitution, since elementary rewrites allow us to use substitutions.

e. $\succ$ is *well founded*, i.e., descending chains $t_1 \succ \cdots$ are always finite. This is because $\mathcal{R}'$ is terminating.

3.14.1 Definition of a Reduction Ordering

These properties motivate the following definitions.

Definition 3.14.1 A *strict partial order* $>$ is an irreflexive and transitive binary relation (on some set).

Definition 3.14.2 Given a language $\mathcal{L}$ of algebras and a set of variables X, a *reduction ordering* on the set of terms $T(X)$ is a strict partial ordering $\succ$ that is compatible, fully invariant, and well founded.

Our preceding observations can now be rephrased as the following proposition:

Proposition 3.14.3 If $\mathcal{R}'$ is a terminating TRS, then $\longrightarrow^+_{\mathcal{R}'}$ is a reduction ordering.

Reduction orderings provide a good context in which to search for a normal form TRS. It is easy to see that such orderings provide good places to look for terminating TRSs. However, it is more subtle to see that they can also lead to normal form TRSs.

Proposition 3.14.4 If $\succ$ is a reduction ordering, and $\mathcal{R}$ is a TRS that is a subset of $\succ$, then $\longrightarrow^+_{\mathcal{R}}$ will be well founded.

Proof. This follows from the fact that if $s \longrightarrow_{\mathcal{R}}^{+} t$ holds, then so does $s \succ t$. $\square$

The fact that we can, in a reasonably straightforward manner, find a large number of reduction orderings has made reduction orderings a standard tool in the study of rewrite systems. Now we look at the original examples of reduction orderings introduced by Knuth and Bendix. We can think of them as weighted lexicographic orderings.

3.14.2 The Knuth–Bendix Orderings

To construct a Knuth–Bendix ordering we first give an order to the function symbols and constant symbols and assign a weight to each of the symbols of the language and to the variables, and then we assign a weight $w(t)$ to each term t by adding together the weights of the symbols in t.

Having done this, under certain constraints, we proceed to inductively define the ordering $\succ$ on the terms. The first requirement for $s \succ t$ to hold is that the weight of s be at least as great as that of t. This is then divided into two cases.

In the first case the weight of s is greater than the weight of t, in which case we require only that each variable occur at least as many times in s as in t.

In the second case, when s and t have the same weight, we require that each variable have the *same* number of occurrences in s and t; then we proceed to compare their first symbols to see which is greater (in the original ordering on the symbols). If the first symbols are the same, then we proceed to compare the arguments of the first symbols. One exceptional case involves the possibility that the last symbol in the ordering of the constant symbols and function symbols is a unary function symbol.

Now we give the detailed description of the steps for obtaining a Knuth–Bendix ordering on $T(X)$. After stating the general version of the Knuth–Bendix orderings we will work out the details of an example, namely, for groups, where the language is $\cdot, ^{-1}, 1$.

Now, let $\mathcal{L}$ be a finite language of algebras, and let $\mathcal{C}$ be the set of constant symbols in $\mathcal{L}$.

- Choose a *linear ordering* $f_1 < \cdots < f_k$ of $\mathcal{L}$. (In this context the f_i can be constant symbols as well as function symbols.)

- Choose a *weight function* $w : \mathcal{L} \rightarrow N$, where $N = \{0, 1, 2, \cdots\}$, on the function and constant symbols satisfying the following constraints:

  $w(c) > 0$ if $c \in \mathcal{C}$, i.e., constant symbols must have positive weight.

  If f_i is *unary* and $w(f_i) = 0$, then $i = k$, i.e., the only time a unary function symbol can have 0 weight is when it is the last symbol f_k in our ordering of the language $\mathcal{L}$.
- Extend the weight function w to the set $T(X)$ of all the $\mathcal{L}$–terms by

  $w(x) = \min\{w(c) : c \in \mathcal{C}\}$ for x a variable; thus all variables have weight equal to the least of the weights of the constant symbols.

  $w(ft_1 \cdots t_n) = w(f) + w(t_1) + \cdots + w(t_n)$. This extends the weight function, using induction over terms, to all other terms.

Before continuing, we define $|t|_x$ to be the *number of occurrences* of the variable x in the term t.

Now we are ready to define the Knuth–Bendix order $\succ$. We inductively define $s \succ t$ to hold if either

Case I: $w(s) > w(t)$ and $|s|_x \geq |t|_x$ for all variables x that occur in s, t, or

Case II: $w(s) = w(t)$ and $|s|_x = |t|_x$ for all variables x, and either

1. f_k is a unary function symbol and $s = f_k \cdots f_k t$, or
2. $s = fs_1 \ldots s_m$ and $t = gt_1 \ldots t_n$, and either
   a. $f > g$ in the ordering of $\mathcal{L}$, or
   b. $f = g$ and $s_1 = t_1, \ldots ,s_{i-1} = t_{i-1}$, and $s_i \succ t_i$, for some i.

Example 3.14.5 Let us repeat the preceding steps for the case where the language is $\{\cdot, {}^{-1}, 1\}$, the language of groups.

First, we need to order $\mathcal{L}$ and choose a weight function.

- Let $1 < \cdot < {}^{-1}$.
- Let $w(\cdot) = w(1) = 1$, and $w({}^{-1}) = 0$. We can let the weight of the inverse function symbol be 0, since it is the last in our ordering of the function symbols.
- Define $w(x) = 1$ for all variables x.

  (Since we have only one constant symbol in our example of groups, namely, 1, whose weight is 1, it follows that all variables will have the weight 1.)

Extend the weight function to all $T(X)$ by the inductive definition:

$w(t^{-1}) = w(^{-1}) + w(t) = w(t)$
$w(t_1 \cdot t_2) = w(\cdot) + w(t_1) + w(t_2) = 1 + w(t_1) + w(t_2)$.

Now, if we total up the information on the weight function for the example of groups, we see that the weight of a term t is the number of constant symbols, i.e., the number of times 1 occurs, plus the number of occurrences of variable symbols in t, plus the number of occurrences of $\cdot$. Thus every symbol in t adds 1 to the weight except the inverse symbols, and they add nothing. So the weight of t is the length of t (i.e., the number of symbols, in prefix form) minus the number of occurrences of the inverse symbol.

In infix form, this says that we count up the number of occurrences of variable symbols, constant symbols and $\cdot$'s. We do not count parentheses or occurrences of the symbol for inverse. So for the term $((x \cdot y) \cdot y^{-1})^{-1}$ the weight is 5.

Now we are ready to define the Knuth–Bendix order $\succ$. We inductively define $s \succ t$ to hold if either

Case I: $w(s) > w(t)$ and $|s|_x \geq |t|_x$ for all variables x that occur in s, t, or

Case II: $w(s) = w(t)$ and $|s|_x = |t|_x$ for all variables x, and either

1. $s = I \cdots It$, where I is the symbol for inverse, or
2. $s = fs_1 \dots s_m$ and $t = gt_1 \dots t_n$, and either
   a. $f = {}^{-1}$ and $g = \cdot$, or
   b. $f = g$ and $s_1 = t_1, \dots, s_{i-1} = t_{i-1}$, and $s_i \succ t_i$, for some i.

Let us consider several terms, with their weights as noted:

Term	1	x	y	x^{-1}	$1 \cdot 1$
Weight	1	1	1	1	3
Term	$x \cdot y$	$(x \cdot y)^{-1}$	$x^{-1} \cdot y^{-1}$	$(x \cdot y) \cdot z$	$x \cdot (y \cdot z)$
Weight	3	3	3	5	5

We have $1 \cdot 1 \succ 1$ by case I. Likewise, $t \succ 1$ for any of these terms t of weight 3. However, $t \not\succ 1$ for any of our terms t of weight 1, since $1 \not\succ 1$, and since for $t \neq 1$ the number of occurrences of each variable is not the same in t and 1.

Now, $x \not\succ y$ as a comparison of x, y puts us in case II, but the number of occurrences of x is not the same in both terms. Likewise, we have $x \not\succ 1$. However, from case II(1) we do have $x^{-1} \succ x$ (because we have selected the inverse operation to be the last symbol in our ordering of the language $\mathcal{L}$).

From case I, $x \cdot y \succ x$. Of particular interest is the fact that $(x \cdot y)^{-1} \succ y^{-1} \cdot x^{-1}$. This comparison is in case II(2a), and ${}^{-1} > \cdot$ by our choice of ordering of the language. (Note that if we had chosen the weight of the function symbol for inverse to be positive, then case I would force $y^{-1} \cdot x^{-1} \succ (x \cdot y)^{-1}$.)

For the weight 5 terms note that $(x \cdot y) \cdot z \succ x \cdot (y \cdot z)$, since this comparison is in case II(2b). The weights are equal and the number of occurrences of each variable is the same for both terms, and the first subterm of the left side, $x \cdot y$, is greater than the first subterm, x, of the right side, i.e., $x \cdot y \succ x$, as we have already observed.

Theorem 3.14.6 Every Knuth–Bendix ordering is a reduction ordering.

Proof. Suppose $\succ$ is defined as already described. We will now verify the properties of a reduction ordering.

a. [$\succ$ *is irreflexive*]

Suppose $s \succ t$. If this falls under case I, then $w(s) > w(t)$, so $s \neq t$. If this falls under case II, then either

1. $s = f_k \cdots f_k t$, so $s \neq t$, or
2. $s = f s_1 \cdots s_m$ and $t = g t_1 \cdots t_n$ with either $f \neq g$ or, for some i, $s_i \neq t_i$. Again $s \neq t$.

b. [$\succ$ *is transitive*]

Suppose $r \succ s$ and $s \succ t$. If at least one of these falls under case I, then $r \succ t$ by case I.

If both of these fall under case II, then if either $r \succ s$ or $s \succ t$ falls under case II(2a), then so does $r \succ t$. Otherwise, we have the following possibilities, where, for example, the possibility [II(1), II(2b)] means $r \succ s$ falls under II(1), and $s \succ t$ falls under II(2b):

[II(1), II(1)] Then $r \succ t$ by II(1).
[II(1), II(2b)] Then $r \succ t$ by either II(2a) or II(2b).
[II(2b), II(1)]Then $r \succ t$ by either II(2a) or II(2b).
[II(2b), II(2b)] Then $r \succ t$ by II(2b).

c. [$\succ$ *is well founded*]

Suppose we have an infinite sequence of terms s_n with $s_1 \succ s_2 \succ \cdots \succ s_n \succ \cdots$.

As the weights are nonincreasing nonnegative integers, it follows that there is an M such that for all $n \geq M$ we have $w(s_n) = w(s_M)$. For the s_n with $n \geq M$ we must have the relation $\succ$ verified by case II. Then the number of occurrences of each variable is the same in each s_n for $n \geq M$. But there are only a finite number of terms we can make using a fixed number of occurrences of each variable. Thus for some $n > m \geq M$ we have $s_m = s_n$. This violates the irreflexivity of $\succ$. Thus every sequence $s_1 \succ \cdots$ is finite.

d. [$\succ$ *has the replacement property*]

If $s_i \succ t$, then by cases I and II(2b), we have the conclusion $f s_1 \cdots s_i \cdots s_n \succ f s_1 \cdots t \cdots s_n$.

e. [$\succ$ *is fully invariant*]

Suppose $s \succ t$ and σ is a substitution. Our first observation is that the number of occurrences of each variable is at least as great in σs as in σt, with equality for the number of occurrences of each variable iff there is equality in the case of s and t.

Then we can conclude that $w(\sigma s) \geq w(\sigma t)$, and $w(\sigma s) > w(\sigma t)$ iff $w(s) > w(t)$.

Now let us look at the various cases that could witness $s \succ t$:

[I] Then case I also applies to $\sigma s \succ \sigma t$.
[II(1)] Then case II(1) also applies to $\sigma s \succ \sigma t$.
[II(2a)] Then case II(2a) also applies to $\sigma s \succ \sigma t$.
[II(2b)] Then case II(2b) also applies to $\sigma s \succ \sigma t$.

This finishes the proof that all Knuth–Bendix orderings are reduction orderings. □

Corollary 3.14.7 The system $\mathcal{R}$ of 10 rewrite rules given in Example 3.13.16 (p. 212) is a normal form TRS for groups.

Proof. Let $\succ$ be the Knuth–Bendix ordering defined in Example 3.14.5 (p. 243). Then we can readily check that for each of the 10 rules $s \longrightarrow t$ we have $s \succ t$. Thus $s \longrightarrow^{+}_{\mathcal{R}} t$ implies $s \succ t$, so we see that $\mathcal{R}$ is terminating (as $\succ$ is well founded).

Then, by application of Theorem 3.13.31 (p. 223), the Critical Pairs Lemma, it follows that $\mathcal{R}$ is a normal form TRS. By Proposition 3.13.19 (p. 213) $\mathcal{R}$ is a normal form TRS for $\mathcal{E}(\mathcal{R})$.

Now we can derive the equations in $\mathcal{E}(\mathcal{R})$ from the axioms $\mathcal{E}$ for groups (Example 3.13.16, p. 212), and conversely. Thus $\mathcal{E}(\mathcal{R}) \vdash s \approx t$ iff $\mathcal{E} \vdash s \approx t$. Consequently, a normal form TRS for $\mathcal{E}(\mathcal{R})$ is also a normal form TRS for $\mathcal{E}$. Thus $\mathcal{R}$ is a normal form TRS for $\mathcal{E}$. □

Exercises

3.14.1 Given the Knuth–Bendix ordering $\succ$ from Example 3.14.5 (p. 243), show that the following are true. (These results are used in Sec. 3.15.)

- **a.** $x^{-1} \cdot (x \cdot y) \succ y$
- **b.** $1^{-1} \cdot y \succ y$
- **c.** $(x^{-1})^{-1} \cdot 1 \succ x$
- **d.** $(x^{-1})^{-1} \cdot y \succ x \cdot y$
- **e.** $x \cdot 1 \succ x$
- **f.** $1^{-1} \succ 1$
- **g.** $(x^{-1})^{-1} \succ x$
- **h.** $x \cdot x^{-1} \succ 1$
- **i.** $x \cdot (y \cdot (x \cdot y)^{-1}) \succ 1$
- **j.** $x \cdot (x^{-1} \cdot y) \succ y$
- **k.** $(x \cdot y)^{-1} \cdot (x \cdot (y \cdot z)) \succ z$
- **l.** $x \cdot (y \cdot ((x \cdot y)^{-1} \cdot z)) \succ z$
- **m.** $x \cdot (y \cdot (z \cdot (x \cdot (y \cdot z))^{-1})) \succ 1$
- **n.** $y \cdot (x \cdot y)^{-1} \succ x^{-1}$
- **o.** $y \cdot ((x \cdot y)^{-1} \cdot z) \succ x^{-1} \cdot z$
- **p.** $x \cdot (y \cdot (z \cdot (x \cdot y))^{-1}) \succ z^{-1}$
- **q.** $(x \cdot y)^{-1} \succ y^{-1} \cdot x^{-1}$

3.14.2 Prove that the Knuth–Bendix orderings $\succ$ that assign distinct weights to distinct constant symbols are *linear orderings* on *ground terms*, i.e., if two different terms s and t do not have any occurrences of variables, then prove $s \succ t$ or $t \succ s$.

What happens if two constant symbols are assigned the same weight?

3.14.3 Prove that all Knuth–Bendix orderings $\succ$ include the *proper superterm* relation, i.e., if t is a *proper* subterm of s, then $s \succ t$. (Reduction orderings that contain the proper superterm relation are called *simplification* orderings.)

3.14.3 Polynomial Orderings

After the introduction of the reduction orderings by Knuth and Bendix other natural reduction orderings were sought out. Among the simpler to explain are the *polynomial orderings*.

Let X be the infinite set of variables $\{x_n : n \geq 1\}$, and let $P[X]$ be the set of polynomials $p(x_1, \dots, x_n)$ with positive integers as coefficients. The relation $\gg$ on the set of polynomials $P[X]$ is given by $p \gg q$ iff

$$p(a_1, \ldots, a_n) > q(a_1, \ldots, a_n) \text{ for } \textit{all} \text{ positive integers } a_1, \ldots, a_n.$$

Example 3.14.8 We have $x + y \gg x$, $(x+y)^2 \gg x^2 + y^2$, etc., but $x^2 + y$ and $x + y^2$ are not comparable by this relation.

Given a language of algebras $\mathcal{L}$ let us associate with each constant symbol c a positive integer q_c, to be thought of as a constant polynomial, and with each n–ary function symbol f a polynomial $q_f(x_1, \ldots, x_n)$ from $P[X]$. Now we use these associations to associate a polynomial p_t to each term t in our language as follows (using definition by induction over terms):

a. $p_{x_i} = x_i$;
b. $p_{f(t_1,\ldots,t_n)} = q_f(p_{t_1}, \ldots, p_{t_n})$.

Then the *polynomial ordering* $\succ$ on $T(X)$ that is associated with our choices (for the constant symbols and function symbols) is defined by

$$s \succ t \qquad \text{iff} \qquad p_s \gg p_t.$$

Example 3.14.9 Consider the language $\mathcal{L} = \{\cdot, {}^{-1}, e\}$ of groups, and attach a polynomial to each symbol of the language as follows:

$$\begin{aligned} q_\cdot(x,y) &= 3xy + x \\ q_{^{-1}}(x) &= x^2 + 1 \\ q_1 &= 1. \end{aligned}$$

Then among the polynomials p_s we have

$$\begin{array}{llll} p_x = x & p_y = y & p_1 = 1 & p_{x\cdot y} = 3xy + x \\ p_{x\cdot 1} = 4x & p_{1\cdot x} = 3x + 1 & p_{x^{-1}} = x^2 + 1 & p_{1^{-1}} = 2. \end{array}$$

This gives a polynomial ordering $\succ$ on the terms $T(X)$. If we let $s(x,y) = (x \cdot y)^{-1}$ and $t(x,y) = y^{-1} \cdot x^{-1}$, then

$$\begin{aligned} p_s(x,y) &= (3xy + x)^2 + 1 &&= 9x^2y^2 + 6x^2y + x^2 + 1 \\ p_t(x,y) &= 3(y^2+1)(x^2+1) + (y^2+1) &&= 3x^2y^2 + 4y^2 + 3x^2 + 4, \end{aligned}$$

and it is clear that for any choice of positive integers a, b we have $p_s(a,b) > p_t(a,b)$, since

$$\begin{aligned} p_s(x,y) - p_t(x,y) &= 6x^2y^2 + 6x^2y - 2x^2 - 4y^2 - 3 \\ &= 2x^2y^2 + 4(x^2y^2 - y^2) + 2(x^2y - x^2) + (4x^2y - 3). \end{aligned}$$

Thus we have $(x \cdot y)^{-1} \succ y^{-1} \cdot x^{-1}$ in this ordering.

Theorem 3.14.10 Given a language $\mathcal{L}$ with its set of terms $T(X)$, every polynomial ordering on $T(X)$ is a reduction ordering.

Proof. Suppose $\succ$ is defined as described previously. We will now verify the properties of a reduction ordering.

a. [$\succ$ *is irreflexive*]

Suppose $s \succ t$. Then $p_s(a_1, \ldots, a_n) > p_t(a_1, \ldots, a_n)$ for every choice of positive integers $a_1, \ldots, a_n$. But then clearly we have $s \neq t$.

b. [$\succ$ *is transitive*]

Suppose $r \succ s \succ t$. From $p_r(a_1, \ldots, a_n) > p_s(a_1, \ldots, a_n)$ and $p_s(a_1, \ldots, a_n) > p_t(a_1, \ldots, a_n)$ follows $p_r(a_1, \ldots, a_n) > p_t(a_1, \ldots, a_n)$, for every sequence $a_1, \ldots, a_n$ of positive integers.

c. [$\succ$ *is well founded*]

Suppose we have an infinite sequence of terms s_n with

$$s_1 \succ s_2 \succ \cdots \succ s_n \succ \cdots,$$

i.e.,

$$p_{s_1} \gg p_{s_2} \gg \cdots \gg p_{s_n} \gg \cdots.$$

A simple observation regarding $\gg$ is that if $p \gg q$, then the variables that occur in q must all appear in p. Thus we have the variables in p_{s_n} all appearing in p_{s_1}, say $x_1, \ldots, x_n$. Now

$$p_{s_1}(1, \ldots, 1) > p_{s_2}(1, \ldots, 1) > \cdots > p_{s_n}(1, \ldots, 1) > \cdots.$$

But it is impossible to have an infinite decreasing sequence of positive integers.

d. [$\succ$ *has the replacement property*]

If $s_i \succ t$, then

$$q_f(p_{s_1}, \ldots, p_{s_i}, \ldots, p_{s_m}) \succ q_f(p_{s_1}, \ldots, p_t, \ldots, p_{s_m}),$$

as $s_i(a_1, \ldots, a_n) > t(a_1, \ldots, a_n)$ for any choice of positive integers $a_1, \ldots, a_n$, and the value of q_f can only increase if any argument is increased. But then, by the inductive definition of the polynomials p_s, we have

$$p_{f(s_1,\ldots,s_i,\ldots,s_m)} \succ p_{f(s_1,\ldots,t,\ldots,s_m)},$$

as desired.

e. [$\succ$ *is fully invariant*]

Suppose $s(x_1, \ldots, x_n) \succ t(x_1, \ldots, x_n)$, and σ is a substitution. As we have

$$\begin{array}{ll} s(x_1, \ldots, x_n) \succ t(x_1, \ldots, x_n) & \text{iff} \\ \qquad p_s(x_1, \ldots, x_n) \gg p_t(x_1, \ldots, x_n), \end{array}$$

we must have $p_s(x_1, \ldots, x_n) \gg p_t(x_1, \ldots, x_n)$. But then it is clear that $p_s(p_{\sigma x_1}, \ldots, p_{\sigma x_n}) \gg p_t(p_{\sigma x_1}, \ldots, p_{\sigma x_n})$, as the values taken on by $p_{\sigma x_1}, \ldots, p_{\sigma x_n}$ will be some sequences of positive integers, and $p_s(a_1, \ldots, a_n) > p_t(a_1, \ldots, a_n)$ for all sequences $a_1, \ldots, a_n$ of positive integers. Now,

$$\begin{array}{rcl} p_{\sigma s} & = & p_s(p_{\sigma x_1}, \ldots, p_{\sigma x_n}) \\ p_{\sigma t} & = & p_t(p_{\sigma x_1}, \ldots, p_{\sigma x_n}), \end{array}$$

and thus $p_{\sigma s} \gg p_{\sigma t}$, so $\sigma s \succ \sigma t$.

This finishes the proof that all polynomial orderings are reduction orderings. □

Corollary 3.14.11 The system $\mathcal{R}$ of 10 rewrite rules given in Example 3.13.16 (p. 212) is a normal form TRS for groups.

Proof. Let $\succ$ be the polynomial ordering defined in Example 3.14.9. Then we can readily check that for each of the 10 rules $s \longrightarrow t$ we have $s \succ t$. Thus $s \longrightarrow^{+}_{\mathcal{R}} t$ implies $s \succ t$, so we see that $\mathcal{R}$ is terminating (as $\succ$ is well founded).

The rest of the proof follows as in Corollary 3.14.7 (p. 246). □

Exercises

3.14.4 Given the polynomial ordering $\succ$ from Example 3.14.9, show that the following are true (corresponding to the rewrite rules for groups given in Example 3.13.16, p. 212).

a. $1^{-1} \succ 1$
b. $x \cdot 1 \succ x$
c. $1 \cdot x \succ x$
d. $(x^{-1})^{-1} \succ x$
e. $x \cdot x^{-1} \succ 1$
f. $x^{-1} \cdot (x \cdot y) \succ y$
g. $x \cdot (x^{-1} \cdot y) \succ y$
h. $(x \cdot y)^{-1} \succ y^{-1} \cdot x^{-1}$
i. $(x \cdot y) \cdot z \succ x \cdot (y \cdot z)$

3.14.5 Let $\succ$ be a polynomial ordering for which, for each function symbol f, $q_f(x_1, \ldots, x_n) \gg x_i$ for each x_i occurring in q_f. Show that $\succ$ includes the *proper superterm* relation, i.e., if t is a proper subterm of s, then $s \succ t$.

3.15 THE KNUTH–BENDIX PROCEDURE

Now we are ready to discuss the Knuth–Bendix procedure, a procedure that tries to convert a finite set of equations $\mathcal{E}$ into a normal form TRS for $\mathcal{E}$.

Given a set of equations $\mathcal{E}$, the first step is to select a reduction ordering $\succ$. To make a good choice seems to be a matter of experience. There are an infinite number of possibilities.

An equation $s \approx t$ is said to be *orientable* (with respect to $\succ$) if either $s \succ t$ or $t \succ s$ holds. To *orient* an equation is to convert it into a rewrite that is compatible with the ordering, i.e., $s \longrightarrow t$ if $s \succ t$, $t \longrightarrow s$ if $t \succ s$. Likewise, we speak of pairs s, t of terms that *can be oriented.*

Let the initial set of term rewrite rules be the result of orienting all the orientable equations in $\mathcal{E}$. (In the original version of Knuth and Bendix it was required that all equations in $\mathcal{E}$ be orientable.) We apply the normal form test of Theorem 3.13.31 (p. 223) to $\mathcal{R}$. If it fails, then there are critical pairs for which we cannot obtain confluence. We use the existing rules to rewrite each critical pair (s', s'') as far as possible, say

$$\begin{array}{lll} s' & \longrightarrow^*_{\mathcal{R}} & t' \\ s'' & \longrightarrow^*_{\mathcal{R}} & t''. \end{array}$$

If $t' \neq t''$, then since we want to have local confluence, it is convenient to add $t' \longrightarrow t''$ or $t'' \longrightarrow t'$ to $\mathcal{R}$. If the pair of terms t', t'' is orientable, then we orient it and add it to $\mathcal{R}$.

We repeatedly interreduce the existing set of rewrite rules, deleting unneeded rules, and repeatedly use the existing rewrite rules to reduce the two sides of the equations that have not been orientable—if some become orientable, then we orient them and add them to the rewrite rules. If some equations reduce to $s \approx s$, we delete them.

> **If at some point there are no equations left, and the existing set of rules $\mathcal{R}$ is a normal form TRS**

(as determined by Theorem 3.13.31, p. 223), then $\mathcal{R}$ is a normal form TRS for $\mathcal{E}$.

3.15.1 Finding a Normal Form TRS for Groups

We will illustrate this procedure with the original example of Knuth and Bendix for groups. We will use the KB reduction ordering $\succ$ for groups given in Example 3.14.5 (p. 243).

The ideas employed in the example should be clear. The use of the terminology *unifier** is to remind us that when forming critical pairs, it is important to first rename the variables so that the head terms of the two rules being considered have no variables in common.

We use the notation

$n. \qquad s \longrightarrow t \qquad\qquad \text{unify}^*\ r_1[k],\ r_2[l]$

in the following, which is read:

> We obtain the rule $s \longrightarrow t$ at the nth step by finding the most general unifier* of r_1 from the kth step and r_2 from the lth step, forming a critical pair (s', s''), reducing the critical pair to get (s, t), and orienting them according to $\succ$.

The following list summarizes the key steps performed by Knuth–Bendix in the group example—after the 32 steps, details of each of the unification* steps are given. The deletion steps throw out redundant rewrite rules. The boldface numbers in the column on the left are the rules that "survive"—there are 10 of them—indeed, the 10 in Example 3.13.16, p. 212.

1	$1 \cdot x \longrightarrow x$	orient equation in $\mathcal{E}$
2	$x^{-1} \cdot x \longrightarrow 1$	orient equation in $\mathcal{E}$
3	$(x \cdot y) \cdot z \longrightarrow x \cdot (y \cdot z)$	orient equation in $\mathcal{E}$
4	$x^{-1} \cdot (x \cdot y) \longrightarrow y$	unify* $x^{-1} \cdot x$ [2], $x \cdot y$ [3]
5	$1^{-1} \cdot y \longrightarrow y$	unify* $1 \cdot x$ [1], $x \cdot y$ [4]
6	$(x^{-1})^{-1} \cdot 1 \longrightarrow x$	unify* $x^{-1} \cdot x$ [2], $x \cdot y$ [4]
7	$(x^{-1})^{-1} \cdot y \longrightarrow x \cdot y$	unify* $x \cdot y$ [3], $(x^{-1})^{-1} \cdot 1$ [6]
8	$x \cdot 1 \longrightarrow x$	unify* $(x^{-1})^{-1} \cdot 1$ [6], $(x^{-1})^{-1} \cdot y$ [7]
9	DELETE 6	by 7,8
10	$1^{-1} \longrightarrow 1$	unify* $x^{-1} \cdot x$ [2], $x \cdot 1$ [8]
11	DELETE 5	by 10,1
12	$(x^{-1})^{-1} \longrightarrow x$	unify* $(x^{-1})^{-1} \cdot y$ [7], $x \cdot 1$ [8]
13	DELETE 7	by 12

14	$x \cdot x^{-1} \longrightarrow 1$	unify* x^{-1} [2], $(x^{-1})^{-1}$ [12]
15	$x \cdot (y \cdot (x \cdot y)^{-1}) \longrightarrow 1$	unify* $(x \cdot y) \cdot z$[3], $x \cdot x^{-1}$ [14]
16	$x \cdot (x^{-1} \cdot y) \longrightarrow y$	unify* $x \cdot y$ [3], $x \cdot x^{-1}$ [14]
17	$(x \cdot y)^{-1} \cdot (x \cdot (y \cdot z)) \longrightarrow z$	unify* $(x \cdot y) \cdot z$ [3], $x \cdot y$ [4]
18	$x \cdot (y \cdot ((x \cdot y)^{-1} \cdot z)) \longrightarrow z$	unify* $(x \cdot y) \cdot z$ [3], $x \cdot (x^{-1} \cdot y)$ [4]
19	$x \cdot (y \cdot (z \cdot (x \cdot (y \cdot z))^{-1})) \longrightarrow 1$	unify* $(x \cdot y) \cdot z$ [3], $y \cdot (x \cdot y)^{-1}$ [15]
20	$y \cdot (x \cdot y)^{-1} \longrightarrow x^{-1}$	unify* $x \cdot y$ [4], $x \cdot (y \cdot (x \cdot y)^{-1})$ [15]
21	$y \cdot ((x \cdot y)^{-1} \cdot z) \longrightarrow x^{-1} \cdot z$	unify* $x \cdot y$[3], $y \cdot (x \cdot y)^{-1}$ [20]
22	DELETE 18	by 21, 16
23	$x \cdot (y \cdot (z \cdot (x \cdot y))^{-1}) \longrightarrow z^{-1}$	unify* $(x \cdot y) \cdot z$ [3], $y \cdot (x \cdot y)^{-1}$ [20]
24	DELETE 19	by 23, 14
25	$(x \cdot y)^{-1} \longrightarrow y^{-1} \cdot x^{-1}$	unify* $x \cdot y$ [4], $y \cdot (x \cdot y)^{-1}$ [20]
26	DELETE 15	by 25, 16, 14
27	DELETE 17	by 25, 3, 4
28	DELETE 18	by 25, 16
29	DELETE 19	by 25, 16, 14
30	DELETE 20	by 25, 16
31	DELETE 21	by 25, 3, 16
32	DELETE 23	by 25, 3, 16

Now we give the details of the preceding unification* steps:

4. The most general unified* form of $x^{-1} \cdot x$ and $x \cdot y$ is $x^{-1} \cdot x$:
   $(x^{-1} \cdot x) \cdot y \longrightarrow 1 \cdot y \longrightarrow y$ by (2), (1)
   $(x^{-1} \cdot x) \cdot y \longrightarrow x^{-1} \cdot (x \cdot y)$ by (3).
5. The most general unified* form of $1 \cdot x$ and $x \cdot y$ is $1 \cdot y$:
   $1^{-1} \cdot (1 \cdot y) \longrightarrow 1^{-1} \cdot y$ by (1)
   $1^{-1} \cdot (1 \cdot y) \longrightarrow y$ by (4).
6. The most general unified* form of $x^{-1} \cdot x$ and $x \cdot y$ is $x^{-1} \cdot x$:
   $(x^{-1})^{-1} \cdot (x^{-1} \cdot x) \longrightarrow (x^{-1})^{-1} \cdot 1$ by (2)
   $(x^{-1})^{-1} \cdot (x^{-1} \cdot x) \longrightarrow x$ by (4).
7. The most general unified* form of $x \cdot y$ and $(x^{-1})^{-1} \cdot 1$ is $(x^{-1})^{-1} \cdot 1$:
   $((x^{-1})^{-1} \cdot 1) \cdot z \longrightarrow (x^{-1})^{-1} \cdot (1 \cdot z) \longrightarrow (x^{-1})^{-1} \cdot z$ by (3), (1)
   $((x^{-1})^{-1} \cdot 1) \cdot z \longrightarrow x \cdot z$ by (6).
8. The most general unified* form of $(x^{-1})^{-1} \cdot y$ and $(x^{-1})^{-1} \cdot 1$ is $(x^{-1})^{-1} \cdot 1$:
   $(x^{-1})^{-1} \cdot 1 \longrightarrow x$ by (6)
   $(x^{-1})^{-1} \cdot 1 \longrightarrow x \cdot 1$ by (7).
10. The most general unified* form of $x^{-1} \cdot x$ and $x \cdot 1$ is $1^{-1} \cdot 1$:
    $1^{-1} \cdot 1 \longrightarrow 1$ by (2)

$1^{-1} \cdot 1 \longrightarrow 1^{-1}$ by (8).

12. The most general unified* form of $(x^{-1})^{-1} \cdot y$ and $x \cdot 1$ is $(x^{-1})^{-1} \cdot 1$:
    $(x^{-1})^{-1} \cdot 1 \longrightarrow x \cdot 1 \longrightarrow x$ by (7), (8)
    $(x^{-1})^{-1} \cdot 1 \longrightarrow (x^{-1})^{-1}$ by (8).
14. The most general unified* form of x^{-1} and $(x^{-1})^{-1}$ is $(x^{-1})^{-1}$:
    $(x^{-1})^{-1} \cdot x^{-1} \longrightarrow 1$ by (2)
    $(x^{-1})^{-1} \cdot x^{-1} \longrightarrow x \cdot x^{-1}$ by (12).
15. The most general unified* form of $(x \cdot y) \cdot z$ and $x \cdot x^{-1}$ is $(x \cdot y) \cdot (x \cdot y)^{-1}$:
    $(x \cdot y) \cdot (x \cdot y)^{-1} \longrightarrow x \cdot (y \cdot (x \cdot y)^{-1})$ by (3)
    $(x \cdot y) \cdot (x \cdot y)^{-1} \longrightarrow 1$ by (14).
16. The most general unified* form of $x \cdot y$ and $x \cdot x^{-1}$ is $x \cdot x^{-1}$:
    $(x \cdot x^{-1}) \cdot z \longrightarrow x \cdot (x^{-1} \cdot z)$ by (3)
    $(x \cdot x^{-1}) \cdot z \longrightarrow 1 \cdot z \longrightarrow z$ by (14), (1).
17. The most general unified* form of $(x \cdot y) \cdot z$ and $x \cdot y$ is $(x \cdot y) \cdot z$:
    $(x \cdot y)^{-1} \cdot ((x \cdot y) \cdot z) \longrightarrow (x \cdot y)^{-1} \cdot (x \cdot (y \cdot z))$ by (3)
    $(x \cdot y)^{-1} \cdot ((x \cdot y) \cdot z) \longrightarrow z$ by (4).
18. The most general unified* form of $(x \cdot y) \cdot z$ and $x \cdot (x^{-1} \cdot y)$ is $(x \cdot y) \cdot ((x \cdot y)^{-1} \cdot z)$:
    $(x \cdot y) \cdot ((x \cdot y)^{-1} \cdot z) \longrightarrow x \cdot (y \cdot ((x \cdot y)^{-1} \cdot z))$ by (3)
    $(x \cdot y) \cdot ((x \cdot y)^{-1} \cdot z) \longrightarrow z$ by (4).
19. The most general unified* form of $(x \cdot y) \cdot z$ and $y \cdot (x \cdot y)^{-1}$ is $(y \cdot z) \cdot (x \cdot (y \cdot z))^{-1}$:
    $x \cdot ((y \cdot z) \cdot (x \cdot (y \cdot z)^{-1})) \longrightarrow x \cdot (y \cdot (z \cdot (x \cdot (y \cdot z)^{-1})))$ by (3)
    $x \cdot ((y \cdot z) \cdot (x \cdot (y \cdot z)^{-1})) \longrightarrow 1$ by (15).
20. The most general unified* form of $x \cdot y$ and $x \cdot (y \cdot (x \cdot y)^{-1})$ is $x \cdot (y \cdot (x \cdot y)^{-1})$:
    $x^{-1} \cdot (x \cdot (y \cdot (x \cdot y)^{-1})) \longrightarrow y \cdot (x \cdot y)^{-1}$ by (4)
    $x^{-1} \cdot (x \cdot (y \cdot (x \cdot y)^{-1})) \longrightarrow x^{-1} \cdot 1 \longrightarrow x^{-1}$ by (15), (6).
21. The most general unified* form of $x \cdot y$ and $y \cdot (x \cdot y)^{-1}$ is $y \cdot (x \cdot y)^{-1}$:
    $(y \cdot (x \cdot y)^{-1}) \cdot z \longrightarrow y \cdot ((x \cdot y)^{-1} \cdot z)$ by (3)
    $(y \cdot (x \cdot y)^{-1}) \cdot z \longrightarrow x^{-1} \cdot z$ by (20).
23. The most general unified* form of $(x \cdot y) \cdot z$ and $y \cdot (x \cdot y)^{-1}$ is $(x \cdot y) \cdot (z \cdot (x \cdot y))^{-1}$:
    $(x \cdot y) \cdot (z \cdot (x \cdot y))^{-1} \longrightarrow x \cdot (y \cdot (z \cdot (x \cdot y))^{-1})$ by (3)
    $(x \cdot y) \cdot (z \cdot (x \cdot y))^{-1} \longrightarrow z^{-1}$ by (20).

25. The most general unified* form of $x \cdot y$ and $y \cdot (x \cdot y)^{-1}$ is $y \cdot (x \cdot y)^{-1}$:
$y^{-1} \cdot (y \cdot (x \cdot y)^{-1}) \longrightarrow (x \cdot y)^{-1}$ by (4)
$y^{-1} \cdot (y \cdot (x \cdot y)^{-1}) \longrightarrow y^{-1} \cdot x^{-1}$ by (20).

3.15.2 A Formalization of the Knuth–Bendix Procedure

Now that we have worked through the Knuth and Bendix example with three group axioms, we want to consider how to proceed in general to convert a system $\mathcal{E}$ of equations into a normal form TRS $\mathcal{R}$.

Knuth and Bendix did not give such a precise formulation of the steps to be followed. We will give a fairly simple way to implement a variation on their procedure.[4]

Fundamental to implementations of the Knuth–Bendix procedure is the necessity of first *making a choice of a reduction ordering* $\succ$ *for* $\mathcal{E}$. We simply assume a choice of $\succ$ has been made for our given finite $\mathcal{E}$.

Let $\mathcal{E}_0 = \mathcal{E}$ and $\mathcal{R}_0 = \emptyset$, and loop as follows to generate a sequence of pairs $\mathcal{E}_i$, $\mathcal{R}_i$. A detailed explanation follows the listing of the eight steps.

Step 1: Input $\mathcal{E}_i$, $\mathcal{R}_i$.
Step 2: Let $\mathcal{E}_{i+1} = \mathcal{E}_i$, $\mathcal{R}_{i+1} = \mathcal{R}_i$.
Step 3: Convert all orientable equations in $\mathcal{E}_i$ to rewrite rules and add them to $\mathcal{R}_{i+1}$; delete the orientable equations from $\mathcal{E}_{i+1}$.
Step 4: Add rules obtained from orientable critical pairs of $\mathcal{R}_i$ to $\mathcal{R}_{i+1}$.
Step 5: Reduce the right–hand sides of rules in $\mathcal{R}_{i+1}$, using the rules in $\mathcal{R}_{i+1}$.
Step 6: Delete the redundant rules in $\mathcal{R}_{i+1}$.
Step 7: Reduce the sides of the equations in $\mathcal{E}_{i+1}$ using $\mathcal{R}_{i+1}$.
Step 8: If $\mathcal{E}_{i+1} = \emptyset$, $\mathcal{R}_{i+1} = \mathcal{R}_i$ and no unorientable critical pair occurs in step 4, then output $\mathcal{R}_i$ and halt; else if $\mathcal{E}_{i+1} \neq \emptyset$ and $\mathcal{R}_{i+1} = \mathcal{R}_i$, output Failure and halt; else output $\mathcal{E}_{i+1}$, $\mathcal{R}_{i+1}$.

Now we give a few comments on the preceding steps.

---

[4]Perhaps the best known precise formulation was given by Huet [**27**] in 1981. A more flexible axiomatic approach was developed by Bachmair and Dershowitz [**2**] in 1994.

Step 3: If $s \approx t$ is an orientable equation in $\mathcal{E}_i$, then it is to be deleted from $\mathcal{E}_{i+1}$, and $t \longrightarrow s$ or $s \longrightarrow t$ is to be added to $\mathcal{R}_{i+1}$.

Step 4: This is where one looks for the failure of $\mathcal{R}_i$ to provide normal forms, that is, one looks for critical pairs (s_1', s_1'') for which there are terminating sequences

$$s_1' \longrightarrow_{\mathcal{R}_i} \cdots \longrightarrow_{\mathcal{R}_i} s'$$

$$s_1'' \longrightarrow_{\mathcal{R}_i} \cdots \longrightarrow_{\mathcal{R}_i} s'',$$

where $s' \neq s''$. Given such a pair, one can remedy the lack of confluence by adding $s' \longrightarrow s''$ to $\mathcal{R}_{i+1}$, provided $s' \succ s''$, or by adding $s'' \longrightarrow s'$ to $\mathcal{R}_{i+1}$, provided $s'' \succ s'$.

Step 5: Now, apply the rules of $\mathcal{R}_{i+1}$ to the right–hand sides of the members of $\mathcal{R}_{i+1}$, thus replacing each member $s \longrightarrow t$ by some $s \longrightarrow t'$ such that t' cannot be further reduced by $\mathcal{R}_{i+1}$.

Step 6: Delete from $\mathcal{R}_{i+1}$ any rule that can be obtained as a sequence of the remaining rules. (The resulting set of independent rules does not depend on the order in which the rules are examined for dependence.)

Step 7: Apply the rules of $\mathcal{R}_{i+1}$ to reduce both sides of each equation $s \approx t$ in $\mathcal{E}_{i+1}$ to obtain equations that can no longer be reduced by $\mathcal{R}_{i+1}$.

Step 8: Two facts are needed to understand why the conditions $\mathcal{E}_{i+1} = \emptyset$, $\mathcal{R}_{i+1} = \mathcal{R}_i$ and no unorientable critical pairs guarantee that $\mathcal{R}_i$ is a complete TRS for $\mathcal{E}$. First, $\mathcal{E}_i \cup \{s \approx t : s \longrightarrow t \in \mathcal{R}_i\}$ has the same equational consequences as $\mathcal{E}$. So if $\mathcal{E}_{i+1} = \emptyset$ then $\{s \approx t : s \longrightarrow t \in \mathcal{R}_{i+1}\}$ axiomatizes the equational consequences of $\mathcal{E}$. Second, as $\mathcal{R}_i$ is confluent on critical pairs, it follows from Theorem 3.13.31 (p. 223) that $\mathcal{R}_i$ is a *normal form* TRS for $\mathcal{E}$.

3.16 HISTORICAL REMARKS

Equations have always been with us, or so it seems. We find them in all areas of mathematics, e.g., number theory (diophantine equations) and geometry (analytic geometry, algebraic geometry). Equations have the power to succinctly and clearly express assertions that require incredible resources to analyze. We have to think only of Fermat's Last Theorem or Burnside's conjecture to gauge the extent to which problems based on simple equations fascinate mathematicians.

Equations have played an influential role in creating new content for mathematics. Solving equations in the real numbers led to the introduction of *complex* numbers. Galois's 1830 work on finding roots of equations in the complex numbers led to the introduction of *fields* and *groups* into mathematics. Boole's application of equations about classes to logic helped establish the study of *classes* as a mathematical subject. By the early part of the twentieth century it had become legitimate mathematics to write down an arbitrary set of equations and investigate their consequences and models.

The logic of equations actually started out as a part of first–order logic, when Gödel proved his completeness theorem in 1930. In particular, this theorem guarantees that one can obtain the equational consequences of a set of equations from the axioms and rules of first–order logic. However, this is overkill. In 1935 Birkhoff proved his completeness theorem for equational logic, a logic just for equations, which consolidated what was essentially known from practice, that one needs only simple rules concerning equations for equational arguments.

Equations were known to be incredibly expressive. Tarski showed that one could use equations from his work on relation algebras[5] to capture any argument in mathematics. By 1953 he had proved that there was no algorithm to determine the equational consequences of his axioms for relation algebras. In 1954 Mal'cev showed that one could find a finite set of equations involving only unary function symbols with the property that there is no algorithm to determine the equational consequences.

A favorite equational problem comes from the study of rings, namely, when is it the case that a ring equation $s \approx t$, added to the ring axioms $\mathcal{R}$, is strong enough to guarantee that the multiplication is commutative, i.e., when does $\mathcal{R} \cup \{s \approx t\} \models x \cdot y \approx y \cdot x$? A beautiful, and nontrivial, result of Jacobson's from 1945 shows that for any integer $n \geq 2$ one has $\mathcal{R} \cup \{x^n \approx x\} \models x \cdot y \approx y \cdot x$. The interesting thing about Jacobson's proof is that he does not find equational derivations but uses a combination of naive set theory and number theory to analyze the rings that satisfy $x^n \approx x$. By

[5]Relation algebras were introduced by Tarski as a continuation of the work on the algebra of relations in Vol. III of Schröder's *Algebra of Logic*.

Birkhoff's completeness theorem there must, for any particular n, be a derivation of the commutative law. This has become a favorite exercise, at least for small values of n, in abstract algebra books, and a favorite test problem for automated equational theorem provers. The algebraist Herstein remarked in the early 1970s that he was confident, given any particular value of n, that he could come up with an equational derivation. But, rather paradoxically, he said that he did not have an algorithm for doing this. John Lawrence (of the University of Waterloo) was the first to put together (in 1990) a simple algorithm to find, given any $n \geq 2$, an equational derivation of $x \cdot y \approx y \cdot x$ from $\mathcal{R} \cup \{x^n \approx x\}$.

It was the realization that equational proofs, including familiar ones, could be highly nontrivial, that seems to have led Knuth to consider using term rewrite systems.[6] Since his pioneering work with Bendix there have been numerous investigations into the limitations of, and extensions of, term rewrite systems. We have mentioned the work on finding reduction orderings. Dershowitz[7] introduced the important and widely used class of *recursive path orderings*.

Considerable work has gone into improving the performance of the unification algorithm. Unfortunately, it can produce most general unifiers that are exponentially larger than the pair of terms to be unified. One example (from [**1**]) is to consider the two terms

$$x_1 + (x_2 + (x_3 + \cdots (x_{n-1} + x_n) \cdots)),$$

$$\begin{aligned}(x_0 + x_0) + ((x_1 + x_1) + ((x_2 + x_2) + \cdots + \\ ((x_{n-2} + x_{n-2}) + (x_{n-1} + x_{n-1})) \cdots)).\end{aligned}$$

Then it is easy to see that the most general unifier μ is given by

$$\begin{array}{lcl} x_1 & \leftarrow & x_0 + x_0 \\ x_2 & \leftarrow & (x_0 + x_0) + (x_0 + x_0) \\ x_3 & \leftarrow & ((x_0 + x_0) + (x_0 + x_0)) + ((x_0 + x_0) + (x_0 + x_0)), \\ & \textit{etc.}, & \end{array}$$

---

[6] Term rewrites are also called *demodulators*, and finding critical pairs can be thought of as *paramodulation* of equations. The general notion of paramodulation (for first–order clauses) is described at the end of the next chapter, on p. 314.

[7] For a comprehensive survey of term rewrite systems consult Dershowitz and Jouannaud's article "Rewrite Systems" in the *Handbook of Theoretical Computer Science*.

so the length of the term $\mu(x_i)$ grows exponentially. This complexity was noticed early on,[8] and soon thereafter compact representations were found for terms, such as by directed acyclic graphs, to obtain fast (i.e., polynomial time) algorithms.

One of the striking accomplishments of automated theorem proving is the settling of Herbert Robbins's conjecture that the following three equations axiomatize Boolean algebras:

$$
\begin{aligned}
&x \vee y \approx y \vee x \\
&(x \vee y) \vee z \approx x \vee (y \vee z) \\
&((x \vee y)' \vee (x \vee y')')' \approx x.
\end{aligned}
$$

(The remaining traditional operations are defined by $x \wedge y = (x' \vee y')'$, $0 = x \wedge x'$, and $1 = x \vee x'$.) These equations are of course true in Boolean algebras, and in 1933 Robbins, a student of E. V. Huntington's, conjectured that they axiomatized Boolean algebras.

In the fall of 1996 EQP, an automated equational theorem prover written by William McCune[9] of Argonne National Laboratory, found a proof for this conjecture. With this discovery an important psychological threshhold had been crossed, namely, the realization that a fully automated theorem prover could find a proof of a mathematical theorem that had eluded many an outstanding mathematician for decades. Gina Kolata, the science reporter for the *New York Times*, wrote a fascinating feature article on this breakthrough.[10] The possibility that machines could *reason* and be *creative* was now a matter of immediate import.

---

[8] For a thorough survey of work on unification see Baader and Siekmann's article "Unification" in the *Handbook of Logic in Artificial Intelligence and Logic Programming*.

[9] His program OTTER, for automated theorem proving, and its companion MACE, for finding small counterexamples, are available from http://www.mcs.anl.gov/home/mccune. For an excellent brief introduction and overview see the first 31 pages of *Automated Deduction in Equational Logic and Cubic Curves* by McCune and Padmanabhan, Lecture Notes in Artificial Intelligence, No. **1095**, Springer–Verlag, 1996. A comprehensive text is *Automated Reasoning: Introduction and Applications*, 2d ed., by Wos, Overbeek, Lusk, and Boyle, McGraw–Hill, 1992.

[10] See the Science Times section of the *New York Times*, Tuesday, December 10, 1996.

The negative results concerning algorithms for equational logic persist in the study of term rewrite systems. There are no algorithms for many of the basic questions about TRSs, e.g., there is no algorithm to determine which TRSs are terminating (Huet and Lankford, 1978), or which are confluent (Bauer and Otto, 1984), or which are normal form TRSs (Huet and Lankford, 1978).

Chapter 4

Predicate Clause Logic

SYMBOLS	$\vee, \neg$
	relation (i.e., predicate) symbols
	function symbols
	constant symbols
	variables

In mathematical practice we make statements, some of which can be analyzed by propositional logic, or equational logic. However, it is often the case that we need to break the statements down further, for example, when we see that they concern *properties*, e.g., the property *of being a perfect square*, *of being a prime number*, *of one number being less than another*, or *of being a continuous function.* It is the necessity to work with such properties (or predicates, or relations) directly that leads us to extend logic languages to include relation symbols.

4.1 FIRST–ORDER LANGUAGES WITHOUT EQUALITY

Equality adds certain complications to working with clauses, so it will not be discussed until Sec. 4.9, starting on p. 307.

Definition 4.1.1 A *first–order language without equality* $\mathcal{L}$ will consist of

- a set $\mathcal{F}$ of *function* symbols $f, g, h, \cdots$ with associated arities;
- a set $\mathcal{R}$ of *relation* symbols $r, r_1, r_2, \cdots$ with associated arities;
- a set $\mathcal{C}$ of *constant* symbols $c, d, e \cdots$;
- a set X of *variables* $x, y, z, \cdots$.

(Any of the sets $\mathcal{F}, \mathcal{R}, \mathcal{C}$ can be empty.)

We have already discussed function symbols and functions in Chapter 3. We want each relation symbol r to have a positive integer, called its *arity*, assigned to it. If the number is n, we say r is n*–ary*. For small n we use the same special names that we use for function symbols, i.e., *unary* ($n = 1$), *binary* ($n = 2$), and *ternary* ($n = 3$). The set $\mathcal{L} = \mathcal{R} \cup \mathcal{F} \cup \mathcal{C}$ is called a *first–order language*.

With our favorite binary relation symbols, like $<$ and $|$, we usually prefer infix notation, e.g., we write $x < y$ instead of $< xy$, and $x|y$ instead of $|xy$.

In mathematical practice we rarely see ternary or higher arity relation symbols—but they are no more difficult for us to analyze than the unary and binary cases.

4.2 INTERPRETATIONS AND STRUCTURES

The obvious interpretation of a relation symbol is as a *relation* on a set.

Definition 4.2.1 If A is a set and n is a positive integer, then an n*–ary relation* r on A is a subset of A^n, i.e., r consists of a collection of n*–tuples* $(a_1, \ldots, a_n)$ of elements of A. Special cases are *unary* ($n = 1$), *binary* ($n = 2$), and *ternary* ($n = 3$) relations.

Definition 4.2.2 An *interpretation* I of the first–order language $\mathcal{L}$ on a set S is a mapping with domain $\mathcal{L}$ such that

- $I(c)$ is an element of S for each constant symbol c in $\mathcal{C}$;
- $I(f)$ is an n–ary function on S for each n–ary function symbol f in $\mathcal{F}$;
- $I(r)$ is an n–ary relation on S for each n–ary relation symbol r in $\mathcal{R}$.

An $\mathcal{L}$*-structure* $\mathbf{S}$ is a pair (S, I), where I is an interpretation of $\mathcal{L}$ on S.

As we see, this is just like our definition of an $\mathcal{L}$–algebra, with the addition of relations. As before, given an $\mathcal{L}$–structure $\mathbf{S} = (S, I)$, we will prefer to write $c^{\mathbf{S}}$ for $I(c)$, and $f^{\mathbf{S}}$ for $I(f)$, and now we want to use $r^{\mathbf{S}}$ for $I(r)$. When we feel confident that the context makes the meaning clear, we will simply write c for $c^{\mathbf{S}}$, f for $f^{\mathbf{S}}$, and r for $r^{\mathbf{S}}$. Also, rather than writing (S, I), we often prefer to write $(S, \mathcal{F}, \mathcal{R}, \mathcal{C})$ or simply to list the symbols in $\mathcal{F}$, $\mathcal{R}$, and $\mathcal{C}$, e.g., the structure $(R, +, \cdot, <, 0, 1)$, the reals with addition, multiplication, less than, and two specified constants. Working backward in this example we find $\mathcal{F} = \{+, \cdot\}$, $\mathcal{R} = \{<\}$, and $\mathcal{C} = \{0, 1\}$.

Example 4.2.3 If $r \in \mathcal{R}$ is a *unary* predicate symbol, then in any $\mathcal{L}$–structure $\mathbf{S}$, the relation $r^{\mathbf{S}}$ is a subset of S. We can picture this as:

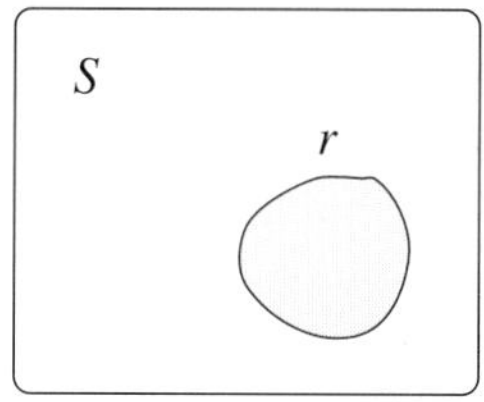

Figure 4.2.4 A unary predicate

Example 4.2.5 If $\mathcal{L}$ consists of a single *binary* relation symbol r, then we call an $\mathcal{L}$–structure a *directed graph*. A small finite directed graph can be conveniently described in three different ways:

- List the ordered pairs in the relation r. A simple example with $S = \{a, b, c, d\}$ is

$$r = \{(a, a), (a, d), (b, b), (b, c), (c, b), (d, b), (d, c)\}.$$

- Use a table. For the same example we have

r	a	b	c	d
a	1	0	0	1
b	0	1	1	0
c	0	1	0	0
d	0	1	1	0

An entry of 1 in the table indicates a pair is in the relation.
- Draw a picture. Again, using the same example:

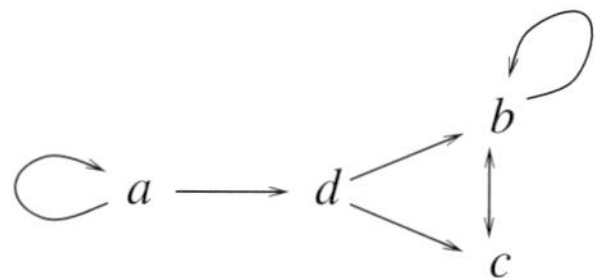

Figure 4.2.6 A directed graph

An interpretation of a language on a small set can be conveniently given by tables.

Example 4.2.7 Let $\mathcal{L} = \{+, -, U, <, k\}$, where $+$ and $<$ are binary, $-$ and U are unary, and k is a constant. Then the following tables give an interpretation of $\mathcal{L}$ on the two–element set $S = \{a, b\}$:

$+$	a	b
a	a	b
b	b	a

$<$	a	b
a	0	1
b	0	0

	$-$
a	b
b	a

	U
a	0
b	1

k is a.

Exercises

4.2.1 Express the following (directed) graphs as (i) sets of ordered pairs, and (ii) by tables.

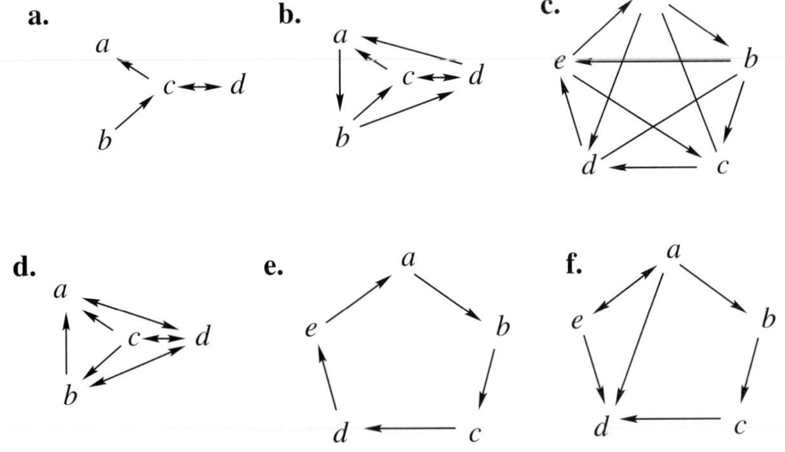

4.2.2 Consider the directed graph **G** given by the binary relation "is less than" on the set of numbers $G = \{0, \ldots, 4\}$.

a. Express **G** as a set of ordered pairs.

b. Give a table for **G**.

c. Draw **G**.

4.2.3 Consider the directed graph **G** given by the binary relation "is greater than or equal to" on the set of numbers $G = \{0, \ldots, 4\}$.

a. Express **G** as a set of ordered pairs.

b. Give a table for **G**.

c. Draw **G**.

4.3 CLAUSES

The definition of a clause in the predicate logic is based on that in the propositional logic, except that we use *atomic formulas* instead of propositional variables. We assume we have a given first–order language $\mathcal{L}$. The notion of a *term* t will be used as defined at the beginning of Chapter 3.

Definition 4.3.1

a. An *atomic formula* A is an expression (i.e., string of symbols) of the form $rt_1 \cdots t_n$, where the t_i are terms, and r is an n–ary relation symbol in the language.

b. A *literal* (or *basic formula*) is either an atomic formula A (a *positive* literal) or a negated atomic formula $\neg \mathsf{A}$ (a *negative* literal).

c. A *clause* C is a finite set of literals

$$\{\mathsf{L}_1, \ldots, \mathsf{L}_n\}.$$

We also use the notation

$$\mathsf{L}_1 \vee \cdots \vee \mathsf{L}_n.$$

Example 4.3.2 If r is a binary relation, f is a binary function symbol, and $0, 1$ are constant symbols in our language, and if x, y, z are variables from X, then

$$rxx,\ ry0,\ rxfxy,\ rffz1yx,\ rfxyfyz,\ \text{etc.}$$

are atomic formulas;

$$rxx,\ \neg rxx,\ rxfxy,\ \neg rxf0y,\ \neg rfxyfyz,\ \text{etc.}$$

are literals; and

$$\{rxx,\ rxf1y,\ \neg rfxyfyz\}$$

is an example of a clause.

Example 4.3.3 The parsing algorithm that we used for terms in Example 3.2.3 (p. 141) can be applied to positive literals (and hence to any literal). Suppose we are given a language $\mathcal{L}$ of structures, and $rs_1 \cdots s_n$ is a string with r a relation symbol, and each s_i is either a function symbol or a constant symbol belonging to this language, or a variable. We define the integer–valued function γ_s on the numbers 0 to n inductively by

$$\gamma_s(0) = 0$$

$$\gamma_s(i+1) = \begin{cases} \gamma_s(i)+1 & \text{if } s_{i+1} \text{ is a variable} \\ \gamma_s(i)+1 & \text{if } s_{i+1} \text{ is a constant symbol} \\ \gamma_s(i)+1-\text{arity}(s_{i+1}) & \text{if } s_{i+1} \text{ is a function symbol.} \end{cases}$$

For $\mathsf{L} = rs_1 \cdots s_n$ *and* γ_s *as just described,* L *is a literal iff*

a. $\gamma_s(i) < \text{arity}(r)$ *for* $i < n$, *and*

b. $\gamma_s(n) = \text{arity}(r)$.

If L *is a literal, then* L *must be of the form* $rt_1 \cdots t_k$, *where* $k =$ arity(r), *for some choice of terms* $t_1, \ldots, t_k$. *We can easily read off the* t_j, *since the first* i *such that* $\gamma_s(i) = j$ *gives the end of the term* t_j.

Let us apply this algorithm to the following situation. Let $\mathcal{L} = \{r, f, g, c\}$ be a language with r a ternary relation symbol, f unary, g binary, and c a constant symbol. We want to determine if $\mathsf{L} = rgxxgcyggxfcx$ is a literal and if so, to find the three subterms t_1, t_2, t_3 such that $rt_1t_2t_3 = rgxxgcyggxfcx$. The following table shows the calculations involved:

i	0	1	2	3	4	5	6	7	8	9	10	11	12
s_i	r	g	x	x	g	c	y	g	g	x	f	c	x
$\gamma(i)$	0	-1	0	1	0	1	2	1	0	1	1	2	3

Thus conditions (a) and (b) of the theorem are fulfilled, as the arity of r is 3, and the value of $\gamma(i)$ is below 3 until the last step, where it is 3. Then t_1 ends at the fourth symbol of L, as that gives the first place where $\gamma(i)$ is 1, t_2 ends at the seventh symbol of L, as that gives the first place where $\gamma(i)$ is 2, and t_3 ends at the last symbol of L. Thus $t_1 = gxx$, $t_2 = gcy$, $t_3 = ggxfcx$.

Predicate clause logic is, like the propositional clause logic, concerned with showing that sets of clauses are *not* satisfiable. Again we want to study this logic from two vantage points:

- (*semantic*) a notion of when a clause is true in a structure;
- (*syntactic*) axioms and rules of inference for deriving clause consequences from clauses.

This is parallel to the study of clauses in the propositional logic. Once we develop these components for clause logic we will show that they are equivalent in the sense that a set of clauses is not satisfiable iff we can derive the empty clause.

Exercises

[f_k is a k–ary function symbol, and r_k is a k–ary relation symbol.]

4.3.1 For each of the following expressions determine if it is (i) a term, (ii) an atomic formula, (iii) a literal, (iv) a clause, or (v) none of the above.

a. $r_1 f_1 x$	**b.** $f_1 r_1 x$	**c.** $\{f_1 x, f_1 y\}$
d. $r_1 r_1 x$	**e.** $\neg r_1 r_1 x$	**f.** $\{r_1 x, \neg f_1 y\}$
g. $r_1 \neg r_1 x$	**h.** $r_2 f_1 x f_1 y$	**i.** $\neg r_2 f_1 x f_1 y$
j. $f_2 r_1 x r_1 y$	**k.** $r_2 f_1 x r_1 y$	**l.** $\{r_2 f_2 xy, \neg r_2 f_2 xy\}$
m. $f_2 f_1 x f_1 y$	**n.** $\neg f_2 f_1 x f_1 y$	**o.** $\{r_2 f_1 x f_2 xy, \neg r_1 f_2 xy, r_3 xyz\}$
p. $\neg r_2 x f_2 yx$	**q.** $\{x, \neg y\}$	**r.** $f_2 \neg r_1 x f_1 y$

4.3.2 In each of the following problems you are given a literal of the form $r_k t_1 \cdots t_k$ or of the form $\neg r_k t_1 \cdots t_k$. Find the terms $t_1, \ldots, t_k$. (You may want to use the parsing algorithm.)

a. $r_2 f_1 f_2 f_1 xv f_2 f_1 f_1 zx$
b. $\neg r_2 f_1 f_2 v f_1 v f_1 f_2 v f_1 w$
c. $\neg r_2 f_1 f_1 f_1 f_3 wzz f_3 v f_3 f_2 uvyzz$
d. $\neg r_3 f_1 f_2 z f_1 y f_1 f_3 yx f_1 u f_2 v f_2 y f_1 y$
e. $\neg r_3 f_3 u f_2 f_3 yw f_2 uwyw f_2 z f_2 f_1 yu f_3 xx f_1 f_1 z$

4.4 SEMANTICS

Now we turn to the "meaning" of terms, literals, and clauses in a structure $\mathbf{S}$, namely,

terms	$t(x_1, \ldots, x_n)$	define	*functions*	$t^{\mathbf{S}} : S^n \to S$
literals	$\mathsf{L}(x_1, \ldots, x_n)$	define	*relations*	$\mathsf{L}^{\mathbf{S}} \subseteq S^n$
clauses	$\mathsf{C}(x_1, \ldots, x_n)$	define	*relations*	$\mathsf{C}^{\mathbf{S}} \subseteq S^n$

We have already explored the notion of *term functions* in Sec. 3.3, and that is exactly how we will give meanings to terms in this chapter, except we now use the notation $t^{\mathbf{S}}$ for $\mathbf{S}$ a first–order structure rather than $t^{\mathbf{A}}$, where $\mathbf{A}$ is an algebra. So it remains only to assign suitable relations to the literals and clauses. We will make extensive use of *vector notation* such as $\vec{x}$ for $(x_1, \ldots, x_n)$, and $t(\vec{x}\,)$ for $t(x_1, \ldots, x_n)$.

Definition 4.4.1 Let $\mathbf{S}$ be a structure.

Atomic formulas:

For $\mathsf{L}(x_1, \ldots, x_n) = rt_1 \cdots t_k$, where $t_i = t_i(x_1, \ldots, x_n)$, $\mathsf{L}^{\mathbf{S}}$ denotes the n–ary relation defined by

$$\mathsf{L}^{\mathbf{S}} = \{(a_1, \ldots, a_n) : (t_1^{\mathbf{S}}(\vec{a}), \ldots, t_k^{\mathbf{S}}(\vec{a})) \in r^{\mathbf{S}}\}.$$

Negated atomic formulas:

For $\mathsf{L} = \neg r t_1 \cdots t_k$,

$$\mathsf{L}^{\mathbf{S}} = \{(a_1, \dots, a_n) : (t_1^{\mathbf{S}}(\vec{a}), \dots, t_k^{\mathbf{S}}(\vec{a})) \notin r^{\mathbf{S}}\}.$$

If a literal $\mathsf{L}^{\mathbf{S}}$ has the tuple $(a_1, \dots, a_n)$ in it, we say that $\mathsf{L}(a_1, \dots, a_n)$ *holds* in $\mathbf{S}$, or L *holds at* $a_1, \dots, a_n$ in $\mathbf{S}$. We also write $\mathbf{S} \models \mathsf{L}(\vec{a})$ [read: $\mathbf{S}$ *satisfies* $\mathsf{L}(\vec{a})$, or $\mathbf{S}$ *models* $\mathsf{L}(\vec{a})$].

Clauses:

For a clause $\mathsf{C}(x_1, \dots, x_n) = \{\mathsf{L}_1(\vec{x}), \dots, \mathsf{L}_k(\vec{x})\}$ the n–ary relation $\mathsf{C}^{\mathbf{S}}$ is defined by

$$\mathsf{C}^{\mathbf{S}} = \{(a_1, \dots, a_n) : \mathsf{L}_i(a_1, \dots, a_n) \text{ holds for some } i,\ 1 \le i \le k\}.$$

If $(a_1, \dots, a_n) \in \mathsf{C}^{\mathbf{S}}$, we say $\mathsf{C}(a_1, \dots, a_n)$ *holds* in $\mathbf{S}$. We also write $\mathbf{S} \models \mathsf{C}(\vec{a})$ [read: $\mathbf{S}$ *satisfies* $\mathsf{C}(\vec{a})$, or $\mathbf{S}$ *models* $\mathsf{C}(\vec{a})$].

For "holds" we also say *is true*.

Example 4.4.2 Let $\mathbf{S}$ be the structure with language $\mathcal{L} = \{f, r\}$ given by the tables

f	a	b
a	a	a
b	a	b

r	a	b
a	0	1
b	0	0

Then rab holds in $\mathbf{S}$, but not rba, i.e., $\mathbf{S} \models rab$ but $\mathbf{S} \not\models rba$.

Let $\mathsf{L}_1 = rfxyfzx$, $\mathsf{L}_2 = \neg rfxyz$, and $\mathsf{C} = \{\mathsf{L}_1, \mathsf{L}_2\}$. Then a combined table for $\mathsf{L}_1, \mathsf{L}_2, \mathsf{C}$ is

					L_1		L_2	C
x	y	z	fxy	fzx	$rfxyfzx$	$rfxyz$	$\neg rfxyz$	$\{rfxyfzx, \neg rfxyz\}$
a	a	a	a	a	0	0	1	1
a	a	b	a	a	0	1	0	0
a	b	a	a	a	0	0	1	1
a	b	b	a	a	0	1	0	0
b	a	a	a	a	0	0	1	1
b	a	b	a	b	1	1	0	1
b	b	a	b	a	0	0	1	1
b	b	b	b	b	0	0	1	1

From this we see that $\mathsf{L}_1(x, y, z)$ holds only at (b, a, b), and $\mathsf{L}_2(x, y, z)$ holds at (a, a, a), (a, b, a), (b, a, a), (b, b, a), (b, b, b); and $\mathsf{C}(x, y, z)$ holds at (a, a, a), (a, b, a), (b, a, a), (b, a, b), (b, b, a), (b, b, b).

Thus $\mathbf{S} \models \mathsf{L}_1(b, a, b)$, $\mathbf{S} \not\models \mathsf{L}_1(a, b, a)$, $\mathbf{S} \models \mathsf{C}(a, a, a)$, and $\mathbf{S} \not\models \mathsf{C}(a, a, b)$.

Definition 4.4.3 [SATISFIABILITY]

- For L a literal we say

$$\mathbf{S} \models \mathsf{L}(x_1, \ldots, x_n),$$

[read: $\mathbf{S}$ *satisfies* L] if for *every* $a_1, \ldots, a_n$ from S we have $\mathsf{L}(a_1, \ldots, a_n)$ holds in $\mathbf{S}$.

- For a clause $\mathsf{C} = \{\mathsf{L}_1, \ldots, \mathsf{L}_k\}$ we say

$$\mathbf{S} \models \mathsf{C}(x_1, \ldots, x_n),$$

[read: $\mathbf{S}$ *satisfies* C] if for every choice of $a_1, \ldots, a_n$ from S we have $\mathsf{C}(a_1, \ldots, a_n)$ holds in $\mathbf{S}$.

- For $\mathcal{S}$ a set of clauses, we say

$$\mathbf{S} \models \mathcal{S},$$

[read: $\mathbf{S}$ *satisfies* $\mathcal{S}$] provided $\mathbf{S}$ satisfies every clause C in $\mathcal{S}$.

- We say $\mathrm{Sat}(\mathcal{S})$, or $\mathcal{S}$ *is satisfiable*, if there is a structure $\mathbf{S}$ such that $\mathbf{S} \models \mathcal{S}$. If this is not the case, we say $\neg\,\mathrm{Sat}(\mathcal{S})$, meaning $\mathcal{S}$ is *not satisfiable*.

A synonym for *satisfies* is *models*. In the preceding items we also say a literal or clause *is true* in $\mathbf{S}$, or *holds* in $\mathbf{S}$, to mean $\mathbf{S}$ *satisfies* the literal or clause. Predicate clause logic, like propositional clause logic, revolves around the study of *not satisfiable*.

Example 4.4.4 If $\mathcal{L}$ has two *unary* relation symbols r_1 and r_2, then the clause

$$\{\neg\, r_1 x,\ \neg\, r_2 x\}$$

is satisfied by a structure $\mathbf{S}$ iff for every $a \in S$ either $\neg\, r_1 a$ or $\neg\, r_2 a$ holds, and this is iff the two sets r_1 and r_2 are disjoint, i.e., $r_1 \cap r_2 = \emptyset$. We can picture this situation as follows:

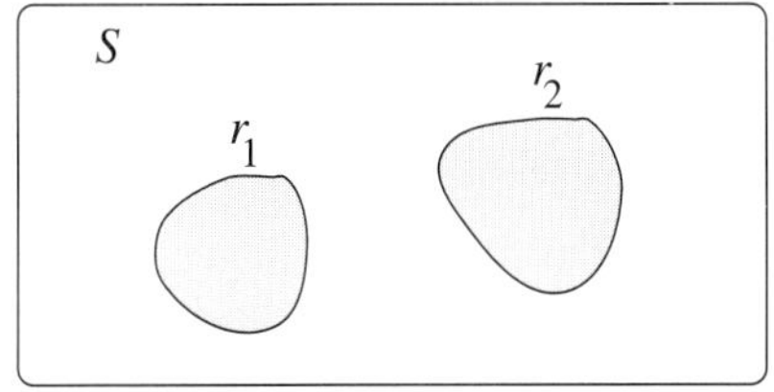

Figure 4.4.5 $\mathbf{S} \models \{\neg\, r_1 x,\ \neg\, r_2 x\}$

Example 4.4.6 If $\mathcal{L}$ has two *unary* relation symbols r_1 and r_2, then the clause

$$\{\neg r_1 x,\ r_2 x\}$$

is satisfied by a structure $\mathbf{S}$ iff the set r_1 is a subset of r_2, i.e., $r_1 \subseteq r_2$. We can picture this situation as follows:

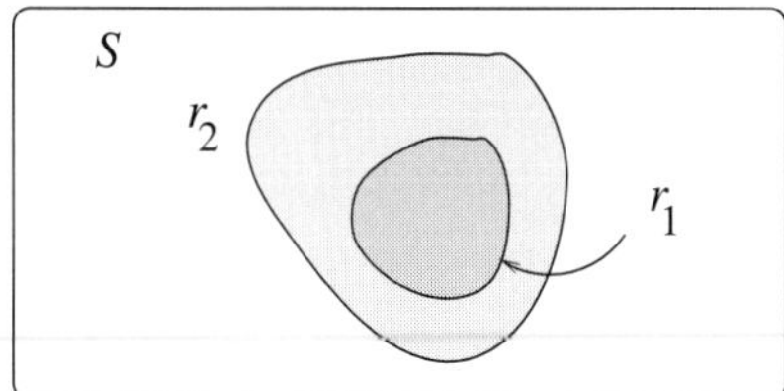

Figure 4.4.7 $\mathbf{S} \models \{\neg r_1 x,\ r_2 x\}$

Example 4.4.8 Suppose $\mathcal{L}$ has a single *binary* predicate r, i.e., we have a language for directed graphs. Let $\mathbf{S}$ be a directed graph.

- $\mathbf{S}$ will satisfy the clause $\{rxx\}$ iff the binary relation r is *reflexive*, i.e., raa holds for every $a \in S$.
- $\mathbf{S}$ will satisfy the clause $\{\neg rxx\}$ iff the binary relation r is *irreflexive*, i.e., raa does not hold for any $a \in S$.
- $\mathbf{S}$ will satisfy the clause $\{\neg rxy,\ ryx\}$ iff the binary relation r is *symmetric*, i.e., rab holds implies rba holds, for any $a, b \in S$.
- $\mathbf{S}$ will satisfy the clause $\{\neg rxy,\ \neg ryz,\ rxz\}$ iff the binary relation r is *transitive*, i.e., if rab and rbc hold then rac holds, for any $a, b, c \in S$.
- A *graph* is an irreflexive, symmetric directed graph.

Graphs are drawn without using directed edges, e.g.,

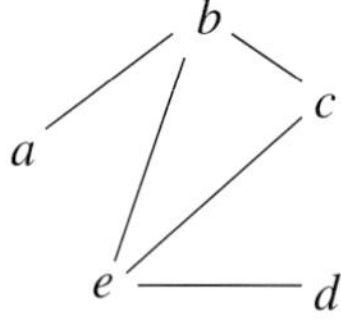

Figure 4.4.9 A graph

Exercises

Let $\mathcal{L} = \{f_3, f_2, f_1, r_3, r_2, r_1\}$, where f_k is a k–ary function symbol and r_k is a k–ary relation symbol. The following tables give an interpretation of $\mathcal{L}$ on the two–element set $S = \{a, b\}$:

			f_3
a	a	a	b
a	a	b	b
a	b	a	a
a	b	b	b
b	a	a	a
b	a	b	a
b	b	a	a
b	b	b	b

f_2	a	b
a	a	b
b	b	b

	f_1
a	b
b	a

			r_3
a	a	a	1
a	a	b	0
a	b	a	0
a	b	b	0
b	a	a	1
b	a	b	1
b	b	a	0
b	b	b	0

r_2	a	b
a	0	1
b	0	0

	r_1
a	1
b	0

4.4.1 Determine if the following literals $\mathsf{L}(x, y, z)$ hold at (a, b, a) in **S**:

a. $r_1 f_2 f_2 x f_2 y f_1 y x$
b. $r_1 f_2 y f_1 f_2 y f_3 y x z$
c. $\neg r_1 f_2 f_3 x f_1 x f_1 x z$
d. $\neg r_1 f_3 x z f_3 x z f_3 x x f_1 y$
e. $r_2 f_2 f_1 x f_2 y f_2 y x f_2 f_3 y f_1 x z f_1 y$

4.4.2 Determine if the following clauses hold at (b, a, b) in **S**:

a. $\mathsf{C}(x, y, z) = \{r_2 f_1 x f_1 y,\ r_1 f_2 z x\}$
b. $\mathsf{C}(x, y, z) = \{r_3 f_1 x f_1 y z,\ \neg r_2 z f_2 y x\}$
c. $\mathsf{C}(x, y, z) = \{\neg r_3 f_1 x f_1 y z,\ \neg r_2 z f_2 y x\}$
d. $\mathsf{C}(x, y, z) = \{r_1 f_3 f_1 y z f_3 x f_1 y x,\ \neg r_1 f_2 f_3 f_2 z y f_1 y z y\}$
e. $\mathsf{C}(x, y, z) = \{\neg r_2 f_2 x y f_3 x y z,\ r_3 f_2 y x f_1 z f_2 x z\}$

4.4.3 Determine if the following clauses hold in **S**:

a. $\{r_2 x f_1 y,\ \neg r_1 f_2 x y,\ \neg r_1 f_1 f_1 x\}$
b. $\{r_2 x f_1 x,\ \neg r_1 f_2 y y,\ r_2 f_1 x x\}$
c. $\{r_3 f_2 z f_1 x x f_1 x,\ r_2 f_1 x y,\ \neg r_2 f_1 x f_1 y\}$
d. $\{\neg r_3 f_1 y f_2 x y f_1 x,\ r_2 x f_1 y,\ \neg r_2 x f_1 f_1 y\}$
e. $\{r_3 f_2 z f_1 x x f_1 x,\ r_2 f_1 x y,\ r_2 f_1 y f_1 y\}$

4.5 REDUCTION TO PROPOSITIONAL LOGIC VIA GROUND CLAUSES, AND THE COMPACTNESS THEOREM FOR CLAUSE LOGIC

In 1915 Löwenheim introduced a method to construct models that essentially used a reduction to a sequence of ground statements. The possibility of using this as a means to prove theorems was not noticed until Skolem gave a lecture on such a procedure in 1928. This idea was developed, from a syntactic perspective, in great detail by Herbrand in his Ph.D. thesis in 1930. This method of proving theorems by examining a sequence of ground statements is commonly referred to as Herbrand's method.

4.5.1 Ground Instances

Suppose a language $\mathcal{L}$ $(= \mathcal{F} \cup \mathcal{R} \cup \mathcal{C})$ is given. We assume that $\mathcal{C} \neq \emptyset$. We want to define the notion of *ground instance*, which is basically a substitution instance without the occurrence of variables.

Definition 4.5.1 Let $\mathbf{A}$ be an algebra in the language $\mathcal{F} \cup \mathcal{C}$. We define the various *ground instances* over $\mathbf{A}$ as follows:

- For a *term* $t(\vec{x}\,)$ a *ground instance* over $\mathbf{A}$ is an element $t^{\mathbf{S}}(\vec{a})$ of A where $\vec{a}$ is from A.

- For an *atomic formula* $rt_1(\vec{x}\,) \cdots t_n(\vec{x}\,)$ a *ground instance* over $\mathbf{A}$ is an expression of the form $rt_1^{\mathbf{S}}(\vec{a}) \cdots t_n^{\mathbf{S}}(\vec{a})$ for $\vec{a}$ from A and is thus of the form $rb_1 \cdots b_n$. For the *negation* of the same formula, $\neg\, rt_1(\vec{x}\,) \cdots t_n(\vec{x}\,)$, we would obtain the *ground instance* $\neg\, rb_1 \cdots b_n$, where $b_i = t_i^{\mathbf{S}}(\vec{a}\,)$.

- A *ground instance* over $\mathbf{A}$ of a *clause* $\{\mathsf{L}_1(\vec{x}\,), \ldots, \mathsf{L}_k(\vec{x}\,)\}$ is obtained by taking some elements $\vec{a}$ from A and forming the clause $\{\mathsf{L}_1(\vec{a}), \ldots, \mathsf{L}_k(\vec{a})\}$, where $\mathsf{L}_i(\vec{a})$ is a ground instance of L_i. This is a clause in the language obtained by adding symbols for the elements of A to our original language and is called a *ground clause* over $\mathbf{A}$.

- Given a set $\mathcal{S}$ of clauses, let $\mathcal{S}/\mathbf{A}$ denote *the set of all ground instances* over $\mathbf{A}$ of all clauses from $\mathcal{S}$.

Example 4.5.2 Let $\mathbf{A} = (\{a,b\}, f)$ be the two–element algebra with the operation f given by

f	a	b
a	a	b
b	b	a

Let $\mathcal{S}$ consist of the single clause

$$\{\neg rxfxy,\ rfxyx\},$$

where r is a binary relation symbol. Then $\mathcal{S}/\mathbf{A}$ is the following collection of ground clauses:

$$\begin{array}{l}\{\neg raf^{\mathbf{A}}(a,a),\ rf^{\mathbf{A}}(a,a)a\}\\ \{\neg raf^{\mathbf{A}}(a,b),\ rf^{\mathbf{A}}(a,b)a\}\\ \{\neg rbf^{\mathbf{A}}(b,a),\ rf^{\mathbf{A}}(b,a)b\}\\ \{\neg rbf^{\mathbf{A}}(b,b),\ rf^{\mathbf{A}}(b,b)b\}.\end{array}$$

We simplify this collection by evaluating the expressions involving an $f^{\mathbf{A}}$ to obtain the following ground clauses over $\mathbf{A}$:

$$\begin{array}{l}\{\neg raa,\ raa\}\\ \{\neg rab,\ rba\}\\ \{\neg rbb,\ rbb\}\\ \{\neg rba,\ rab\}.\end{array}$$

Example 4.5.3 Let $\mathbf{P} = (\mathrm{P}, +, \times, 1)$, the positive integers, and let $\mathcal{S}$ consist of the two clauses $\{r1\}$ and $\{\neg rx,\ r(x+1)\}$, where r is a unary relation symbol. Then $\mathcal{S}/\mathbf{P}$ is the set consisting of the ground clauses

$$\{r1\},\ \{\neg r1,\ r2\},\ \{\neg r2,\ r3\},\ \ldots.$$

Exercises

Given the language $\mathcal{L} = \{f, r, 0\}$ with both f and r *binary*, let $\mathbf{A}$ be the algebra for the language $\{f, 0\}$ defined by the table

f	a	b
a	a	a
b	a	b

0 is a.

4.5.1 For each of the following literals and clauses find all the ground instances over $\mathbf{A}$.

a. rxy **b.** $rx0$ **c.** $rfxyz$ **d.** $rfxyfzx$
e. $\neg rfxyz$ **f.** $rffxyfyzy$ **g.** $\{rxfyz,\ rfxyz\}$
h. $\{ryfxfyz,\ \neg rzfyy\}$ **i.** $\{rxfyz,\ \neg rzfyy,\ \neg rfxxx\}$
j. $\{\neg rfxxfyx,\ \neg rffyyxy,\ \neg rxfxx\}$

4.5.2 Satisfiable over an Algebra

Given that our objective is to study Sat($\mathcal{S}$), we can think of choosing an algebra $\mathbf{A}$ as a halfway step to finding a model of $\mathcal{S}$—namely, we have then made a choice of the universe, and a choice of the interpretation of the constant symbols and the function symbols. We need only to interpret the relation symbols.

Definition 4.5.4 Let $\mathcal{L}$ and $\mathcal{L}'$ be two languages with $\mathcal{L} \subseteq \mathcal{L}'$ and with both languages having the same arities for the common symbols in $\mathcal{L}$. We say $\mathcal{L}$ is a *reduct* of $\mathcal{L}'$, and $\mathcal{L}'$ is an *expansion* of $\mathcal{L}$.

Let $\mathcal{L}'$ be an expansion of the language $\mathcal{L}$. Let I be an interpretation of $\mathcal{L}$ on a given set S and let I' be an interpretation of $\mathcal{L}'$ on the same set. If I and I' agree on $\mathcal{L}$, i.e., $I(f) = I'(f)$, $I(r) = I'(r)$, $I(c) = I'(c)$, for all $f, r, c \in \mathcal{L}$, then we say $\mathbf{S} = (S, I)$ is a *reduct* of $\mathbf{S}' = (S, I')$, and $\mathbf{S}' = (S, I')$ is an *expansion* of $\mathbf{S} = (S, I)$.

Example 4.5.5 Let N be the set of nonnegative integers. Then the structure $\mathbf{N}' = (N, +, \times, <, 0, 1)$ is an expansion of the structure $\mathbf{N} = (N, +, \times, 0)$, and $\mathbf{N}$ is a reduct of $\mathbf{N}'$.

Definition 4.5.6 A set $\mathcal{S}$ of clauses is *satisfiable over an algebra* $\mathbf{A} = (A, I)$ if it is possible to expand I to an interpretation I' of the relation symbols on A such that the resulting expansion of the algebra $\mathbf{A}$ to the structure $\mathbf{S} = (A, I')$ satisfies $\mathcal{S}$.

Example 4.5.7 Let $\mathcal{L}$ have a single *unary* function symbol f and a single *binary* relation symbol r. Consider the collection $\mathcal{S}$ of clauses:

$$\{rxfx,\ rfxx\},\ \{\neg rxfx,\ \neg rfxx\},\ \{\neg rxx\}.$$

We picture two monounary algebras $\mathbf{A}_1$ and $\mathbf{A}_2$ as follows:

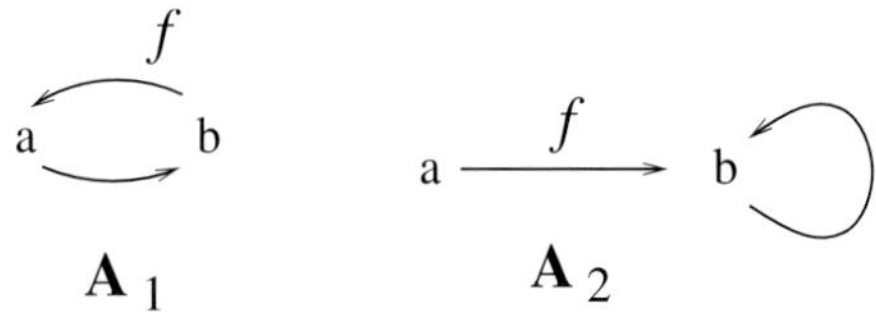

Figure 4.5.8 Two monounary algebras

$\mathcal{S}$ is satisfiable over the first algebra, $\mathbf{A}_1$, by interpreting r as the binary relation $\{(a, b)\}$, i.e., its directed graph is simply

$$a \longrightarrow b$$

Figure 4.5.9 A directed graph on $\{a, b\}$

But $\mathcal{S}$ is not satisfiable over the second algebra, $\mathbf{A}_2$.

Lemma 4.5.10 Given a set $\mathcal{S}$ of clauses, we have $\mathrm{Sat}(\mathcal{S})$ iff there is an algebra $\mathbf{A}$ such that $\mathcal{S}$ is satisfiable over $\mathbf{A}$.

Proof. If $\mathrm{Sat}(\mathcal{S})$, say $\mathbf{S} = (S, I')$ satisfies $\mathcal{S}$, let $\mathbf{A}$ be the algebra obtained by taking the interpretation I' and restricting it to the function symbols and constant symbols to obtain the interpretation I of $\mathcal{F} \cup \mathcal{C}$ on S. Then $\mathbf{A} = (S, I)$ is an algebra, and $\mathcal{S}$ is satisfiable over $\mathbf{A}$, namely, we use the given I' to expand $\mathbf{A}$ back to $\mathbf{S}$.

For the converse, if $\mathcal{S}$ is satisfiable over $\mathbf{A} = (A, I)$, say by the expansion $\mathbf{S} = (A, I')$, then $\mathrm{Sat}(\mathcal{S})$, since $\mathbf{S} \models \mathcal{S}$. $\square$

Definition 4.5.11 A set $\mathcal{G}$ of ground clauses is *p–satisfiable* (*propositionally satisfiable*) if $\mathcal{G}$ is satisfiable as a set of propositional clauses, where the atomic formulas $r\vec{a}$ are treated as the propositional variables.

Example 4.5.12 Let $\mathcal{L} = \{f, r\}$, where f and r are both binary, and let $\mathbf{A}$ be the algebra (A, f) given by

f	a	b
a	a	b
b	b	b

Let $\mathcal{G}$ be the collection of ground clauses consisting of

$$\begin{array}{l}\{rab,\ \neg\, rba\}\\ \{\neg\, raa,\ \neg\, rbb\}\\ \{\neg\, rab,\ rba\}\\ \{\neg\, rab,\ \neg\, raa\}.\end{array}$$

Let us treat this as a collection of propositional clauses, where the atomic ground formulas are the propositional variables. Let us rename the atomic ground formulas with single letters as follows:

atomic ground formula	raa	rab	rba	rbb
letter name	P	Q	R	S

Then the preceding ground clauses become

$$\begin{array}{l}\{Q,\ \neg\, R\}\\ \{\neg\, P,\ \neg\, S\}\\ \{\neg\, Q,\ R\}\\ \{\neg\, Q,\ \neg\, P\}.\end{array}$$

We can easily check that this collection of propositional clauses is satisfied by

P	Q	R	S
0	1	1	0

.

Thus the collection $\mathcal{G}$ of ground clauses is p–satisfiable.

Now we come to a fundamental observation.

Lemma 4.5.13 Let $\mathcal{G}$ be a set of ground clauses over $\mathbf{A}$. Then $\mathcal{G}$ is satisfiable over $\mathbf{A}$ iff $\mathcal{G}$ is p–satisfiable.

Proof. If $\mathcal{G}$ is satisfiable over $\mathbf{A}$, say $\mathbf{S}$ is an expansion that satisfies $\mathcal{G}$, then each atomic ground formula is either true or false in $\mathbf{S}$, and this assignment of truth values shows $\mathcal{G}$ is p–satisfiable.

Conversely, if $\mathcal{G}$ is p–satisfiable, then we choose a truth evaluation of the atomic ground formulas (as propositional variables) that makes $\mathcal{G}$ true. We extend the algebra $\mathbf{A}$ to a structure $\mathbf{S}$ by requiring that $r\vec{a}$ hold precisely when the truth evaluation of $r\vec{a}$ is 1. Then $\mathbf{S}$ satisfies $\mathcal{G}$. □

Example 4.5.14 If we return to the algebra $\mathbf{A}$ in Example 4.5.12 and expand it by interpreting r according to the truth evaluation given in Example 4.5.12, i.e., by

raa	rab	rba	rbb
0	1	1	0

i.e., by

r	a	b
a	0	1
b	1	0

,

then we have a structure $\mathbf{S}$ such that $\mathbf{S} \models \mathcal{G}$.

Lemma 4.5.15 Let $\mathcal{S}$ be a set of clauses and let $\mathbf{A}$ be an algebra. Then

$$\begin{aligned} \mathcal{S} \text{ is satisfiable over } \mathbf{A} \quad & \text{iff} \quad \mathcal{S}/\mathbf{A} \text{ is } p\text{–satisfiable} \\ & \text{iff} \quad \text{every finite } \mathcal{G} \subseteq \mathcal{S}/\mathbf{A} \\ & \qquad \text{is } p\text{–satisfiable.} \end{aligned}$$

Proof. It is a straightforward application of the definitions to show $\mathcal{S}$ is satisfiable over $\mathbf{A}$ iff $\mathcal{S}/\mathbf{A}$ is p–satisfiable. This allows us to invoke the compactness theorem (Theorem 2.8.1, p. 75) for propositional logic. □

Exercises

4.5.2 Let $\mathbf{P} = (\mathrm{P}, +, \times, 1)$, the positive integers with the usual operations, and let $\mathcal{S}$ consist of the two clauses $\{r1\}$ and $\{\neg rx, r(x+1)\}$, where r is a unary relation symbol. Prove that $\mathcal{S}$ is satisfiable in only one way over $\mathbf{P}$.

4.5.3 Let $\mathbf{A}$ be the algebra given by the table

f	a	b
a	a	a
b	a	b

and let r_k be a k–ary relation symbol.

For each of the following sets $\mathcal{G}$ of ground clauses over $\mathbf{A}$ determine if $\mathcal{G}$ is p–satisfiable. If so, find an expansion $\mathbf{S}$ of $\mathbf{A}$ such that $\mathbf{S} \models \mathcal{G}$.

a.

$$\begin{array}{l} \{r_1a,\ \neg r_1b\} \\ \{r_1b,\ \neg r_2ab\} \\ \{r_2ab,\ \neg r_2ba\} \\ \{\neg r_1a,\ \neg r_2aa\}. \end{array}$$

b.

$$\begin{array}{l} \{\neg r_2ab,\ \neg r_2bb,\ r_1a\} \\ \{\neg r_2ba,\ r_2aa,\ \neg r_2ab,\ \neg r_1a\} \\ \{r_2ab,\ r_2ba,\ r_1b\} \\ \{\neg r_1a,\ \neg r_2aa\}. \end{array}$$

4.5.3 The Herbrand Universe

Definition 4.5.16 Given a language $\mathcal{L} = \mathcal{F} \cup \mathcal{R} \cup \mathcal{C}$, the set of *ground terms* $T_\mathcal{C}$ for the language is the set of terms that have no variables. $T_\mathcal{C}$ is called the *Herbrand universe* for the language $\mathcal{L}$.

We note that $T_\mathcal{C} = \mathcal{C}$ iff there are no function symbols, i.e., $\mathcal{F} = \emptyset$, or if $\mathcal{C} = \emptyset$. Otherwise $T_\mathcal{C} \setminus \mathcal{C}$ is infinite.

Example 4.5.17 Suppose our language has a binary function symbol f and two constants $0, 1$. Then the following ground terms will be in the Herbrand universe:

$$0,\ 1,\ f00,\ f01,\ f10,\ f11,\ f0f00, \text{ etc.}$$

Now we give a natural interpretation I of the constant symbols of $\mathcal{C}$ and the function symbols of $\mathcal{F}$ on the Herbrand universe to create the algebra $\mathbf{T}_\mathcal{C}$ as follows:

- $I(c) = c$.
- $I(f)(t_1, \dots, t_n) = ft_1 \cdots t_n$.

As we shall see, the Herbrand universe provides an analog of the two–element algebra of connectives in the propositional calculus, i.e., it provides a place to check for satisfiability. The basic theorem says that a set of clauses is satisfiable iff every finite set of ground clauses over the Herbrand universe is satisfiable.

4.5.4 Growth of the Herbrand Universe

In Löwenheim's and Skolem's constructions of the Herbrand universe $T_\mathcal{C}$ for a language, $T_\mathcal{C}$ is built up in stages $S_0, S_1, \cdots$ defined by

$$\begin{aligned} S_0 &= \mathcal{C} \\ S_{n+1} &= S_n \cup \{ft_1 \cdots t_k : f \in \mathcal{F},\, t_i \in S_n,\, \text{arity}(f) = k\}. \end{aligned}$$

We note that S_n is the collection of *ground terms with tree depth at most* n.

Example 4.5.18 Let $\mathcal{L} = \{f, r, 0\}$, where f and r are binary. Then we have

$$\begin{aligned} S_0 &= \{0\} \\ S_1 &= \{0, f00\} \\ S_2 &= \{0, f00, f0f00, ff000, ff00f00\}. \end{aligned}$$

We can find a recursion formula for the size of S_n by noting that the terms in $S_n \setminus S_{n-1}$ are of the form ft_1t_2, where $t_1, t_2 \in S_{n-1}$, and at least one of the t_i has tree depth $n-1$, i.e., is in $S_{n-1} \setminus S_{n-2}$. Letting $H(n)$ be $|S_n|$, the size of S_n, we have $H(0) = 1$, $H(1) = 2$, and for $n > 1$

$$\begin{aligned} H(n) &= |S_{n-1}| + |S_n \setminus S_{n-1}| \\ &= H(n-1) + (H(n-1)^2 - H(n-2)^2). \end{aligned}$$

Given a (finite) set $\mathcal{S}$ of clauses[1] Löwenheim considered extensions of potential models of $\mathcal{S}$ on S_i to S_{i+1}; equivalently, Skolem considered the satisfiability of the sequence $\mathcal{S}/S_0, \mathcal{S}/S_1, \cdots$ of sets of ground instances of $\mathcal{S}$.

[1] Both Löwenheim and Skolem worked with arbitrary quantifier–free formulas (see Chapter 5), and not just clauses.

Now let us look at the feasibility of this procedure for any given language $\mathcal{L} = \mathcal{F} \cup \mathcal{R} \cup \mathcal{C}$. Let m_0 be the size of $\mathcal{C}$, and let m_i be the number of function symbols in $\mathcal{F}$ of rank i. Letting $H(n)$ be the size of the nth stage S_n in the construction of the Herbrand universe, we have

$$\begin{aligned} H(0) &= m_0 \\ H(1) &= m_0 + \sum_{i>0} m_i \cdot m_0^i \\ H(n+1) &= H(n) + \sum_{i>0} m_i \cdot [H(n)^i - H(n-1)^i]. \end{aligned}$$

Next, suppose our set $\mathcal{S}$ of clauses uses the set $\mathcal{R}$ of relation symbols, say with ρ_i relation symbols of rank i. Let $V(n)$ be the size of $\mathcal{R}/S_n$, the number of atomic ground formulas obtained from $\mathcal{R}$ using the ground terms S_n. Then we have

$$V(n) = \sum_{i>0} \rho_i \cdot H(n)^i.$$

Now, let us consider a simple case where $\mathcal{C} = \{a\}$, $\mathcal{F} = \{+\}$ (where $+$ is a binary operation), and $\mathcal{R} = \{<\}$ (where $<$ is a binary relation). Then the functions $H(n)$ and $V(n)$ grow as in the table:

n	$H(n)$	$V(n)$
0	1	1
1	2	4
2	5	25
3	26	676
4	677	458,329

Determining the p–satisfiability of $\mathcal{S}/S_n$ is a propositional question involving clauses in $V(n)$ propositional variables. To tackle this by hand with truth tables would, in practice, limit us to the first two stages, S_0, S_1, and we could not hope to look at more than the third stage, S_2. Unfortunately, interesting mathematical questions, when phrased as showing a set of clauses is not satisfiable, usually require far more than the first few stages of the Herbrand universe, and in general, $\mathcal{F}$ will have a generous supply of function symbols of rank greater than 2. Thus the reduction to ground terms is more of a theoretical tool than a practical tool for theorem proving.

Exercises

4.5.4 Find S_4 for $\mathcal{L} = \{f, r, c, d\}$, where f is a unary function symbol, r is a ternary relation symbol, and c, d are constant symbols.

4.5.5 Find S_3 for $\mathcal{L} = \{f, g, r, c\}$, where f, g are unary function symbols, r is a binary relation symbol, and c is a constant symbol.

4.5.6 For $\mathcal{L} = \{f, r, c, d\}$, where f is a unary function symbol, r is a ternary relation symbol, and c, d are constant symbols, find a recursion formula for $H(n)$, the number of elements in S_n.

4.5.7 For $\mathcal{L} = \{f, g, r, c\}$, where f, g are unary function symbols, r is a binary relation symbol, and c is a constant symbol, find a recursion formula for $H(n)$, the number of elements in S_n.

4.5.8 Let $\mathcal{L} = \{f, r, c\}$, where f is a unary function symbol, r is a unary relation symbol, and c is a constant symbol. Let $\mathcal{S}$ consist of the single clause $\{rfx, \neg rfy\}$.

a. Find $\mathcal{S}/S_4$.
b. Determine if $\mathcal{S}/S_4$ is p–satisfiable.

4.5.9 Let $\mathcal{L} = \{f, r, c\}$, where f is a binary function symbol, r is a unary relation symbol, and c is a constant symbol. Let $\mathcal{S}$ consist of the single clause $\{rfxy, \neg rfyx\}$.

a. Find $\mathcal{S}/S_2$.
b. Determine if $\mathcal{S}/S_2$ is p–satisfiable.

4.5.10 Let $\mathcal{L} = \{f, r, c\}$, where f is a unary function symbol, r is a binary relation symbol, and c is a constant symbol. Let $\mathcal{S}$ consist of the single clause $\{rfxy, \neg rxfy\}$.

a. Find $\mathcal{S}/S_2$.
b. Determine if $\mathcal{S}/S_2$ is p–satisfiable.

4.5.5 Satisfiability over the Herbrand Universe

One of the remarkable properties for checking the satisfiability of a set of clauses is that we can completely standardize the algebra part, and then we need only find suitable interpretations of the relation symbols. We write p–Sat($\mathcal{G}$) to assert that $\mathcal{G}$ is p–satisfiable.

Theorem 4.5.19 Given a set $\mathcal{S}$ of clauses we have

$$\begin{array}{lll} \text{Sat}(\mathcal{S}) & \text{iff} & \mathcal{S} \text{ is satisfiable over } \mathbf{T}_\mathcal{C} \\ & \text{iff} & p\text{–Sat}(\mathcal{S}/\mathbf{T}_\mathcal{C}) \\ & \text{iff} & p\text{–Sat}(\mathcal{G}) \text{ for all finite } \mathcal{G} \subseteq \mathcal{S}/\mathbf{T}_\mathcal{C}. \end{array}$$

Proof. The equivalence of the second, third, and fourth assertions is simply an application of Lemma 4.5.15, and from p–Sat$(\mathcal{S}/\mathrm{T}_\mathcal{C})$ we have Sat$(\mathcal{S})$ by Lemma 4.5.10 (p. 275).

Now suppose Sat$(\mathcal{S})$ holds, say $\mathbf{S} \models \mathcal{S}$. Let $\mathbf{T}_\mathcal{C} = (T_\mathcal{C}, I)$.

Let g be a mapping defined inductively on the ground terms $T_\mathcal{C}$ to the set S by

- $\mathsf{g}(c) = c^{\mathbf{S}}$;
- $\mathsf{g}(ft_1 \cdots t_n) = f^{\mathbf{S}}(\mathsf{g}t_1, \ldots, \mathsf{g}t_n)$.

Then we define an interpretation I' on $T_\mathcal{C}$ by

- $I'(c) = c$;
- $I'(f)(t_1, \ldots, t_n) = ft_1 \cdots t_n$;
- $I'(r)(t_1, \ldots, t_n)$ holds iff $r(\mathsf{g}t_1, \ldots, \mathsf{g}t_n)$ holds.

Thus I' extends the interpretation I used for the constant and function symbols in $\mathbf{T}_\mathcal{C}$.

Now let $\mathbf{T}'$ be the structure $(T_\mathcal{C}, I')$. We claim that $\mathbf{T}' \models \mathcal{S}$. To see this let $\mathsf{C} = \{\mathsf{L}_1(\vec{x}), \ldots, \mathsf{L}_k(\vec{x})\}$ be a clause in $\mathcal{S}$. Given $\vec{t}$ from $T_\mathcal{C}$ we have, from the fact that $\mathbf{S} \models \mathsf{C}$, that $\mathsf{C}(\mathsf{g}t_1, \ldots, \mathsf{g}t_n)$ holds in $\mathbf{S}$. Then, for some i, $\mathsf{L}_i(\mathsf{g}t_1, \ldots, \mathsf{g}t_n)$ holds in $\mathbf{S}$, so $\mathsf{L}_i(t_1, \ldots, t_n)$ holds in $\mathbf{T}'$. But then $\mathbf{T}' \models \mathsf{C}$.

As I' extends I, we see that $\mathcal{S}$ is satisfiable over $\mathbf{T}_\mathcal{C}$. □

Thus to check a set $\mathcal{S}$ of clauses for satisfiability we can work over the Herbrand universe $T_\mathcal{C}$ with the operations defined in a natural manner, namely, as in $\mathbf{T}_\mathcal{C}$; we *need only to see if it is possible to interpret the relation symbols to obtain a model of* $\mathcal{S}$.

Remark 4.5.20 When we refer to ground instances, clauses, etc., without naming a particular algebra $\mathbf{A}$, it is understood that we are working over $\mathbf{T}_\mathcal{C}$, the Herbrand universe with the natural interpretation I.

Corollary 4.5.21 A set $\mathcal{S}$ of clauses is not satisfiable iff there is a finite set of ground instances of $\mathcal{S}$ that is not p–satisfiable.

Proof. This is immediate from Theorem 4.5.19. □

Example 4.5.22 The following set of clauses is not satisfiable:

$$\{\neg rxy,\ rfxfy\}$$
$$\{\neg rxy,\ \neg rxz,\ ryz\}$$
$$\{rffxx\}$$
$$\{rfffxx\}$$
$$\{\neg rfxx\}.$$

We give a proof using ground clauses, namely, we consider the following ground instances (one for each clause):

$$\{\neg rffaa,\ rfffafa\}$$
$$\{\neg rfffafa,\ \neg rfffaa,\ rfaa\}$$
$$\{rffaa\}$$
$$\{rfffaa\}$$
$$\{\neg rfaa\}.$$

Translating these into propositional clauses we have

$$\{\neg P,\ Q\}$$
$$\{\neg Q,\ \neg R,\ S\}$$
$$\{P\}$$
$$\{R\}$$
$$\{\neg S\},$$

which we saw earlier in Example 2.10.10 (p. 102) and proved to be unsatisfiable by resolution. Writing out that same proof using the ground clauses would give the following:

1.	$\{\neg rfaa\}$	given
2.	$\{\neg rfffafa,\ \neg rfffaa,\ rfaa\}$	given
3.	$\{\neg rfffafa,\ \neg rfffaa\}$	resolution 1,2
4.	$\{rfffaa\}$	given
5.	$\{\neg rfffafa\}$	resolution 3,4
6.	$\{\neg rffaa,\ rfffafa\}$	given
7.	$\{\neg rffaa\}$	resolution 5,6
8.	$\{rffaa\}$	given
9.	$\{\ \}$	resolution 7,8.

Remark 4.5.23 One is tempted to extend Theorem 4.5.19 to include clauses *with equality*, e.g., the clause $\{x < y, x \approx y, y < x\}$. However, there is a problem with "pulling back from **S** to **T** using **g**" in the proof of Theorem 4.5.19, since we would want equality to pull back to equality, and this would force **g** to be one–to–one. If we consider the unit clause $\{fx \approx 0\}$, then it is clearly satisfiable by a

one–element algebra but *cannot* be satisfied on the infinite Herbrand universe for this language. Thus Theorem 4.5.19 fails to hold if we add equality to our language. One way around this problem is to *axiomatize equality* and add this to the set of clauses—we will do this in Sec. 4.9.

Exercises

4.5.11 Let $\mathcal{L} = \{f, r, 0\}$, where f, r are both *unary*. Each of the following collections of clauses is not satisfiable. Find, in each case, a set of ground instances of the clauses that is not p–satisfiable, and prove that the set of ground instances is not p–satisfiable.

a. $\{r0\},\ \{\neg rx,\ rfx\},\ \{\neg rff0,\ \neg rfff0\}$
b. $\{r0\},\ \{\neg rx,\ rfx\},\ \{\neg rf0,\ \neg rfff0\}$
c. $\{r0\},\ \{\neg rx,\ \neg ry\}$

4.5.12 Let $\mathcal{L} = \{f, r, 0\}$, where f is *unary* and r is *binary*. Each of the following collections of clauses is not satisfiable. Find, in each case, a set of ground instances of the clauses that is not p–satisfiable, and prove that the set of ground instances is not p–satisfiable.

a. $\{\neg rxy,\ \neg ryx\},\ \{rfxfy\}$
b. $\{\neg rxy,\ \neg ryx\},\ \{rxfx\},\ \{rfxx\}$
c. $\{\neg rxy,\ \neg ryx\},\ \{rxfx\},\ \{rffxfx\}$
d. $\{\neg rxy,\ \neg ryz,\ rxz\},\ \{\neg rxffx\},\ \{rxfx\}$
e. $\{\neg rxx\},\ \{\neg rxy,\ rfxy\},\ \{rxfx\}$

4.5.6 Compactness for Predicate Clause Logic without Equality

Theorem 4.5.24 A set $\mathcal{S}$ of clauses is satisfiable iff every finite subset is satisfiable.

Proof. The direction ($\Longrightarrow$) is trivial. For the converse ($\Longleftarrow$) suppose $\mathcal{S}$ is not satisfiable. Then by Corollary 4.5.21 we see that some finite set $\mathcal{G}$ of ground instances is not p–satisfiable. We choose a finite $\mathcal{S}_0 \subseteq \mathcal{S}$ such that $\mathcal{G} \subseteq \mathcal{S}_0/\mathbf{T}_{\mathcal{C}}$. Then $\mathcal{S}_0/\mathbf{T}_{\mathcal{C}}$ is not p–satisfiable, and hence by Theorem 4.5.19 $\mathcal{S}_0$ is not satisfiable. □

4.6 RESOLUTION

The efficient application of the method of reduction to ground clauses to prove a set of clauses unsatisfiable requires careful pattern matching, called unification, to obtain clauses with complementary literals. Analyzing this procedure leads to the fact that we can *bypass* the propositional calculus (i.e., the reduction to ground clauses) and carry out the proof of unsatisfiability directly with the original clauses.

4.6.1 Substitution

Given a substitution

$$\sigma = \begin{pmatrix} x_1 \leftarrow t_1 \\ \vdots \\ x_n \leftarrow t_n \end{pmatrix}$$

and a literal $\mathsf{L}(x_1, \ldots, x_n)$, we write $\sigma\mathsf{L}$ or $\mathsf{L}(t_1, \ldots, t_n)$ for the result of applying the substitution σ to L. Given a clause

$$\mathsf{C} = \mathsf{C}(x_1, \ldots, x_n) = \{\mathsf{L}_1(x_1, \ldots, x_n), \cdots, \mathsf{L}_k(x_1, \ldots, x_n)\},$$

we write $\sigma\mathsf{C}$ or $\mathsf{C}(t_1, \ldots t_n)$ for the result of applying σ to C to obtain the clause $\{\sigma\mathsf{L}_1, \ldots, \sigma\mathsf{L}_k\}$.

Example 4.6.1 Consider the clause $\mathsf{C} = \{\neg rxfx, \neg rfxy\}$. Then if σ is the substitution

$$\sigma = \begin{pmatrix} x \leftarrow gxfz \\ y \leftarrow fgyx \end{pmatrix},$$

we have $\sigma\mathsf{C} = \{\neg rgxfzfgxfz, \neg rfgxfzfgyx\}$.

Now we prove that substitution applied to clauses preserves satisfaction, as is the case with equations.

Theorem 4.6.2 [SUBSTITUTION THEOREM] If $\mathbf{S}$ is a structure and C is a clause, then for any substitution σ,

$$\mathbf{S} \models \mathsf{C} \quad \text{implies} \quad \mathbf{S} \models \sigma\mathsf{C}.$$

Proof. Let $\mathsf{C} = \mathsf{C}(x_1, \ldots, x_k)$. Then for every choice of elements $b_1, \ldots, b_k$ from S we know that $\mathsf{C}(b_1, \ldots, b_k)$ holds in $\mathbf{S}$. As the k–element sequences $t_1^{\mathbf{S}}(a_1, \ldots, a_n), \ldots, t_k^{\mathbf{S}}(a_1, \ldots, a_n)$ from S are just some of the sequences $b_1, \ldots, b_k$, it follows that for any sequence

$a_1, \ldots, a_n$ from S we have $\mathsf{C}(t_1(\vec{a}), \ldots, t_k(\vec{a}))$ holding. But then $\mathsf{C}(t_1, \ldots, t_k)$ holds in $\mathbf{S}$. □

The *complement* $\overline{\mathsf{L}}$ of a literal L is defined much as it was in the propositional calculus, namely, we convert an atomic formula A to a negated atomic formula $\neg\,\mathsf{A}$, and vice versa.

4.6.2 Opp–Unification

Resolution for clauses looks similar to that for propositional logic except that *we can first use a substitution.* If a substitution σ is applied to a clause $\mathsf{C} = \{\mathsf{L}_1, \ldots, \mathsf{L}_k\}$, then the resulting clause is $\sigma\mathsf{C} = \{\sigma\mathsf{L}_1, \ldots, \sigma\mathsf{L}_k\}$. As a result of substitution, several literals may collapse into a single literal.

Example 4.6.3 Let $\mathsf{C} = \{rxy,\ rxz,\ \neg\, rzx\}$. Applying the substitution

$$\sigma = \begin{pmatrix} x \leftarrow w \\ z \leftarrow y \end{pmatrix}$$

yields the clause

$$\sigma\mathsf{C} = \{rwy,\ rwy,\ \neg\, ryw\} = \{rwy,\ \neg\, ryw\}.$$

Definition 4.6.4 An *opp–unifier* (*opposite unifier*) of a pair of clauses $\mathsf{C}'', \mathsf{D}''$ is a pair of substitutions σ_1, σ_2 such that

$$\sigma_1\mathsf{C}'' = \{\mathsf{L}\}$$
$$\sigma_2\mathsf{D}'' = \{\overline{\mathsf{L}}\},$$

where L is a literal. This says *all* the literals in C'' become L under the substitution σ_1, and *all* the literals in D'' become $\neg\,\mathsf{L}$ under σ_2. If an opp–unifier exists, then we say the clauses are *opp–unifiable*.

Example 4.6.5 Let $\mathsf{C}'' = \{rxfz,\ rxffy\}$ and $\mathsf{D}'' = \{\neg\, rf0ffx\}$. Then the pair of substitutions σ_1, σ_2 given by

$$\sigma_1 = \begin{pmatrix} x \leftarrow f0 \\ z \leftarrow fy \end{pmatrix} \quad \text{and} \quad \sigma_2 = (x \leftarrow y)$$

is an opp–unifier of the pair of clauses $\mathsf{C}'', \mathsf{D}''$; indeed, $\sigma_1\mathsf{C}'' = \{rf0ffy\}$ and $\sigma_2\mathsf{D}'' = \{\neg\, rf0ffy\}$.

4.6.3 Resolution

Robinson needed only a single rule of inference, namely, the following *resolution of clauses* for $\mathsf{C} = \mathsf{C}' \cup \mathsf{C}''$ and $\mathsf{D} = \mathsf{D}' \cup \mathsf{D}''$:

Resolution of Clauses

$$\frac{\mathsf{C}' \cup \mathsf{C}'',\ \mathsf{D}' \cup \mathsf{D}''}{\sigma_1 \mathsf{C}' \cup \sigma_2 \mathsf{D}'},$$

where σ_1, σ_2 is an opp–unifier of $\mathsf{C}'', \mathsf{D}''$, i.e.,

$$\sigma_1 \mathsf{C}'' = \{\mathsf{L}\}$$
$$\sigma_2 \mathsf{D}'' = \{\overline{\mathsf{L}}\},$$

with L a literal and $\overline{\mathsf{L}}$ its complement.

Definition 4.6.6 A *derivation* using resolution of a clause C from a set $\mathcal{S}$ of clauses is a sequence of clauses $\mathsf{C}_1, \ldots, \mathsf{C}_n$ such that each C_i is either a member of $\mathcal{S}$ or results from applying resolution to two previous clauses in the sequence, and the last clause C_n is the clause C. We write $\mathcal{S} \vdash \mathsf{C}$ (read: C is *derivable from* $\mathcal{S}$) if there is such a derivation.

Example 4.6.7 Now we give a predicate logic resolution derivation of the empty clause from the clauses in Example 4.5.22 (p. 282):

1.	$\{\neg rfxx\}$	given
2.	$\{\neg rxy,\ \neg rxz,\ ryz\}$	given
3.	$\{\neg rfffxfx,\ \neg rfffxx\}$	resolution 1,2
4.	$\{rfffxx\}$	given
5.	$\{\neg rfffxfx\}$	resolution 3,4
6.	$\{\neg rxy,\ rfxfy\}$	given
7.	$\{\neg rffxx\}$	resolution 5,6
8.	$\{rffxx\}$	given
9.	$\{\,\}$	resolution 7,8.

To justify these resolution steps consider the following.

In step 3 let σ_1 be the identity map, applied to (1), and apply

$$\sigma_2 = \begin{pmatrix} x \leftarrow fffx \\ y \leftarrow fx \\ z \leftarrow x \end{pmatrix}$$ to (2).

In step 5 let both σ_1 and σ_2 be the identity maps.

In step 7 let σ_1 be the identity map, applied to (5), and apply $\sigma_2 = \begin{pmatrix} x \leftarrow ffx \\ y \leftarrow x \end{pmatrix}$ to (6).

In step 9 let both σ_1 and σ_2 be the identity maps.

Exercises

4.6.1 Let $\mathcal{L} = \{f, r, 0\}$, where f, r are both unary. Each of the following collections of clauses is not satisfiable. Find, in each case, a derivation of the empty clause using resolution.

a. $\{r0\}$, $\{\neg rx,\ rfx\}$, $\{\neg rff0\}$
b. $\{r0\}$, $\{\neg rx,\ rfx\}$, $\{\neg rf0,\ \neg rfff0\}$
c. $\{r0\}$, $\{\neg rx,\ rfx\}$, $\{\neg rff0,\ \neg rfff0,\ \neg rffff0\}$
d. $\{\neg rx,\ \neg rffx\}$, $\{rfx\}$

4.6.2 Let $\mathcal{L} = \{f, r, 0\}$, where f is unary and r is binary. Each of the following collections of clauses is not satisfiable. Find, in each case, a derivation of the empty clause using resolution.

a. $\{\neg rxy,\ \neg ryx\}$, $\{rfxy\}$
b. $\{\neg rxy,\ \neg ryx\}$, $\{rffxfy\}$
c. $\{\neg rxy,\ \neg ryx\}$, $\{rfxfx\}$
d. $\{\neg rxy,\ \neg ryx\}$, $\{rxfx\}$, $\{\neg rxy,\ \neg ryz,\ rxz\}$, $\{rfffxfx\}$
e. $\{\neg rxy,\ \neg ryz,\ rxz\}$, $\{\neg rfxfffx\}$, $\{rxfx\}$

4.6.4 Soundness and Completeness of Resolution

Theorem 4.6.8 A set $\mathcal{S}$ of clauses is not satisfiable iff there is a derivation of the empty clause by resolution.

Proof. (*Soundness*): Suppose $\mathbf{S} \models \mathsf{C}' \cup \mathsf{C}''$ and $\mathbf{S} \models \mathsf{D}' \cup \mathsf{D}''$ and that σ_1, σ_2 is an opp–unifier of $\mathsf{C}'', \mathsf{D}''$ with $\sigma_1 \mathsf{C}'' = \{\mathsf{L}\}$ and $\sigma_2 \mathsf{D}'' = \{\overline{\mathsf{L}}\}$. From Theorem 4.6.2 we have $\mathbf{S}$ satisfies $\sigma_1(\mathsf{C}' \cup \mathsf{C}'')$ and $\sigma_2(\mathsf{D}' \cup \mathsf{D}'')$, so $\mathbf{S}$ satisfies $\sigma_1 \mathsf{C}' \cup \sigma_2 \mathsf{C}''$ and $\sigma_2 \mathsf{D}' \cup \sigma_2 \mathsf{D}''$. But then $\mathbf{S}$ satisfies both $\sigma_1 \mathsf{C}' \cup \{\mathsf{L}\}$ and $\sigma_2 \mathsf{D}' \cup \{\overline{\mathsf{L}}\}$.

For any elements $\vec{a}$ from S, one of $\mathsf{L}(\vec{a})$ and its complement $\overline{\mathsf{L}}(\vec{a})$ does not hold in $\mathbf{S}$, let us say the first. But then $(\sigma_1 \mathsf{C}')(\vec{a})$ must hold in $\mathbf{S}$, as $\mathbf{S}$ satisfies $\sigma_1 \mathsf{C}' \cup \{\mathsf{L}\}$. This implies $(\sigma_1 \mathsf{C}' \cup \sigma_2 \mathsf{D}')(\vec{a})$ holds in $\mathbf{S}$. Consequently, $\mathbf{S} \models \sigma_1 \mathsf{C}' \cup \sigma_2 \mathsf{D}'$. This gives the soundness of the resolution, i.e., resolution preserves satisfiability. Thus we cannot derive the empty clause from a satisfiable set of premisses.

(*Completeness*): Now we turn to the other direction of the proof. Suppose $\mathcal{S}$ is an unsatisfiable set of clauses.

Case 1: There are constant symbols, i.e., $\mathcal{C} \neq \emptyset$. By Theorem 4.5.19 (p. 281) we see that the ground instances of $\mathcal{S}$ are not p–satisfiable, and hence using Lemma 4.5.13 (p. 276) and the completeness of resolution in the propositional calculus we can find a propositional derivation of the empty clause from the ground clauses.

Since the conversion of clauses into ground clauses is nothing more than an instance of substitution, the propositional derivation can easily be converted into a derivation in the predicate clause logic, namely, we replace each step that is a ground instance of $\mathcal{S}$ with the clause from $\mathcal{S}$ that gave rise to the ground instance. The other steps of the derivation are left unchanged.

Case 2: If the original language of $\mathcal{S}$ has no constant symbols, then adding a constant symbol c allows us to proceed as before to obtain a resolution derivation of the empty clause from ground clauses. We convert this into a derivation using resolution in the predicate clause logic. Then we take such a derivation and replace the constant symbol c with a variable x not appearing in the derivation to obtain a resolution derivation of the empty clause in the original language. □

4.7 THE UNIFICATION OF LITERALS

When carrying out resolution we would like to make the steps as efficient as possible. One obvious place to look is the choice of the substitution in resolution—why not make it as simple or, more precisely, as general as possible. We will start by recalling the definition of unifiers of terms, and then we will define unifiers of literals.

4.7.1 Unifying Pairs of Literals

Definition 4.7.1 Given two terms t_1 and t_2 we say that a substitution σ is a *unifier* of t_1, t_2 if

$$\sigma t_1 = \sigma t_2.$$

Given two literals L_1 and L_2 we say that a substitution σ is a *unifier* of $\mathsf{L}_1, \mathsf{L}_2$ if

$$\sigma \mathsf{L}_1 = \sigma \mathsf{L}_2.$$

If a unifier exists, we say the literals (or terms) are *unifiable*.

The following list collects together some simple facts about unifying literals.

a. If the literals L_1 and L_2 are unifiable, then either both are positive literals, or both are negative literals, since substitutions take positive literals to positive literals, and negative literals to negative literals.

b. L_1 and L_2 are unifiable iff $\neg\mathsf{L}_1$ and $\neg\mathsf{L}_2$ are unifiable. Indeed, $\sigma\mathsf{L}_1 = \sigma\mathsf{L}_2$ iff $\sigma(\neg\mathsf{L}_1) = \sigma(\neg\mathsf{L}_2)$.

c. If L_1 and L_2 are unifiable, then in the positive case,

$$\begin{aligned} \mathsf{L}_1 &= rs_1\cdots s_n \\ \mathsf{L}_2 &= rt_1\cdots t_n, \end{aligned} \tag{48}$$

and in the negative case

$$\begin{aligned} \mathsf{L}_1 &= \neg rs_1\cdots s_n \\ \mathsf{L}_2 &= \neg rt_1\cdots t_n, \end{aligned} \tag{49}$$

i.e., both L_1 and L_2 involve *the same relation symbol* r, and thus *the same number of terms* (where n is the arity of r).

In either case, a substitution σ is a unifier of L_1 and L_2 iff σ is a unifier of the pairs $(s_1, t_1), \ldots, (s_n, t_n)$, i.e., $\sigma s_1 = \sigma t_1, \ldots,$ $\sigma s_n = \sigma t_n$. Thus the unification of a pair of literals reduces to the unification of finitely many pairs of terms, a topic discussed in Sec. 3.12.6.

Example 4.7.2 Let $\mathsf{L}_1 = rxfyz$ and $\mathsf{L}_2 = rfzyw$, where f and r are both *binary*. Then the substitution

$$\sigma = \begin{pmatrix} x \leftarrow ffyyy \\ y \leftarrow y \\ z \leftarrow fyy \\ w \leftarrow fyfyy \end{pmatrix}$$

is a unifier of L_1 and L_2, and $\sigma(\mathsf{L}_i) = rffyyyfyfyy$ for $i = 1, 2$.

Actually this choice is rather special, since the substitution

$$\sigma' = \begin{pmatrix} x \leftarrow ffyyz \\ y \leftarrow z \\ z \leftarrow fyy \\ w \leftarrow fzfyy \end{pmatrix}$$

is also a unifier, and $\sigma'(\mathsf{L}_i) = rffyyzfzfyy$ for $i = 1, 2$ We can obtain $\sigma(\mathsf{L}_i)$ from this by applying the substitution $z \leftarrow y$. So we say σ' is a *more general unifier* of $\mathsf{L}_1, \mathsf{L}_2$ than σ. However we cannot obtain $\sigma'(\mathsf{L}_i)$ from $\sigma(\mathsf{L}_i)$ by substitution. Consequently, we say σ' is a *strictly* more general unifier of $\mathsf{L}_1, \mathsf{L}_2$ than σ.

And even this choice is rather special, since the substitution

$$\sigma'' = \begin{pmatrix} x \leftarrow fzy \\ y \leftarrow y \\ z \leftarrow z \\ w \leftarrow fyz \end{pmatrix}$$

is a unifier, and $\sigma''(\mathsf{L}_i) = rfzyfyz$. We can obtain $\sigma'(\mathsf{L}_i)$ from this by applying the substitution

$$\tau = \begin{pmatrix} y \leftarrow z \\ z \leftarrow fyy \end{pmatrix},$$

but we cannot obtain $\sigma''(\mathsf{L}_i)$ from $\sigma'(\mathsf{L}_i)$ by substitution. So σ'' is a strictly more general unifier of $\mathsf{L}_1, \mathsf{L}_2$ than σ'.

Exercises

Consider the following pairs of literals, where r is *binary* and f is *unary*:

$$\text{1.} \begin{cases} rxy \\ ryz \end{cases} \quad \text{2.} \begin{cases} rfxfy \\ ryfz \end{cases} \quad \text{3.} \begin{cases} rxfy \\ ryfz \end{cases} \quad \text{4.} \begin{cases} rxffy \\ rfyfz \end{cases}$$

$$\text{5.} \begin{cases} rfxy \\ ryz \end{cases} \quad \text{6.} \begin{cases} rffxy \\ rfyz \end{cases} \quad \text{7.} \begin{cases} rfxfy \\ rfyfz \end{cases} \quad \text{8.} \begin{cases} rffxfy \\ rfyfz. \end{cases}$$

4.7.1 Determine which of the preceding pairs of literals are unified by the following substitutions:

$$\text{a.} \begin{pmatrix} x \leftarrow y \\ y \leftarrow z \end{pmatrix} \text{b.} \begin{pmatrix} y \leftarrow fx \\ z \leftarrow fx \end{pmatrix} \text{c.} \begin{pmatrix} y \leftarrow x \\ z \leftarrow x \end{pmatrix} \text{d.} \begin{pmatrix} x \leftarrow fy \\ z \leftarrow fy \end{pmatrix}.$$

4.7.2 The Unification Algorithm for Pairs of Literals

In Example 4.7.2 we cannot find a more general unifier of $\mathsf{L}_1, \mathsf{L}_2$ than σ'', and this is not an isolated case. If the literals $\mathsf{L}_1, \mathsf{L}_2$ have a unifier, then they have a *most general unifier*, i.e., a unifier μ such that for any unifier σ we have $\sigma = \tau\mu$ for some substitution τ. Thus every unifier is a special case of a most general unifier.

The algorithm that we presented in Sec. 3.12 to unify terms works in essentially the same manner (with the same proof) to find most general unifiers for literals.

We define the *critical subterms* s', t' of two strings of terms $s_1 \cdots s_n$ and $t_1 \cdots t_n$ exactly as for the case with just two terms s_1 and t_1 (see

Chapter 3, p. 192). We will use the critical subterm condition (CSC) as defined in Definition 3.12.7 on p. 192.

Unification algorithm for $\mathsf{L}_1, \mathsf{L}_2$:

If $\mathsf{L}_1, \mathsf{L}_2$ are not in the form

$$\begin{pmatrix} \mathsf{L}_1 & = & rs_1 \cdots s_n \\ \mathsf{L}_2 & = & rt_1 \cdots t_n \end{pmatrix} \quad \text{or} \quad \begin{pmatrix} \mathsf{L}_1 & = & \neg rs_1 \cdots s_n \\ \mathsf{L}_2 & = & \neg rt_1 \cdots t_n \end{pmatrix},$$

  then return NOT UNIFIABLE

else

  let μ be the identity substitution
  let $S = s_1 \cdots s_n$
  let $T = t_1 \cdots t_n$
  WHILE $S \neq T$
    find the critical subterms s', t'
    if the CSC fails
      then return NOT UNIFIABLE
    else with $\{s', t'\} = \{x, p\}$
      apply $(x \leftarrow p)$ to both S and T, i.e.,
        $S \leftarrow (x \leftarrow p)S$ and $T \leftarrow (x \leftarrow p)T$
      apply $(x \leftarrow p)$ to μ, i.e.,
        $\mu \leftarrow (x \leftarrow p)\mu$
  ENDWHILE
  return (μ)

In the following, r is a *binary* relation symbol, f is a *unary* function symbol, and g is a *binary* function symbol.

Example 4.7.3 Consider $\neg rfxx$ and $rfyy$:

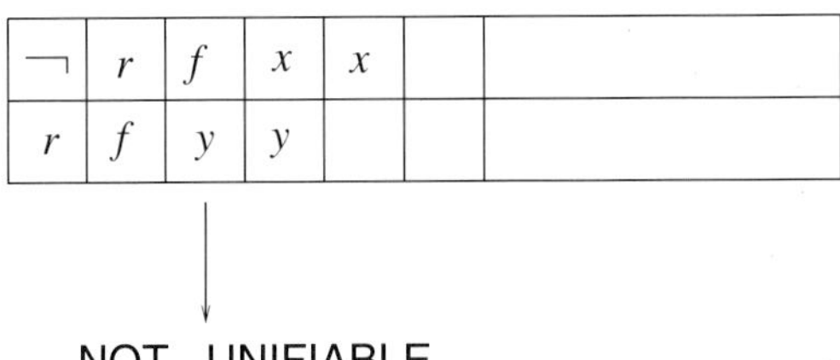

Figure 4.7.4 Applying the unification algorithm to the literals $\neg rfxx$ and $rfyy$

One literal is positive and the other is negative, so they cannot be unified.

Example 4.7.5 Consider $rfxy$ and $sufv$:

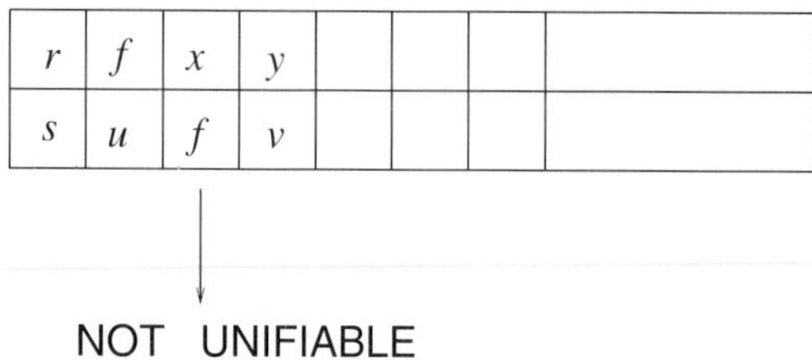

Figure 4.7.6 Applying the unification algorithm to the literals $rfxy$ and $sufv$

The literals have different relation symbols, so they cannot be unified.

Example 4.7.7 Consider $rfgxxy$ and $rxfy$:

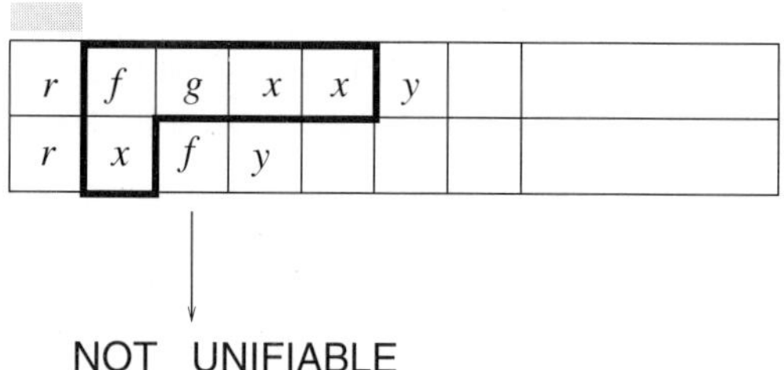

Figure 4.7.8 Applying the unification algorithm to the literals $rfgxxy$ and $rxfy$

The pair of critical subterms x and $fgxx$ violates the CSC. Thus these literals are not unifiable.

Example 4.7.9 Consider $rgyygyy$ and rxx:

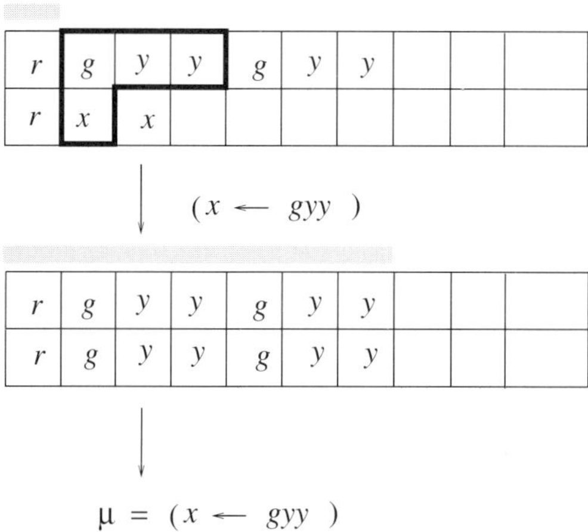

Figure 4.7.10 Applying the unification algorithm to the literals $rgyygyy$ and rxx

The literals are unifiable, and the most general unifier is $\mu = (x \leftarrow gyy)$.

Example 4.7.11 Consider $rxfy$ and $rgzyw$:

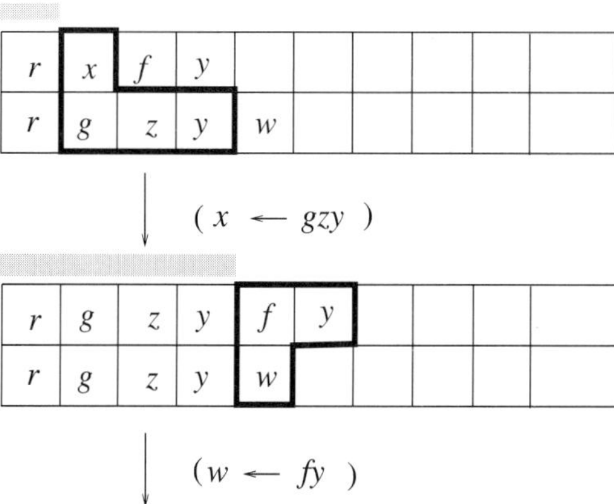

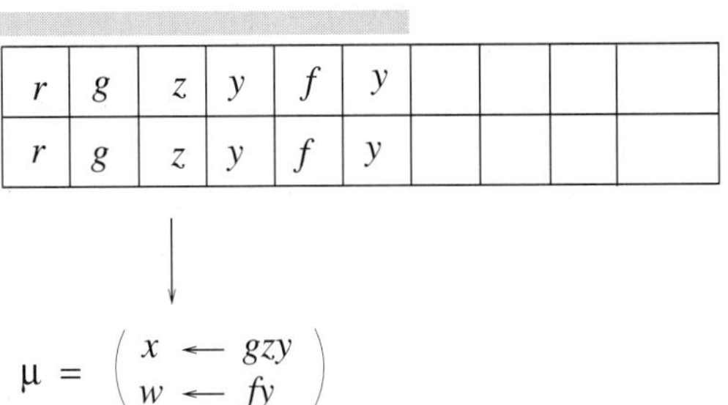

Figure 4.7.12 Applying the unification algorithm to the literals $rxfy$ and $rgzyw$

The literals are unifiable, and the most general unifier is

$$\mu = \begin{pmatrix} x \leftarrow gzy \\ w \leftarrow fy \end{pmatrix}.$$

Exercises

In the following, r_k denotes a k–ary relation symbol, and f_k a k–ary function symbol.

4.7.2 Determine if the following pairs of literals are unifiable, and if so, find the most general unifier.

- **a.** r_1x and r_2xy
- **b.** r_2xy and r_2uv
- **c.** r_2f_1xy and r_2f_1yx
- **d.** r_2f_2xyz and r_2uf_2vw
- **e.** $r_2f_1xf_2yz$ and r_2uf_2vw
- **f.** $r_3f_1f_1wf_1xf_1f_1u$ and $r_3f_1zf_1f_1wf_1y$

4.7.3 Most General Unifiers of Finitely Many Literals

Definition 4.7.13 A *unifier* for $\mathsf{L}_1, \ldots, \mathsf{L}_m$, a finite collection of literals, is a substitution σ such that $\sigma\mathsf{L}_1 = \cdots = \sigma\mathsf{L}_m$. If such a unifier exists, we say $\mathsf{L}_1, \ldots, \mathsf{L}_m$ are *unifiable*.

The unification algorithm for finitely many literals $\mathsf{L}_1, \ldots, \mathsf{L}_m$ works essentially as for finitely many terms, as given in Sec. 3.12.6.

Theorem 4.7.14 Let $\mathsf{L}_1, \ldots, \mathsf{L}_m$ be literals. If they are unifiable then they have a most general unifier, i.e., a unifier μ such that for any unifier σ there is a substitution τ such that $\sigma = \tau\mu$.

If finitely many literals $\mathsf{L}_1, \ldots, \mathsf{L}_m$ are unifiable, they must all be positive or all be negative, and all must involve the same relation symbol r, and thus all have the same number of terms following the occurrence of the relation symbol. Thus in the positive case we would have

$$\begin{aligned} \mathsf{L}_1 &= rt_{11} \cdots t_{1n} \\ &\vdots \\ \mathsf{L}_m &= rt_{m1} \cdots t_{mn}, \end{aligned} \tag{50}$$

and in this case a substitution σ would be a unifier of $\mathsf{L}_1, \ldots, \mathsf{L}_m$ iff

$$\sigma t_{1j} = \cdots = \sigma t_{mj} \qquad \text{for } 1 \leq j \leq n.$$

(A similar discussion applies to the case that the literals are all negative.) We will give two versions of the unification algorithm—the first one concentrates on the rows of the following grid, and the second one on the columns:

t_{11}	$\ldots$	t_{1n}
$\vdots$		$\vdots$
t_{m1}	$\ldots$	t_{mn}

Figure 4.7.15 Rows to be unified

```
First version of the unification algorithm for L_1, ..., L_m:

If L_1, ..., L_m are not in the form
  L_1 = r t_11 ··· t_1n              L_1 = ¬ r t_11 ··· t_1n
   ⋮                  or              ⋮
  L_m = r t_m1 ··· t_mn,             L_m = ¬ r t_m1 ··· t_mn,
    then return NOT UNIFIABLE
else
 let μ be the identity substitution
 FOR i = 1 to m − 1
     if μL_i and μL_{i+1} are not unifiable
         then return NOT UNIFIABLE
     else let μ_0 be the most general unifier of μL_i and μL_{i+1}
         apply μ_0 to μ, i.e.,
             μ ← μ_0 μ
 ENDFOR
 return (μ)
```

```
Second version of the unification algorithm for L_1, ..., L_m:

If L_1, ..., L_n are not in the form
  L_1 = r t_11 ··· t_1n              L_1 = ¬ r t_11 ··· t_1n
   ⋮                  or              ⋮
  L_m = r t_m1 ··· t_mn,             L_m = ¬ r t_m1 ··· t_mn,
    then return NOT UNIFIABLE
else
 let μ be the identity substitution
 FOR j = 1 to n
     if μt_1j, ..., μt_mj are not unifiable
         then return NOT UNIFIABLE
     else let μ_0 be the most general unifier of μt_1j, ..., μt_mj
         apply μ_0 to μ, i.e.,
             μ ← μ_0 μ
 ENDFOR
 return (μ)
```

There are variations of the preceding algorithms that we can use, namely, we can handle all the literals simultaneously and, at any step, choose critical subterms from any two distinct literals, as illustrated in the following example.

Example 4.7.16 Apply the unification algorithm to the following three literals, where r is binary and f is unary:

$$rffxfy$$
$$rfyfffz$$
$$rfffzffx$$

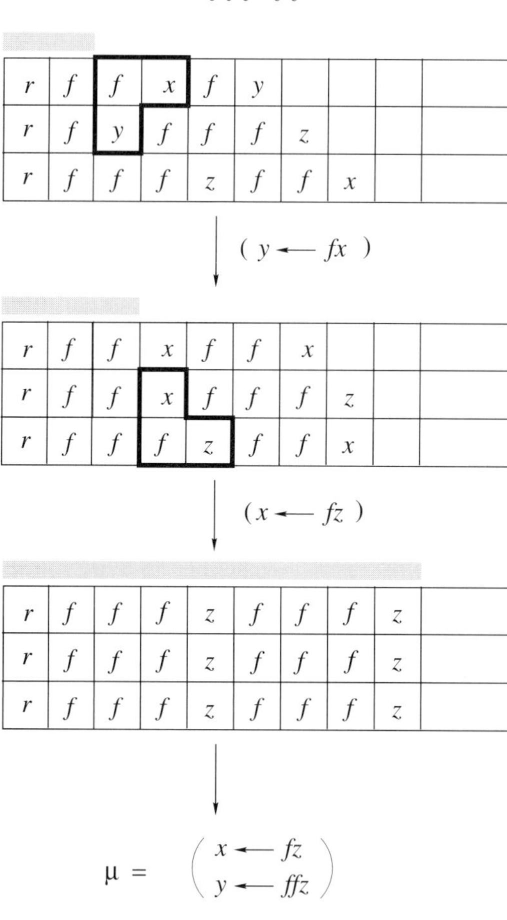

Figure 4.7.17 The unification algorithm for the literals $rffxfy$, $rfyfffz$, and $rfffzffx$

Thus the three literals are unifiable, and the most general unifier is given by

$$\mu = \begin{pmatrix} x \leftarrow fz \\ y \leftarrow ffz \end{pmatrix}.$$

Exercises

4.7.3 In the following, r is binary and f is unary. For each of the following collections of three literals determine if they are unifiable, and if so, find the most general unifier.

a. $\neg rfxfy$	b. $rfxffy$	c. $rffxy$
$\neg rfyffz$	$rfyffz$	$rfyz$
$\neg rfzffx$	$rfzffx$	$rfzffx$
d. $\neg rffxy$	e. $rffxffy$	f. $rffxffy$
$\neg rfyz$	$ryfz$	$ryfz$
$\neg rfzfx$	$rzfx$	rzx

4.8 RESOLUTION WITH MOST GENERAL OPP–UNIFIERS

Now that we have developed the unification algorithm for literals, we return to our study of resolution. We can do resolution theorem proving using only most general opp–unifiers. This is the way to implement efficient pattern matching.

4.8.1 Most General Opp–Unifiers

Definition 4.8.1 A *most general opp–unifier* for a pair of clauses $\mathsf{C}'', \mathsf{D}''$ is an opp–unifier μ_1, μ_2 such that any other opp–unifier σ_1, σ_2 can be obtained by applying a further substitution to the pair μ_1, μ_2, i.e., there exists τ such that $(\sigma_1, \sigma_2) = (\tau\mu_1, \tau\mu_2)$.

The next lemma ties together the definition of unification for literals with the definition of opp–unification for clauses.

Lemma 4.8.2 Suppose σ_1, σ_2 is an opp–unifier of the pair of clauses C'', D'', say $\sigma_1 C'' = \{L\}$, $\sigma_2 D'' = \{\overline{L}\}$. If C'' and D'' have no variables in common, then there is a substitution σ such that $\sigma C'' = \{L\}$, $\sigma D'' = \{\overline{L}\}$.

Proof. We simply choose σ such that it equals σ_1 on the variables of C'', σ_2 on the variables of D''. □

Lemma 4.8.3 Suppose C'', D'' are opp–unifiable clauses. Then they have a most general opp–unifier.

Proof. Since the pair of clauses is opp–unifiable, the literals in each clause are unifiable. Let μ_1 be the most general unifier of the literals in C'' (such exists by Theorem 4.7.14, p. 295), and let μ_2 be the most general unifier of the literals in D''. *Without loss of generality we can assume that the variables of $\mu_1 C''$ and $\mu_2 D''$ are disjoint.* Now let μ be a most general unifier of $\mu_1 C''$ and $\mu_2 \overline{D''}$ (where $\overline{D''}$ means to replace the literals in D'' by their complements). Then $\mu\mu_1 C'' = \mu\mu_2 \overline{D''}$, so $\mu\mu_1, \mu\mu_2$ will be an opp–unifier of C'', D''.

We now claim that $\mu\mu_1, \mu\mu_2$ is a most general opp–unifier of C'', D''. To see this, suppose σ_1, σ_2 is any opp–unifier of C'', D''. Then we have

$$\begin{aligned} \sigma_1 &= \tau_1 \mu_1 \\ \sigma_2 &= \tau_2 \mu_2 \end{aligned}$$

for suitable τ_1, τ_2 (as the μ_i are most general unifiers), and thus

$$\sigma_1 C'' = \sigma_2 \overline{D''}$$

leads to

$$\tau_1 \mu_1 C'' = \tau_2 \mu_2 \overline{D''}.$$

By Lemma 4.8.2 we can assume $\tau_1 = \tau_2 = \tau^*$. As τ^* is a unifier of $\mu_1 C''$ and $\mu_2 \overline{D''}$, there must exist τ such that $\tau^* = \tau\mu$, as μ is a most general unifier. This gives

$$\begin{aligned} \sigma_1 &= \tau\mu\mu_1 \\ \sigma_2 &= \tau\mu\mu_2, \end{aligned}$$

so, indeed, $\mu\mu_1, \mu\mu_2$ is more general than σ_1, σ_2. □

4.8.2 An Opp–Unification Algorithm

Now, a simple modification of the algorithm for literals will give us an opp–unification algorithm for clauses. Let $\mathsf{C}''(x_1, \ldots, x_n)$ and $\mathsf{D}''(x_1, \ldots, x_n)$ be two clauses for which we want to determine whether or not they can be opp–unified, and if so, to find a most general opp–unifier. The first step is to rename the variables in D'' so that they are disjoint from the variables in C'', say we rename $x_1, \ldots, x_n$ to $x'_1, \ldots, x'_n$ in D''. Then any opp–unifier σ_1, σ_2 of the pair of clauses $\mathsf{C}''(x_1, \ldots, x_n), \mathsf{D}''(x'_1, \ldots, x'_n)$ can be thought of as a single substitution σ, where $\sigma = \sigma_1$ on the variables $x_1, \ldots, x_n$, and $\sigma = \sigma_2$ on $x'_1, \ldots, x'_n$.

Now, the opp–unification algorithm is as follows, where the notation $\overline{\mathsf{D}''}$ means to replace each literal in D'' by its complement.

> **An opp–unification algorithm for $\mathsf{C}'', \mathsf{D}''$:**
>
> **If** $\mathsf{C}''(x_1, \ldots, x_n) \cup \overline{\mathsf{D}''}(x'_1, \ldots, x'_n)$ is not
> a unifiable set of literals
>
> **then** return NOT OPP–UNIFIABLE
>
> **else**
>
> let μ be the most general unifier of the set of literals
> $\mathsf{C}''(x_1, \ldots, x_n) \cup \overline{\mathsf{D}''}(x'_1, \ldots, x'_n)$
>
> let $\sigma_1 = \begin{pmatrix} x_1 \leftarrow \mu x_1 \\ \vdots \\ x_n \leftarrow \mu x_n \end{pmatrix}, \quad \sigma_2 = \begin{pmatrix} x_1 \leftarrow \mu x'_1 \\ \vdots \\ x_n \leftarrow \mu x'_n \end{pmatrix}$
>
> return(σ_1, σ_2).

Example 4.8.4 For r a *binary* relation symbol and f a *unary* function symbol let

$$\begin{aligned} \mathsf{C}''(x, y, z) &= \{rfxfy,\ rzy,\ rfyfz\} \\ \mathsf{D}''(x, y, z) &= \{\neg rxfy,\ \neg rfyx\}. \end{aligned}$$

We will use u, v, w rather than x', y', z' for our change of variables in D''. Then applying the opp–unification algorithm, we need to analyze the unifiability of the literals in $\mathsf{C}''(x, y, z) \cup \overline{\mathsf{D}''}(u, v, w)$, which gives the following collection of five literals:

$$rfxfy,\ rzy,\ rfyfz,\ rufv,\ rfvu.$$

Applying the unification algorithm for literals:

r	f	x	f	y					
r	z	y							
r	f	y	f	z					
r	u	f	v						
r	f	v	u						

↓ (z ← fy)

r	f	x	f	y					
r	f	y	y						
r	f	y	f	f	y				
r	u	f	v						
r	f	v	u						

↓ (u ← fv)

r	f	x	f	y					
r	f	y	y						
r	f	y	f	f	y				
r	f	v	f	v					
r	f	v	f	v					

↓ (v ← y)

r	f	x	f	y					
r	f	y	y						
r	f	y	f	f	y				
r	f	y	f	y					
r	f	y	f	y					

↓ (x ← y)

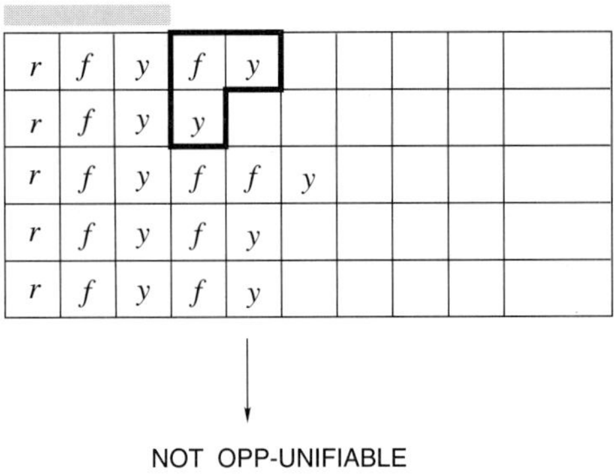

r	f	y	f	y	
r	f	y	y		
r	f	y	f	f	y
r	f	y	f	y	
r	f	y	f	y	

↓

NOT OPP-UNIFIABLE

Figure 4.8.5 The opp–unification algorithm for the clauses $\{rfxfy,\ rzy,\ rfyfz\}$, $\{\neg rxfy,\ \neg rfyx\}$

Thus the clauses C'', D'' are not opp–unifiable.

Example 4.8.6 Now let

$$\begin{aligned} \mathsf{C}''(x,y,z) &= \{rfxfy,\ rfyfz\} \\ \mathsf{D}''(x,y,z) &= \{\neg rxfy,\ \neg rfyx\}. \end{aligned}$$

We need to analyze the unifiability of the four literals in the clause $\mathsf{C}''(x,y,z) \cup \overline{\mathsf{D}''}(u,v,w)$, i.e., of the four literals:

$$rfxfy,\ rfyfz,\ rufv,\ rfvu.$$

Applying the unification algorithm for literals:

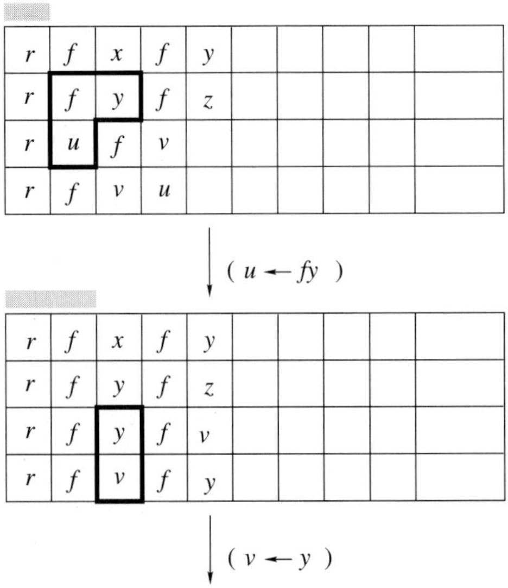

r	f	x	f	y
r	f	y	f	z
r	u	f	v	
r	f	v	u	

↓ (u ← fy)

r	f	x	f	y
r	f	y	f	z
r	f	y	f	v
r	f	v	f	y

↓ (v ← y)

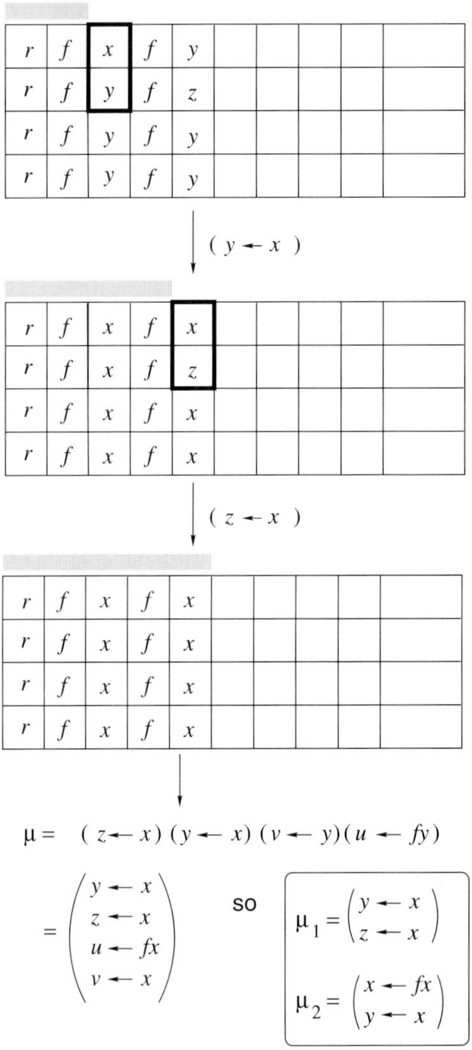

r	f	x	f	y					
r	f	y	f	z					
r	f	y	f	y					
r	f	y	f	y					

$\downarrow (y \leftarrow x)$

r	f	x	f	x					
r	f	x	f	z					
r	f	x	f	x					
r	f	x	f	x					

$\downarrow (z \leftarrow x)$

r	f	x	f	x					
r	f	x	f	x					
r	f	x	f	x					
r	f	x	f	x					

$\downarrow$

$$\mu = (z \leftarrow x)\,(y \leftarrow x)\,(v \leftarrow y)(u \leftarrow fy)$$

$$= \begin{pmatrix} y \leftarrow x \\ z \leftarrow x \\ u \leftarrow fx \\ v \leftarrow x \end{pmatrix} \qquad \text{so} \qquad \mu_1 = \begin{pmatrix} y \leftarrow x \\ z \leftarrow x \end{pmatrix} \quad \mu_2 = \begin{pmatrix} x \leftarrow fx \\ y \leftarrow x \end{pmatrix}$$

Figure 4.8.7 The opp–unification algorithm for the clauses $\{rfxfy,\ rfyfz\},\ \{\neg rxfy,\ \neg rfyx\}$

Thus the clauses C'', D'' are opp–unifiable, and the most general opp–unifier is given by

$$\mu_1 = \begin{pmatrix} y \leftarrow x \\ z \leftarrow x \end{pmatrix} \qquad \mu_2 = \begin{pmatrix} x \leftarrow fx \\ y \leftarrow x \end{pmatrix}.$$

4.8.3 Resolution and Most General Opp–Unifiers

Now, if we have a derivation of the empty clause from a set of clauses, we can go through the derivation, starting at the beginning, and successively replace each application of resolution with a resolution step using a most general opp–unifier. This procedure is detailed in the following lemma.

Lemma 4.8.8 [LIFTING LEMMA] Suppose $\mathsf{C}_1, \ldots, \mathsf{C}_n$ is a resolution derivation from a set $\mathcal{S}$ of clauses. Then there is a resolution derivation $\mathsf{D}_1, \ldots, \mathsf{D}_n$ from $\mathcal{S}$ using only most general opp–unifiers, and there are substitutions $\nu_1, \ldots, \nu_n$ such that $\nu_i \mathsf{D}_i = \mathsf{C}_i$ for $i = 1, \ldots, n$. In particular, if C_n is the empty clause, so is D_n.

Proof. We proceed by induction on n.

Ground case: For $n = 1$ we choose $\mathsf{D}_1 = \mathsf{C}_1$, and ν_1 is the identity map. (No resolutions, and hence no opp–unifiers, are used if $n = 1$.)

Induction hypothesis: Suppose the Lifting Lemma holds for resolution derivations of length less than n.

Induction step: We are given a resolution derivation $\mathsf{C}_1, \ldots, \mathsf{C}_n$. We choose $\mathsf{D}_1, \ldots, \mathsf{D}_{n-1}$ and $\nu_1, \ldots, \nu_{n-1}$ as guaranteed by the induction hypothesis.

Case I: If $\mathsf{C}_n \in \mathcal{S}$, then let $\mathsf{D}_n = \mathsf{C}_n$, and let ν_n be the identity map.

Case II: If $\mathsf{C}_n \notin \mathcal{S}$, then suppose C_n is the result of a resolution step applied to $\mathsf{C}_i, \mathsf{C}_j$. Let $\mathsf{C}_i = \mathsf{C}'_i \cup \mathsf{C}''_i$, $\mathsf{C}_j = \mathsf{C}'_j \cup \mathsf{C}''_j$, and σ_i, σ_j be an opp–unifier of $(\mathsf{C}''_i, \mathsf{D}''_j)$ such that $\mathsf{C}_n = \sigma_i \mathsf{C}'_i \cup \sigma_j \mathsf{C}'_j$. Let $\mathsf{D}_i = \mathsf{D}'_i \cup \mathsf{D}''_i$ with $\nu_i \mathsf{D}'_i = \mathsf{C}'_i$ and $\nu_i \mathsf{D}''_i = \mathsf{C}''_i$; and let $\mathsf{D}_j = \mathsf{D}'_j \cup \mathsf{D}''_j$ with $\nu_j \mathsf{D}'_j = \mathsf{C}'_j$ and $\nu_j \mathsf{D}''_j = \mathsf{C}''_j$. Then $\mathsf{D}''_i, \mathsf{D}''_j$ is opp–unifiable, namely, by $\sigma_i \nu_i, \sigma_j \nu_j$. Let μ_i, μ_j be a most general opp–unifier of $\mathsf{D}''_i, \mathsf{D}''_j$. Then there is a τ such that $(\sigma_i \nu_i, \sigma_j \nu_j) = (\tau \mu_i, \tau \mu_j)$. Let $\mathsf{D}_n = \mu_i \mathsf{D}'_i \cup \mu_j \mathsf{D}'_j$, a resolution step with most general opp–unifiers applied to $\mathsf{D}_i, \mathsf{D}_j$. Then

$$\begin{aligned} \tau \mathsf{D}_n &= \tau \mu_i \mathsf{D}'_i \cup \tau \mu_j \mathsf{D}'_j \\ &= \sigma_i \nu_i \mathsf{D}'_i \cup \sigma_j \nu_j \mathsf{D}'_j \\ &= \sigma_i \mathsf{C}'_i \cup \sigma_j \mathsf{C}'_j \\ &= \mathsf{C}_n. \end{aligned}$$

So we can let $\nu_n = \tau$.

This completes the induction proof. □

Example 4.8.9 The following is a predicate logic resolution derivation of the empty clause from the clauses in Example 4.5.22 (p. 282), using most general opp–unifiers.

1.	$\{\neg rfxx\}$	given
2.	$\{\neg rxy,\ \neg rxz,\ ryz\}$	given
3.	$\{\neg rxfy,\ \neg rxy\}$	resolution 1,2
4.	$\{rfffxx\}$	given
5.	$\{\neg rfffxfx\}$	resolution 3,4
6.	$\{\neg rxy,\ rfxfy\}$	given
7.	$\{\neg rffxx\}$	resolution 5,6
8.	$\{rffxx\}$	given
9.	$\{\ \}$	resolution 7,8

4.8.4 Soundness and Completeness with Most General Opp–Unifiers

Now we can prove the version of completeness and soundness for resolution theorem proving that is the most popular, namely, the one using only most general opp–unifiers.

Theorem 4.8.10 [J. A. ROBINSON] A set $\mathcal{S}$ of clauses is not satisfiable iff there is a derivation of the empty clause by resolution using only most general opp–unifiers.

Proof. From the completeness theorem for resolution, Theorem 4.6.8 (p. 287), we know that a set $\mathcal{S}$ of clauses is not satisfiable iff there is a derivation of the empty clause by resolution. Given a derivation of the empty clause, we simply apply the Lifting Lemma to obtain a derivation using most general opp–unifiers. □

Example 4.8.11 With the clauses

$$\begin{aligned} \mathsf{C} &= \{\neg rxx,\ rfzx,\ rzy,\ rfxfy,\ rfyfz\} \\ \mathsf{D} &= \{rxffz,\ \neg rffyx,\ \neg rxfy,\ \neg rfyx\}, \end{aligned}$$

where r is *binary* and f is *unary*, determine if it is possible to do a resolution step with

$$\begin{aligned} \mathsf{C}'' &= \{rzy,\ rfxfy,\ rfyfz\} \\ \mathsf{D}'' &= \{\neg rxfy,\ \neg rfyx\}. \end{aligned}$$

This is just the $\mathsf{C}'', \mathsf{D}''$ from Example 4.8.4 (p. 300), and there we saw that $\mathsf{C}'', \mathsf{D}''$ are not opp–unifiable. Thus no such resolution step is possible, i.e.,

$$\frac{\{\overbrace{\neg rxx,\ rfzx,}^{\mathsf{C}'}\ \overbrace{rzy,\ rfxfy,\ rfyfz}^{\mathsf{C}''}\},\ \{\overbrace{rxffz,\ \neg rffyx,}^{\mathsf{D}'}\ \overbrace{\neg rxfy,\ \neg rfyx}^{\mathsf{D}''}\}}{\textit{(No resolution is possible using opp–unification of } C'', D''.)}$$

Example 4.8.12 With the same clauses C, D as in the previous example, namely,

$$\begin{aligned} \mathsf{C} &= \{\neg rxx,\ rfzx,\ rzy,\ rfxfy,\ rfyfz\} \\ \mathsf{D} &= \{rxffz,\ \neg rffyx,\ \neg rxfy,\ \neg rfyx\}, \end{aligned}$$

determine if it is possible to do a resolution step with

$$\begin{aligned} \mathsf{C}'' &= \{rfxfy,\ rfyfz\} \\ \mathsf{D}'' &= \{\neg rxfy,\ \neg rfyx\}. \end{aligned}$$

This is just the $\mathsf{C}'', \mathsf{D}''$ from Example 4.8.6 (p. 302), and there we saw that opp–unification was possible, and the most general opp–unifiers are

$$\mu_1 = \begin{pmatrix} y \leftarrow x \\ z \leftarrow x \end{pmatrix} \qquad \mu_2 = \begin{pmatrix} x \leftarrow fx \\ y \leftarrow x \end{pmatrix}.$$

Thus for

$$\begin{aligned} \mathsf{C}' &= \{\neg rxx,\ rfzx,\ rzy\} \\ \mathsf{D}' &= \{rxffz,\ \neg rffyx\}, \end{aligned}$$

we have the valid resolution

$$\frac{\mathsf{C},\ \mathsf{D}}{\mu_1\mathsf{C}' \cup \mu_2\mathsf{D}'}\ ,\ \text{i.e.,}\ \frac{\mathsf{C},\ \mathsf{D}}{\{\neg rxx,\ rfxx,\ rxx\} \cup \{rfxffz,\ \neg rffxfx\}}.$$

Thus, in summary, we have the following resolution using most general opp–unifiers:

$$\frac{\{\overbrace{\neg rxx,\ rfzx,\ rzy,}^{\mathsf{C}'}\ \overbrace{rfxfy,\ rfyfz}^{\mathsf{C}''}\},\ \{\overbrace{rxffz,\ \neg rffyx,}^{\mathsf{D}'}\ \overbrace{\neg rxfy,\ \neg rfyx}^{\mathsf{D}''}\}}{\{\underbrace{\neg rxx,\ rfxx,\ rxx,}_{\mu_1\mathsf{C}'}\ \underbrace{rfxffz,\ \neg rffxfx}_{\mu_2\mathsf{D}'}\}}$$

Exercises

4.8.1 Let $\mathcal{L} = \{f, r, 0\}$, where f, r are both unary. Give resolution derivations of the empty clause for each of the following collections of clauses, using only most general opp–unification.

a. $\{r0\},\ \{\neg rx,\ rfx\},\ \{\neg rff0\}$
b. $\{r0\},\ \{\neg rx,\ rfx\},\ \{\neg rff0,\ \neg rfff0,\ \neg rffff0\}$
c. $\{r0\},\ \{\neg rx,\ \neg ry\}$

4.8.2 Let $\mathcal{L} = \{f, r, 0\}$, where f is unary and r is binary. Give resolution derivations of the empty clause for each of the following collections of clauses, using only most general opp–unification.

a. $\{\neg rxy,\ \neg ryx\},\ \{rfxy\}$
b. $\{\neg rxy,\ \neg ryx\},\ \{rfxfy\}$
c. $\{\neg rxy,\ \neg ryx\},\ \{rfxfx\}$
d. $\{\neg rxy,\ \neg ryx\},\ \{rxfx\},\ \{rffxfx\}$
e. $\{\neg rxy,\ \neg ryz,\ rxz\},\ \{\neg rfxfffx\},\ \{rxfx\}$
f. $\{\neg rxx\},\ \{\neg rxy,\ rfxy\},\ \{rxfx\}$

4.9 ADDING EQUALITY TO THE LANGUAGE

SYMBOLS	$\approx$
	$\vee, \neg$
	relation (i.e., predicate) symbols
	function symbols
	constant symbols
	variables

When we add equality to our symbols, we add another possibility to the atomic formulas.

Definition 4.9.1 Expressions of the form $s(\vec{x}) \approx t(\vec{x})$ are also atomic formulas, where s and t are terms. Any such atomic formula or its negation will be added to our list of *literals*.

As in the case of equational logic, we define satisfaction for equality by:

Definition 4.9.2 $s(\vec{x}) \approx t(\vec{x})$ *holds at* $\vec{a}$ in $\mathbf{S}$ means $s^{\mathbf{S}}(\vec{a}) = t^{\mathbf{S}}(\vec{a})$. In this case we say simply that $s(\vec{a}) \approx t(\vec{a})$ holds in $\mathbf{S}$.

With this addition we define $\mathbf{S} \models \mathcal{S}$, and $\mathrm{Sat}(\mathcal{S})$, for $\mathcal{S}$ a set of clauses, as in Definition 4.4.3 (p. 269). However, with equality we can no longer use the Herbrand universe as a generalized truth table to test for satisfiability without making some modifications to our procedure. In the next section we will look at one way to handle the addition of equality to our symbols.

To translate an equational argument into the setting of clause logic we use the following, where $s \not\approx t$ means $\neg(s \approx t)$.

Theorem 4.9.3 An equational argument $\mathcal{S} \therefore s \approx t$ (i.e., $\mathcal{S} \models s \approx t$) is valid iff

$$\neg\,\mathrm{Sat}(\mathcal{S} \cup \{s(\vec{c}\,) \not\approx t(\vec{c}\,)\}),$$

where $\vec{c}$ is a sequence of constant symbols that do not appear in the original argument.

Proof. This follows from observing that the argument $\mathcal{S} \therefore s \approx t$ is valid iff for every algebra $\mathbf{A}$ satisfying $\mathcal{S}$ we have $\mathbf{A}$ satisfying $s \approx t$, and the latter condition means that for every $\vec{a}$ from A we have $s(\vec{a}\,) \approx t(\vec{a}\,)$ holding in $\mathbf{A}$. But this is equivalent to saying that we cannot find a model of $\mathcal{S} \cup \{s(\vec{c}\,) \not\approx t(\vec{c}\,)\}$. □

Example 4.9.4 Translating the equational argument

$$x \cdot y \approx x \qquad \therefore x \cdot (y \cdot z) \approx (x \cdot y) \cdot z$$

into clause logic gives the equivalent assertion

$$\neg\,\mathrm{Sat}(\{x \cdot y \approx x\},\ \{a \cdot (b \cdot c) \not\approx (a \cdot b) \cdot c\}).$$

4.10 REDUCTION TO PROPOSITIONAL LOGIC

We will extend the scope of the reduction in Sec. 4.5 by axiomatizing equality and then treating it like any other binary relation symbol. So let $\mathcal{S}$ be a set of clauses in the language $\mathcal{L} = \mathcal{R} \cup \mathcal{F} \cup \mathcal{C}$ with equality. Let $\equiv$ be a new binary relation symbol. First, we give axioms for $\equiv$ so it behaves like equality. It will be convenient to write $s \not\equiv t$ instead of $\neg(s \equiv t)$.

4.10.1 Axiomatizing Equality

Definition 4.10.1 Let $\mathrm{Ax}_{\equiv}$ be the set of clauses given by

$\{x \equiv x\}$	
$\{x \not\equiv y,\ y \equiv x\}$	
$\{x \not\equiv y,\ y \not\equiv z,\ x \equiv z\}$	
$\{x_1 \not\equiv y_1, \ldots, x_n \not\equiv y_n,\ f\vec{x} \equiv f\vec{y}\,\}$	for f an n–ary function symbol in $\mathcal{F}$
$\{x_1 \not\equiv y_1, \ldots, x_n \not\equiv y_n,\ \neg\, r\vec{x},\ r\vec{y}\,\}$	for r an n–ary relation symbol in $\mathcal{R}$.

Remark 4.10.2 Horn clauses are defined in the predicate setting as in the propositional setting, namely, a clause is a *Horn clause* if there is at most one positive literal in the clause. Thus we see that the clauses in $\mathrm{Ax}_{\equiv}$ are Horn clauses.

Next, we replace equality ($\approx$) with our new binary relation ($\equiv$).

Definition 4.10.3

a. Given a clause C define the clause $\mathsf{C}_{\equiv}$ to be the result of replacing the occurrences of $\approx$ in C with $\equiv$.

b. Given a set $\mathcal{S}$ of clauses define the set $\mathcal{S}_{\equiv}$ of clauses to be the set of $\mathsf{C}_{\equiv}$ for $\mathsf{C} \in \mathcal{S}$.

Example 4.10.4 For the language $\mathcal{L} = \{f, r\}$, where f is *unary* and r is *binary*, we can formulate $\mathrm{Ax}_{\equiv}$ as the five clauses

$$\begin{array}{l} \{x \equiv x\} \\ \{x \not\equiv y,\ y \equiv x\} \\ \{x \not\equiv y,\ y \not\equiv z,\ x \equiv z\} \\ \{x \not\equiv y,\ fx \equiv fy\} \\ \{x_1 \not\equiv y_1,\ x_2 \not\equiv y_2,\ \neg r x_1 x_2,\ r y_1 y_2\}. \end{array}$$

4.10.2 The Reduction

Now we are ready for the reduction—we assume $\mathcal{C} \neq \emptyset$ (if not, we simply add a constant symbol to the language).

Theorem 4.10.5 Given a set $\mathcal{S}$ of clauses we have

$\mathrm{Sat}(\mathcal{S})$	iff	$\mathrm{Sat}(\mathcal{S}_{\equiv} \cup \mathrm{Ax}_{\equiv})$
	iff	$p\text{–}\mathrm{Sat}\left((\mathcal{S}_{\equiv} \cup \mathrm{Ax}_{\equiv})/\mathrm{T}_{\mathcal{C}}\right)$
	iff	$p\text{–}\mathrm{Sat}(\mathcal{G})$ for all finite $\mathcal{G} \subseteq (\mathcal{S}_{\equiv} \cup \mathrm{Ax}_{\equiv})/\mathrm{T}_{\mathcal{C}}$.

Proof. We need only to establish the first "iff" since the remaining parts are covered by Theorem 4.5.19 (p. 281).

If $\mathrm{Sat}(\mathcal{S})$, say $\mathbf{S} \models \mathcal{S}$, then let $\mathbf{S}'$ be the structure $\mathbf{S}$ augmented with a binary relation $\equiv$ that is precisely the identity relation. Then clearly, $\mathbf{S}' \models \mathcal{S}_{\equiv} \cup \mathrm{Ax}_{\equiv}$, and thus we have $\mathrm{Sat}(\mathcal{S}_{\equiv} \cup \mathrm{Ax}_{\equiv})$.

Conversely, suppose that $\mathrm{Sat}(\mathcal{S}_{\equiv} \cup \mathrm{Ax}_{\equiv})$ holds, say $\mathbf{S} \models \mathcal{S}_{\equiv} \cup \mathrm{Ax}_{\equiv}$.

From the truth of the clauses $\mathrm{Ax}_\equiv$ we see that $\equiv$ is actually an equivalence relation on S with the following property: if $a_i \equiv b_i$, for all i, then $f(a_1, \ldots, a_n) \equiv f(b_1, \ldots, b_n)$. Now we will employ the same construction that we used to prove Birkhoff's completeness theorem.

Let S' be the set of equivalence classes $[a]$ on S using $\equiv$. We define the structure $\mathbf{S}'$ on S' as follows:

- $f([s_1], \ldots, [s_n]) = [f^{\mathbf{S}}(s_1, \ldots, s_n)]$ for f an n–ary function symbol in $\mathcal{F}$.
- $r([s_1], \ldots, [s_n])$ holds iff $r^{\mathbf{S}}(s_1, \ldots, s_n)$ holds for r an n–ary relation symbol $\mathcal{R}$.
- $c = [c^{\mathbf{S}}]$ for c a constant symbol in $\mathcal{C}$.

The fact that $\mathbf{S} \models \mathrm{Ax}_\equiv$ guarantees that the preceding definitions are consistent, and we have $[s_1] = [s_2]$ iff $s_1 \equiv s_2$. Consequently, the satisfiability of the clauses $\mathcal{S}_\equiv \cup \mathrm{Ax}_\equiv$ by $\mathbf{S}$ gives the satisfiability of the clauses $\mathcal{S}$ by $\mathbf{S}'$ (using essentially the argument in the last paragraph of the proof of Theorem 4.5.19, p. 281). Thus we have $\mathrm{Sat}(\mathcal{S})$. $\square$

Applying Corollary 4.5.21 (p. 281) gives the following.

Corollary 4.10.6 A set $\mathcal{S}$ of clauses is not satisfiable iff there is a finite set of ground instances of $\mathcal{S}_\equiv \cup \mathrm{Ax}_\equiv$ that is not p–satisfiable.

Example 4.10.7 To show $x \cdot y \approx x \models x \cdot (y \cdot z) \approx (x \cdot y) \cdot z$ using clause logic we first translate the argument into the nonsatisfiability of clauses as in Example 4.9.4 (p. 308), namely, we want to show

$$\neg\,\mathrm{Sat}(\{x \cdot y \approx x\},\ \{a \cdot (b \cdot c) \not\approx (a \cdot b) \cdot c\}).$$

Applying Theorem 4.10.5 we want to show that the following set of clauses $\mathcal{S}_\equiv \cup \mathrm{Ax}_\equiv$ is not satisfiable:

$$\begin{array}{l}
\{x \cdot y \equiv x\} \\
\{a \cdot (b \cdot c) \not\equiv (a \cdot b) \cdot c\} \\
\{x \equiv x\} \\
\{x \not\equiv y,\ y \equiv x\} \\
\{x \not\equiv y,\ y \not\equiv z,\ x \equiv z\} \\
\{x_1 \not\equiv y_1,\ x_2 \not\equiv y_2,\ x_1 \cdot x_2 \equiv y_1 \cdot y_2\}.
\end{array}$$

By Theorem 4.10.5 it suffices to find finitely many ground instances that are not p-satisfiable. Let us look at the following ground instances (that we found by "trial and error") of the above clauses:

Clause	Ground instances
$\{x \cdot y \equiv x\}$	$\{(a \cdot b) \cdot c \equiv a \cdot b\}$
	$\{a \cdot b \equiv a\}$
	$\{a \cdot (b \cdot c) \equiv a\}$
$\{a \cdot (b \cdot c) \not\equiv (a \cdot b) \cdot c\}$	$\{a \cdot (b \cdot c) \not\equiv (a \cdot b) \cdot c\}$
$\{x \not\equiv y,\ y \equiv x\}$	$\{(a \cdot b) \cdot c \not\equiv a,\ a \equiv (a \cdot b) \cdot c\}$
$\{x \not\equiv y,\ y \not\equiv z,\ x \equiv z\}$	$\{(a \cdot b) \cdot c \not\equiv a \cdot b,\ a \cdot b \not\equiv a,\ (a \cdot b) \cdot c \equiv a\}$
	$\{a \cdot (b \cdot c) \not\equiv a,\ a \not\equiv (a \cdot b) \cdot c,\ a \cdot (b \cdot c) \equiv (a \cdot b) \cdot c\}$

(51)

If we use the following renaming of the atomic ground formulas,

$$\begin{aligned}
P:&\quad (a \cdot b) \cdot c \equiv a \cdot b\\
Q:&\quad a \cdot b \equiv a\\
R:&\quad a \cdot (b \cdot c) \equiv a\\
S:&\quad (a \cdot b) \cdot c \equiv a\\
T:&\quad a \equiv (a \cdot b) \cdot c\\
U:&\quad a \cdot (b \cdot c) \equiv (a \cdot b) \cdot c,
\end{aligned}$$

then the ground instances in (51) become

$$\begin{array}{llll}
\{P\} & \{Q\} & \{R\} & \{\neg U\}\\
\{\neg S,\ T\} & & & \\
\{\neg P,\ \neg Q,\ S\} & & & \\
\{\neg R,\ \neg T,\ U\}. & & &
\end{array}$$

This collection of propositional clauses is easily seen to be unsatisfiable. This means that the equational argument at the beginning of this example is a valid argument.

Exercises

4.10.1 Convert the following equational arguments into assertions about the nonsatisfiability of a collection of clauses using $\equiv$, and then use resolution on ground instances to prove the arguments are valid.

a. $x \cdot y \approx y \qquad \therefore x \cdot (y \cdot z) \approx (x \cdot y) \cdot z.$

b. $x \cdot y \approx x \qquad \therefore x \cdot (y \cdot z) \approx x \cdot z.$

c. $x \cdot y \approx u \cdot v \qquad \therefore x \cdot (y \cdot z) \approx (x \cdot y) \cdot z.$

d. $ffffx \approx x, fffx \approx fx \qquad \therefore ffx \approx x.$

e. $fx \approx fy, ffx \approx x \qquad \therefore x \approx y.$

4.10.3 Compactness for Clause Logic with Equality

Theorem 4.10.8 A set of clauses is satisfiable iff every finite subset is satisfiable.

Proof. One direction is trivial, so let us assume the set $\mathcal{S}$ of clauses is not satisfiable. By Corollary 4.10.6 there is a finite set $\mathcal{G}$ of ground instances of $\mathcal{S}_{\equiv} \cup \mathrm{Ax}_{\equiv}$ that is not p–satisfiable. Choose a finite $\mathcal{S}_0 \subseteq \mathcal{S}$ such that $\mathcal{G} \subseteq ((\mathcal{S}_0)_{\equiv} \cup \mathrm{Ax}_{\equiv})/\mathbf{T}_{\mathcal{C}}$. Then $((\mathcal{S}_0)_{\equiv} \cup \mathrm{Ax}_{\equiv})/\mathbf{T}_{\mathcal{C}}$ is not p–satisfiable, so by Corollary 4.10.6 $\mathcal{S}_0$ is not satisfiable. □

Example 4.10.9 A simple, and somewhat typical, application of compactness is the following. Let $\mathcal{S}$ be a set of clauses. Suppose $\mathcal{S}$ has arbitrarily large finite models, i.e., for every integer n there is a model $\mathbf{S}_n$ of $\mathcal{S}$ with S_n having at least n elements. Then $\mathcal{S}$ has an infinite model.

To see this, simply consider the set of clauses

$$\mathcal{S}' = \mathcal{S} \cup \{c_m \not\approx c_n : m, n \geq 1, m \neq n\},$$

where the constant symbols c_n do not appear in $\mathcal{S}$. Then, using the $\mathbf{S}_n$ mentioned, we see that any finite subset of $\mathcal{S}'$ has a model, so by compactness $\mathcal{S}'$ also has a model, say $\mathbf{S}'$. But in a model of $\mathcal{S}'$ distinct constant symbols c_n must be interpreted differently. Thus $\mathbf{S}'$ is infinite. Let $\mathbf{S}$ be the structure obtained by restricting the interpretation I' of $\mathbf{S}'$ to the original language. Since $\mathbf{S}' \models \mathcal{S}'$, it follows that $\mathbf{S} \models \mathcal{S}$.

4.10.4 Soundness and Completeness

Next, we have the following version of the soundness and completeness results in Theorems 4.6.8 (p. 287) and 4.8.10 (p. 305).

Theorem 4.10.10 A set $\mathcal{S}$ of clauses is not satisfiable iff there is a derivation of the empty clause by resolution [using only most general opp–unifiers] from $\mathcal{S}_{\equiv} \cup \mathrm{Ax}_{\equiv}$.

Example 4.10.11 Let us apply resolution to show that the equational argument

$$x \cdot y \approx x \qquad \therefore x \cdot (y \cdot z) \approx (x \cdot y) \cdot z$$

is valid. The conversion into clause logic in Example 4.10.7 (p. 310) gave the following set of clauses for $\mathcal{S}_{\equiv} \cup \mathrm{Ax}_{\equiv}$:

$$\begin{array}{l} \{x \cdot y \equiv x\} \\ \{a \cdot (b \cdot c) \not\equiv (a \cdot b) \cdot c\} \\ \{x \equiv x\} \\ \{x \not\equiv y,\ y \equiv x\} \\ \{x \not\equiv y,\ y \not\equiv z,\ x \equiv z\} \\ \{x_1 \not\equiv y_1,\ x_2 \not\equiv y_2,\ x_1 \cdot x_2 \equiv y_1 \cdot y_2\}. \end{array}$$

We would like to show that they are not satisfiable. The following resolution proof uses most general opp–unifiers:

1.	$\{x \cdot y \equiv x\}$	given
2.	$\{a \cdot (b \cdot c) \not\equiv (a \cdot b) \cdot c\}$	given
3.	$\{x \not\equiv y,\ y \equiv x\}$	given
4.	$\{x \not\equiv y,\ y \not\equiv z,\ x \equiv z\}$	given
5.	$\{a \cdot (b \cdot c) \not\equiv y,\ y \not\equiv (a \cdot b) \cdot c\}$	$2, 4$
6.	$\{a \not\equiv (a \cdot b) \cdot c\}$	$1, 5$
7.	$\{(a \cdot b) \cdot c \not\equiv a\}$	$3, 6$
8.	$\{(a \cdot b) \cdot c \not\equiv y,\ y \not\equiv a\}$	$4, 7$
9.	$\{a \cdot b \not\equiv a\}$	$1, 8$
10.	$\{\ \}$	$1, 9.$

Exercises

4.10.2 Convert the following equational arguments into assertions about the nonsatisfiability of a collection of clauses using $\equiv$, and then use resolution on clauses with most general opp–unification to prove the arguments are valid.

a. $x \cdot y \approx y \qquad \therefore x \cdot (y \cdot z) \approx (x \cdot y) \cdot z$
b. $x \cdot y \approx x \qquad \therefore x \cdot (y \cdot z) \approx x \cdot z$
c. $x \cdot y \approx u \cdot v \qquad \therefore x \cdot (y \cdot z) \approx (x \cdot y) \cdot z$
d. $ffffx \approx x, fffx \approx fx \qquad \therefore ffx \approx x$
e. $fx \approx fy, ffx \approx x \qquad \therefore x \approx y$

4.11 HISTORICAL REMARKS

The reduction to ground statements was one of the first techniques tried by the early (1950s) theorem provers. One would set out some way to enumerate the ground statements and, starting with the first ground statement, keep adding new ones until one had an unsatisfiable set (or one gave up). By Lemma 4.5.13 one could use the techniques of propositional logic to determine the satisfiability of such finite sets.

A 1960 paper by Davis and Putnam introduced the focus on *clauses.* The same year Prawitz advocated the use of *unification*, an idea he and Kanger discovered in the late 1950s, without knowing that Herbrand had worked with unification in 1930.

J. A. Robinson developed the clause logic using resolution with most general unifiers in 1963. (Putnam was one of his thesis advisors, but Robinson did not learn of Putnam's work on automated theorem proving until after he had finished his graduate work.) His paper on resolution with most general unifiers was submitted for publication in 1963 and evidently spent more than a year resting on a referee's desk. It finally appeared in print in 1965. Knuth discovered the unification algorithm in his work with Bendix on term rewriting systems. He presented his results at a conference in 1967. The Knuth–Bendix paper was published in 1970.

Larry Wos and George Robinson, both of Argonne National Laboratory, introduced *paramodulation*

$$\frac{\mathsf{C} \cup \{s \approx t\},\ \mathsf{D}(\cdots s' \cdots)}{\sigma[\mathsf{C} \cup \mathsf{D}(\cdots t \cdots)]}, \text{ where } \sigma s = \sigma s',$$

around 1970 for clause logic with equality. Their completeness result is

> A set $\mathcal{S}$ of clauses is not satisfiable iff one can derive the empty clause from $\mathcal{S}$ using resolution and paramodulation (using most general unifiers).

William McCune implemented these rules in his well–known automated theorem prover OTTER, which we mentioned, with important literature and Web site references, on p. 259.

When working with equality one cannot use the Herbrand universe to test for satisfiability of a set of clauses, as we noted. One does have a test algebra, at least in the case that one is working with Horn clauses, from the theoretical point of view. Given a set $\mathcal{S}$ of Horn clauses, one can take the equivalence relation $\sim$ to be the set of all pairs (s, t) of ground terms such that $\mathcal{S} \models s \approx t$ and define an algebra $\mathbf{A}$ on the equivalence classes of terms, essentially as in the proof of Birkhoff's theorem. This is called the *initial* algebra, or the *free* algebra generated by the constants.

Part II

Logic with Quantifiers

Chapter 5

First–Order Logic: Introduction, and Fundamental Results on Semantics

SYMBOLS	connectives ($\vee, \wedge, \rightarrow, \leftrightarrow, \neg$)
	quantifiers ($\forall, \exists$)
	function symbols
	equality ($\approx$)
	relation symbols
	constant symbols
	variables

Now, we turn to the most complicated logic we will treat in this text, namely, first–order logic. First–order logic is the simplest logic that allows us to easily and naturally express a truly wide range of important concepts and assertions in mathematics. As we shall see in Sec. 5.13, any sentence F in first–order logic can easily be translated into a finite set $\mathcal{C}$ of clauses such that F is true iff $\mathcal{C}$ is not satisfiable. Thus clause logic is just as expressive as first–order logic. Still, in order to find clauses for a mathematical assertion it is considered more natural to first translate the assertion into a first–order sentence. So for computer scientists, the first–order logic is often a convenient intermediate stage in the translation of a mathematical problem into the clause logic. We will see examples of how this is done in Sec. 5.13.

The new ingredient in first–order logic is the use of the quantifiers $\forall$ and $\exists$. After a brief look at the syntax, our initial treatment of first–order logic will be rather informal, using the natural numbers and graphs to illustrate the basic ideas. The immediate goal is to gain some experience in expressing mathematical concepts and assertions in first–order logic. Then, starting with Sec. 5.7, we develop the general results, building up to the use of skolemization to translate mathematical problems into clause logic, which we will use to give a simple proof of the compactness theorem for first–order logic.

5.1 THE SYNTAX OF FIRST–ORDER LOGIC

First–order languages were introduced at the beginning of Chapters 3 and 4. We need only to add formulas and sentences. These concepts are defined in this section.

Definition 5.1.1 *Formulas* for a language $\mathcal{L}$ are defined inductively as follows:

a. There are two kinds of *atomic* formulas:

$(s \approx t)$, where s and t are terms, and

$(rt_1 \cdots t_n)$, where r is an n–ary relation symbol and $t_1, \cdots, t_n$ are terms.

b. If F is a formula, then so is $(\neg \mathsf{F})$.

c. If F and G are formulas, then so are $(\mathsf{F} \vee \mathsf{G})$, $(\mathsf{F} \wedge \mathsf{G})$, $(\mathsf{F} \rightarrow \mathsf{G})$, $(\mathsf{F} \leftrightarrow \mathsf{G})$.

d. If F is a formula and x is a variable, then $(\forall x\, \mathsf{F})$ and $(\exists x\, \mathsf{F})$ are formulas.

Notational Convention 5.1.2 We will usually drop outer parentheses, and we adopt the previous precedence conventions for connectives. Quantifiers bind more strongly than any of the connectives. Thus $\forall x \forall y\, (rxy) \rightarrow (rxy)$ means $(\forall x (\forall y\, (rxy))) \rightarrow (rxy)$.

Definition 5.1.3 We define the *subformulas* of a formula F as follows:

a. The only subformula of an atomic formula F is F itself.

b. The subformulas of $\neg \mathsf{F}$ are $\neg \mathsf{F}$ itself and all the subformulas of F.

c. The subformulas of $\mathsf{F} \,\square\, \mathsf{G}$ are $\mathsf{F} \,\square\, \mathsf{G}$ itself and all the subformulas of F and all the subformulas of G (where $\square$ is any of the binary connectives $\vee$, $\wedge$, $\rightarrow$, $\leftrightarrow$).

d. The subformulas of $\forall x\, \mathsf{F}$ are $\forall x\, \mathsf{F}$ itself and all the subformulas of F.

e. The subformulas of $\exists x\, \mathsf{F}$ are $\exists x\, \mathsf{F}$ itself and all the subformulas of F.

Definition 5.1.4 An occurrence of a variable x in a formula F is *bound* if the occurrence is in a subformula of the form $\forall x\, \mathsf{G}$ or of the form $\exists x\, \mathsf{G}$ (such a subformula is called the *scope* of the quantifier that begins the subformula). Otherwise the occurrence of the variable is said to be *free*. A formula with no free occurrences of variables is called a *sentence*.

Given a bound occurrence of x in F, we say that x is *bound by* an occurrence of a quantifier Q if (i) the occurrence of Q quantifies the variable x, and (ii) subject to this constraint the scope of this occurrence of Q is the smallest in which the given occurrence of x occurs.

Example 5.1.5 It is easier to explain scope, and quantifiers that bind variables, with a diagram:

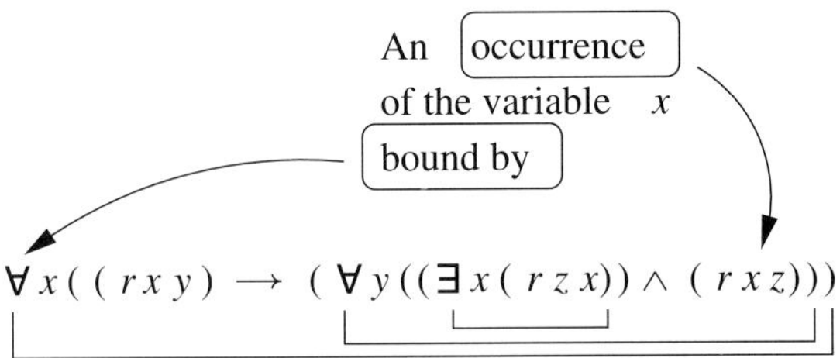

Scopes of quantifiers are underlined

Figure 5.1.6 The scope of a quantifier

The following figure indicates all the bound and free variables in the previous formula:

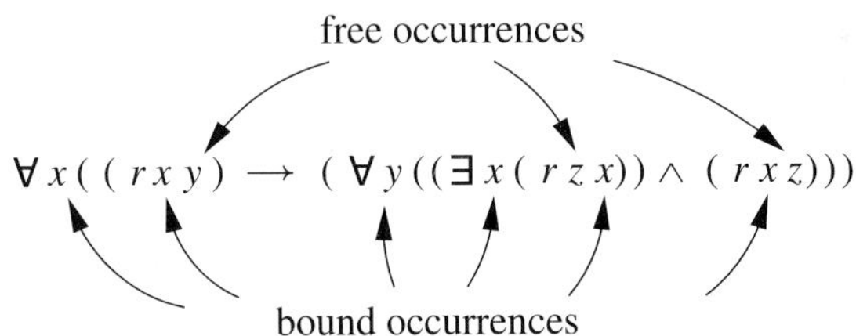

Figure 5.1.7 Free and bound occurrences of variables

Notational Convention 5.1.8 If F is a formula, we use the notation $\mathsf{F}(x_1, \ldots, x_n)$ to indicate the free variables are *among* $x_1, \ldots, x_n$.

Exercises

In the following set of problems f_n is an n–ary function symbol, and r_n is an n–ary relation symbol, for $n = 1, 2, 3$.

5.1.1 Which of the following formulas are atomic? Find all subformulas.

a. $\forall y(r_2 f_1 xy)$
b. $\neg\,(r_2 y f_3 zyx)$
c. $r_3 f_3 yzzy f_1 z$
d. $(r_2 yz) \wedge (r_3 zz f_3 zxy)$
e. $\exists z\big((r_1 z) \leftrightarrow (r_1 f_2 yz)\big)$
f. $\big(\exists x(r_3 f_2 xyx f_2 yy)\big) \leftrightarrow (r_2 zy)$
g. $\exists y\big((r_2 y f_3 f_1 x f_2 xxy) \wedge \forall x(r_1 x)\big)$
h. $(r_2 zx) \to \Big((r_2 f_1 xy) \to \big((r_1 z) \to (r_1 f_1 x)\big)\Big)$

5.1.2 Indicate the bound and free occurrences of variables in the following formulas, and the scopes of the occurrences of the quantifiers. If a variable is bound, indicate the occurrence of the quantifier that binds it.

a. $\big(\exists y(r_3 f_2 xyy f_2 yz)\big) \leftrightarrow (r_2 xy)$
b. $\forall y \exists y\big((r_2 xz) \vee \exists z(r_2 xz)\big)$
c. $\big((r_2 xx) \wedge \forall x(r_2 xz)\big) \to \exists x(r_3 zx f_1 z)$
d. $\exists x\Big(\big((r_2 zz) \to \exists x(r_1 x)\big) \wedge \exists y\big(\neg\,(r_1 z)\big)\Big)$
e. $\forall x\Bigg(\forall z\Big(\big((r_1 y) \wedge \neg\,(r_2 zy)\big) \vee \exists y(r_3 yxz)\Big) \wedge (r_2 yz)\Bigg)$
f. $\big(\exists y(r_3 xyz)\big) \to \exists x\Big(\big(\forall x(r_3 zyy)\big) \to (r_1 f_3 xyx)\Big)$
g. $\forall x\Bigg((r_3 z f_1 yy) \leftrightarrow \forall y\Big(\big(\exists z(r_3 yz f_3 xyx)\big) \vee (r_1 f_3 xx f_3 yxx)\Big)\Bigg)$

5.2 FIRST–ORDER SYNTAX FOR THE NATURAL NUMBERS

N denotes the structure $(N, +, \cdot, <, 0, 1)$, i.e., the set N of natural numbers $\{0, 1, 2, \cdots\}$ with the familiar operations $+, \cdot$, the relation $<$, and the two constants $0, 1$. The language of **N** is $\mathcal{L} = \{+, \cdot, <, 0, 1\}$. Throughout this section we will be working with the traditional *infix* notation.

In this language the *atomic* formulas look like either $(s \approx t)$ or $(s < t)$.

Example 5.2.1 The following are atomic formulas:

$$(0 < 0),\ (1 < 0),\ (x < 0),\ (x \cdot (y+z) < x \cdot z),$$
$$(x \cdot (y+1) < x \cdot x + y \cdot z).$$

Example 5.2.2 The following are formulas:

$$((x < y) \to (x + x < y + y)),$$
$$\Big(\forall x \,\big((x \cdot (y+1) < x \cdot x + y \cdot z) \to (\exists y \,(y \cdot y < x + z))\big)\Big).$$

Example 5.2.3 With the usual conventions the formulas in the previous two examples could be written as follows:

$$0 < 0,\ 1 < 0,\ x < 0,\ x \cdot (y+z) < x \cdot z,\ x \cdot (y+1) < x \cdot x + y \cdot z,$$
$$(x < y) \to (x + x < y + y)$$
$$\forall x \,\big((x \cdot (y+1) < x \cdot x + y \cdot z) \to \exists y \,(y \cdot y < x + z)\big).$$

Example 5.2.4 The subformulas of the formula $\big(\forall x \,(x \cdot (y+1) < x \cdot x + y \cdot z)\big) \to (\exists y \,(y \cdot y < x + z))$ are

$\big(\forall x \,(x \cdot (y+1) < x \cdot x + y \cdot z)\big) \to (\exists y \,(y \cdot y < x + z))$
$\forall x \,(x \cdot (y+1) < x \cdot x + y \cdot z)$
$x \cdot (y+1) < x \cdot x + y \cdot z$
$\exists y \,(y \cdot y < x + z)$
$y \cdot y < x + z.$

Exercises

5.2.1 Determine which of the following are terms in the language $\{+, \cdot, <, 0, 1\}$, using conventional infix notation. If one has a term, find all subterms.

a. $x \vee y$

b. $\exists x$

c. $\exists x \vee \exists y$

d. $x + (y + z)$

e. $x + (y - z)$

f. $x < x + 1$

g. $(x + (y \cdot y)) \cdot (y + (z + x \cdot z))$

h. $(x + (y \cdot y)) < (y + (z + (x \cdot z)))$

5.2.2 Determine which of the following are formulas. For the formulas, determine if they are atomic, if they are sentences, and find all subformulas.

a. $1 \cdot 1 < 1 + 1$

b. $1 \cdot 1 \approx 1 + 1$

c. $(1 + 1) \cdot (x + y)$

d. $(1 + 1 < 1) \approx 0$

e. $(x + x \approx (1 + 1) \cdot x)\forall x$

f. $(x \cdot (x \approx y) \cdot z) \leftrightarrow \big(1 < (x + (y \cdot z))\big)$

g. $\neg \forall z\Big(\big((z \approx x) \wedge (y < x)\big) \vee \exists y \exists z(z \approx (z \cdot y))\Big)$

5.2.3 Indicate the bound and free occurrences of variables in the following formulas, and the scopes of the occurrences of the quantifiers. If a variable is bound, indicate the occurrence of the quantifier that binds it.

a. $\neg \Big(\forall x(x \approx y) \leftrightarrow \exists x\big((y \approx x) \wedge (x < z)\big)\Big)$

b. $\Big(\neg \forall y\big((y < x) \rightarrow \neg (z < y)\big)\Big) \leftrightarrow \forall z(z \approx y)$

c. $\forall y\Big(\big(z < (x \cdot y)\big) \rightarrow \big((z \approx z) \wedge \forall x(y < x)\big)\Big)$

d. $\Big(\forall z\big((y \approx z) \vee \exists z(x \approx x \cdot z)\big)\Big) \rightarrow (x < z)$

e. $\Bigg(\forall x \neg \Big(\exists x(x < y) \vee \neg \big((x < y) \wedge \exists x(y < x)\big)\Big)\Bigg) \wedge (z \approx z)$

5.3 THE SEMANTICS OF FIRST–ORDER SENTENCES IN N

In this section we will look informally at how to express statements about the natural numbers as first–order sentences and how to interpret first–order sentences F in $\mathbf{N}$. We will use the traditional symbols 2,3, etc. as abbreviations for the terms $1 + 1$, $(1 + 1) + 1$, etc., as well as for the actual natural numbers 2,3, etc., and we will use the quantifiers as we have previously, namely, $\forall x$ is read "for all x," and $\exists x$ is read "there exists an x."

Example 5.3.1 Now, let us consider the following first–order sentences:

a. $3 < 5$

b. $2 + 2 < 3$

c. $\forall x \exists y \, (x < y)$

d. $\exists y \forall x \, (x < y)$

e. $\forall x \exists y \, (y < x)$

f. $\forall x \, \big((0 < x) \to \exists y \, (y \cdot y \approx x)\big)$

g. $\forall x \exists y \, \big((y \cdot y) \cdot y \approx x\big)$

h. $\forall x \forall y \, \Big((x < y) \to \exists z \, \big((x < z) \wedge (z < y)\big)\Big)$.

Let us translate these sentences into English and determine which are true in **N**.

(a) is an atomic sentence that says "three is less than five." True in **N**.

(b) is also an atomic sentence, which says "four is less than three." False in **N**.

(c) says that "for every number there is a larger number." True in **N**.

(d) says that "there is a number that is larger than every other number." False in **N**.

(e) says that "for every number there is a smaller number." False in **N**.

(f) says that "every positive number is a square." False in **N**.

(g) says that "every number is a cube." False in **N**.

(h) says that "if one number is less than another, then there is a number properly between the two." False in **N**.

Now let us practice translating English statements into first–order sentences, but first, we give a useful definition.

Definition 5.3.2 Let F_i be first–order formulas, where $1 \leq i \leq n$. We will use the shorthand notation

$$\bigwedge_{1 \leq i \leq n} \mathsf{F}_i$$

to mean the same as the notation

$$\mathsf{F}_1 \wedge \cdots \wedge \mathsf{F}_n.$$

Both stand for the first–order formula that we would obtain by writing out all the details, e.g.,

$$\bigwedge_{1\leq i\leq 4} \mathsf{F}_i$$

means

$$\mathsf{F}_1 \wedge \mathsf{F}_2 \wedge \mathsf{F}_3 \wedge \mathsf{F}_4.$$

Actually, we do need to put parentheses in to obtain a formula, but since the interpretation of $\wedge$ is associative, all legal ways of inserting parentheses lead to formulas that express the same thing. It is only when we are discussing syntactic issues that we need to be cautious about inserting parentheses.

Likewise, we will use the notations

$$\bigvee_{1\leq i\leq n} \mathsf{F}_i$$

and

$$\mathsf{F}_1 \vee \cdots \vee \mathsf{F}_n.$$

However, we do not have similar notations for $\rightarrow$ and $\leftrightarrow$. (For "$\rightarrow$" the reason is simply that it is not associative—one needs to know where to put parentheses!)

Proposition 5.3.3 Given a first–order formula $\mathsf{F}(x)$ we can find first–order sentences to say

- **a.** There is at least one number such that $\mathsf{F}(x)$ is true in $\mathbf{N}$.
- **b.** There are at least two numbers such that $\mathsf{F}(x)$ is true in $\mathbf{N}$.
- **c.** There are at least n numbers (for fixed n) such that $\mathsf{F}(x)$ is true in $\mathbf{N}$.
- **d.** There are infinitely many numbers that make $\mathsf{F}(x)$ true in $\mathbf{N}$.
- **e.** There is at most one number such that $\mathsf{F}(x)$ is true in $\mathbf{N}$.
- **f.** There are at most two numbers such that $\mathsf{F}(x)$ is true in $\mathbf{N}$.
- **g.** There are at most n numbers (for fixed n) such that $\mathsf{F}(x)$ is true in $\mathbf{N}$.
- **h.** There are only finitely many numbers that make $\mathsf{F}(x)$ true in $\mathbf{N}$.

Proof. We can use the following:

- **a.** $\exists x\, \mathsf{F}(x)$
- **b.** $\exists x \exists y \left(\neg\,(x \approx y) \wedge \mathsf{F}(x) \wedge \mathsf{F}(y)\right)$

c. $\exists x_1 \cdots \exists x_n \Big(\big(\bigwedge_{1 \le i < j \le n} \neg (x_i \approx x_j) \big) \wedge \big(\bigwedge_{1 \le i \le n} \mathsf{F}(x_i) \big) \Big)$

d. $\forall x \exists y \, ((x < y) \wedge \mathsf{F}(y))$

e. $\forall x \forall y \, ((\mathsf{F}(x) \wedge \mathsf{F}(y)) \rightarrow (x \approx y))$

f. $\forall x \forall y \forall z \Big((\mathsf{F}(x) \wedge \mathsf{F}(y) \wedge \mathsf{F}(z)) \rightarrow ((x \approx y) \vee (x \approx z) \vee (y \approx z)) \Big)$

g. $\forall x_1 \cdots \forall x_{n+1} \Big(\big(\bigwedge_{1 \le i \le n+1} \mathsf{F}(x_i) \big) \rightarrow \big(\bigvee_{1 \le i < j \le n+1} (x_i \approx x_j) \big) \Big)$

h. $\exists x \forall y \, (\mathsf{F}(y) \rightarrow (y < x))$.

□

To better understand what we can express with first–order sentences we need to introduce "definable relations." Given a first–order formula $\mathsf{F}(x_1, \ldots, x_k)$ we say F is "true" at a k–tuple $(a_1, \ldots, a_k)$ of natural numbers if the expression $\mathsf{F}(a_1, \ldots, a_k)$ is a true statement about the natural numbers. (A mathematically more precise definition of this is given in Sec. 5.7.)

Example 5.3.4 Let $\mathsf{F}(x, y)$ be the formula $x < y$. Then F is true at $(1, 2)$ but false at $(2, 1)$ in **N**. Indeed, F is true at (a, b) iff a is less than b.

Example 5.3.5 Let $\mathsf{F}(x, y)$ be $\exists z \, (x \cdot z \approx y)$. The F is true at (a, b) iff a divides b, written $a|b$, e.g., F is true at $(2, 6)$ but not at $(2, 5)$. [*Note:* We allow $a|0$ for any a, including $a = 0$.]

Definition 5.3.6 For $\mathsf{F}(x_1, \ldots, x_k)$ a formula let $\mathsf{F}^{\mathbf{N}}$ be the set of k–tuples $(a_1, \ldots, a_k)$ of natural numbers for which $\mathsf{F}(a_1, \ldots, a_k)$ is true in **N**. We say $\mathsf{F}^{\mathbf{N}}$ is *the relation on N defined by* F.

Definition 5.3.7 A k–ary relation $r \subseteq N^k$ is *definable* in **N** if there is a formula $\mathsf{F}(x_1, \ldots, x_k)$ such that $r = \mathsf{F}^{\mathbf{N}}$.

Remark 5.3.8 If $k = 1$ in Definition 5.3.7, then we usually think of r as a subset of N, and we say that the *subset is definable* in **N**.

Remark 5.3.9 Not all relations on N are definable. One reason is that there are uncountably many relations and only countably many definable relations.

Example 5.3.10 The relations described by

a. x is an even number

b. x divides y

c. x is prime

d. $x \equiv y$ modulo n

are definable in **N** by

a. $\exists y\,(x \approx y+y)$
b. $\exists z\,(x \cdot z \approx y)$
c. $(1<x) \wedge \forall y\,\Big((y|x) \to ((y \approx 1) \vee (y \approx x))\Big)$
d. $\exists z\,\big((x \approx y+n\cdot z) \vee (y \approx x+n\cdot z)\big)$.

Now that we have the concept of definable relations to work with, we can use it to express more interesting statements about numbers.

Definition 5.3.11 We will adopt the following abbreviations:

a. $x|y$ for the formula $\exists z\,(x \cdot z \approx y)$
b. $prime(x)$ for the formula

$$(1<x) \wedge \forall y\,\Big((y|x) \to ((y \approx 1) \vee (y \approx x))\Big).$$

Remark 5.3.12 Note that in the definition of $prime(x)$ we have used the previous abbreviation. To properly write $prime(x)$ as a first–order formula we need to replace that abbreviation; doing so gives us

$$(1<x) \wedge \forall y\,\Big(\exists z\,(y \cdot z \approx x) \to ((y \approx 1) \vee (y \approx x))\Big)$$

to obtain $prime(x)$. Note that abbreviations are *not* a feature of first–order logic, but rather they are a tool used by people to discuss first–order logic. Why do we use abbreviations? Without them, writing out the first–order sentences that we find interesting would fill up lines with tedious, hard–to–read symbolism.

Remark 5.3.13 Thus $(u+1)|(u\cdot u+1)$ is an abbreviation for $\exists z\,((u+1)\cdot z \approx u\cdot u+1)$. This seems simple enough. Now let us write out $z|2$ to obtain $\exists z\,(z \cdot z \approx 1+1)$. Unfortunately, this last formula does not define the set of elements in N that divide 2. It is a first–order sentence that is simply false in **N**—the square root of 2 is not a natural number. We have stumbled onto one of the subtler points of first–order logic, namely, we must *be careful with substitution*.

The remedy for defining "z divides 2" is to use another formula, like $\exists w\,(x \cdot w \approx y)$, for "$x$ divides y." We obtain such a formula by simply renaming the bound variable z in the formula for $x|y$. With this formula we can correctly express "z divides 2" by $\exists w\,(z\cdot w \approx 2)$.

The danger in using abbreviations in first–order logic is that we forget the names of the bound variables in the abbreviation. Our choice of a solution to this dilemma is to add a $\star$ to the abbreviation when we want to alert the reader to the necessity for renaming the bound variables that overlap with the variables in the term to be substituted into the abbreviation. For example, $prime^\star(y+z)$ alerts the reader to the need to change the formula for $prime(x)$, say to $(1 < x) \wedge \forall v \Big((v|x) \to ((v \approx 1) \vee (v \approx x))\Big)$, so that when we substitute $y + z$ for x in the formula, no new occurrence of y or z becomes bound. Thus we could express $prime(y + z)$ by

$$(1 < x) \wedge \forall v \Big(((y+z)|^\star x) \to (((y+z) \approx 1) \vee ((y+z) \approx x))\Big).$$

Example 5.3.14 The following statements:

a. The relation "divides" is transitive.

b. There are an infinite number of primes.

c. All primes except 2 are odd.

d. There are an infinite number of pairs of primes that differ by the number 2. (*Twin Prime Conjecture*)

e. All even numbers greater than two are the sum of two primes. (*Goldbach's Conjecture*)

f. There are an infinite number of primes of the form $7n + 3$.

g. If a number greater than one is composite, then it has a prime factor no larger than the square root of the number.

can be expressed by the following first–order sentences:

a. $\forall x \forall y \forall z \Big(((x|y) \wedge (y|^\star z)) \to (x|^\star z)\Big)$

b. $\forall x \exists y \left((x < y) \wedge prime^\star(y)\right)$

c. $\forall x \Big((prime(x) \wedge \neg(x \approx 2)) \to \exists y\,(x \approx 2 \cdot y + 1)\Big)$

d. $\forall x \exists y \left((x < y) \wedge prime^\star(y) \wedge prime^\star(y+2)\right)$

e. $\forall x \Big(((2|x) \wedge (2 < x)) \to \exists y \exists z\, (prime^\star(y) \wedge prime(z) \wedge$
$(x \approx y + z))\Big)$

f. $\forall x \exists y \Big((x < y) \wedge prime^\star(y) \wedge (\exists z\,(y \approx 7 \cdot z + 3))\Big)$

g. $\forall x \Bigg((\neg prime(x) \wedge (1 < x)) \to \Big(\exists y\,(prime^\star(y) \wedge (y|x) \wedge$
$(y \cdot y < x + 1))\Big)\Bigg).$

Exercises

5.3.1 Translate the following sentences into English.

a. $\forall x\,(\exists y\,(x \approx y + y) \vee \exists y\,(x \approx y + y + 1))$

b. $\forall x\,\Big((\exists y\,(x \approx 2 \cdot y) \wedge \exists y\,(x \approx 3 \cdot y)) \rightarrow \exists y\,(x \approx 6 \cdot y)\Big)$

c. $\forall x\,\Big(\neg\,(x \approx 0) \rightarrow \forall y\,((y|x) \rightarrow ((y < x) \vee (y \approx x)))\Big)$

5.3.2 Find first–order formulas that express the following relations:

a. *gcd(x,y,z)* is to hold iff z is the greatest common divisor of x and y.

b. *lcm(x,y,z)* is to hold iff z is the least common multiple of x and y.

c. *congr(x,y,z)* is to hold iff x is congruent to y modulo z.

d. *power_of_p(x)* is to hold iff x is a power of the (fixed) prime p.

e. *prime_divisors_are_p,q(x)* is to hold iff x is of the form $p^m \cdot q^n$, where p and q are (fixed) primes.

f. *Pythagorean(x,y,z)* is to hold iff (x, y, z) is a Pythagorean triple, i.e., the three numbers are the lengths of the sides of a right triangle, with z being the length of the hypotenuse.

g. *rem(x,y,z)* is to hold iff z is the remainder of dividing x by y.

h. *of_form_[2ⁿ + 1](x)* is to hold iff x is a number of the form $2^n + 1$.

i. *sum_1_to_n(x)* is to hold iff x can be expressed as $1+2+\cdots+n$, for *some* n.

j. *sum_squares_1_to_n(x)* is to hold iff x can be expressed as $1^2 + 2^2 + \cdots + n^2$, for *some* n.

k. *sum_2_squares(x)* is to hold iff x can be written as the sum of two squares.

5.3.3 Translate the following English statements into first–order sentences. (You may use the definitions in Exercise 5.3.2 as well as the ones in Sec. 5.3.)

a. Every natural number is the sum of four squares.(*Lagrange*)

b. a plus a is twice a, for any a.

c. If a prime number p divides $a \cdot b$ and p does not divide a, then p divides b.

d. If c divides $a \cdot b$ and the greatest common divisor of c and a is one, then c divides b.

e. If a and b are distinct, then there are infinitely many numbers c for which the greatest common divisor of $a+c$ and $b+c$ is equal to one.

f. The fourth power of any number, minus nine, is not a prime.

g. For every positive integer n there is a prime number between n and $2n$. (*Bertrand's Postulate*)

h. If the greatest common divisor of two numbers is one, then some multiple of the first, minus one, is divisible by the second.

i. If d is not a perfect square, then *Pell's* equation $x^2 - d \cdot y^2 = 1$ has an infinite number of solutions (in the natural numbers).

j. If (a, b, c) is a *Pythagorean triple* (i.e., the length of the sides of some right triangle, with c being the length of the hypotenuse) and the greatest common divisor of a and b is one, then one of a and b is even.

k. There are infinitely many Pythagorean triples.

l. For any $k \neq 0$, *Bachet's* equation $y^2 = x^3 + k$ has only finitely many solutions. (*Mordell's Theorem*)

m. A prime number is the sum of two squares iff it is equal to 2 or is congruent to one modulo four.

n. There are infinitely many primes that can be written as the sum of two squares.

5.4 OTHER NUMBER SYSTEMS

Our first–order language $\mathcal{L} = \{+, \cdot, <, 0, 1\}$ can just as easily be used to study other number systems, in particular, the integers $\mathbf{Z} = (Z, +, \cdot, <, 0, 1)$, the rationals $\mathbf{Q} = (Q, +, \cdot, <, 0, 1)$, and the reals $\mathbf{R} = (R, +, \cdot, <, 0, 1)$. However, first–order sentences that are true in one can be false in another.

Example 5.4.1 Consider the following first–order sentences:

a. $\forall x \exists y\, (x < y)$

b. $\forall y \exists x\, (x < y)$

c. $\forall x \forall y\, \Big((x < y) \rightarrow \exists z\, \big((x < z) \wedge (z < y) \big) \Big)$

d. $\forall x \exists y\, \big((0 < x) \rightarrow (x \approx y \cdot y) \big)$

e. $\exists x \forall y\, (x < y)$.

Then we have

	N	**Z**	**Q**	**R**
a.	true	true	true	true
b.	false	true	true	true
c.	false	false	true	true
d.	false	false	false	true
e.	false	false	false	false

We arrive at the simple conclusion that the notion of truth depends on the structure being considered!

Exercises

5.4.1 Translate the following English statements into first–order sentences:

a. No number is the square root of 2.

b. Every cubic polynomial with leading coefficient one has a root.

c. Every nonzero element has a multiplicative inverse.

5.4.2 Find first–order formulas in the language $\mathcal{L} = \{+, \cdot, <, 0, 1\}$ that express the following relations in the indicated number system:

a. *minus(x,y)* holds in **R** iff y is equal to $-x$.

b. *inverse(x,y)* holds in **R** iff x has a multiplicative inverse y.

c. *gcd(x,y,z)* holds in **Z** iff z is the greatest common divisor of x and y.

d. *distinct(x,y,u,v)* holds in **R** iff (x, y) and (u, v) are distinct points in the plane $\mathbf{R}^2$.

e. *on_the_line(x,y,u,v,w,z)* holds in **R** iff the points (x, y), (u, v) are distinct and the point (w, z) is on the line determined by (x, y) and (u, v).

f. *gen_pos_3(x,y,u,v,w,z)* holds in **R** iff the three points (x, y), (u, v), (w, z) are in general position, i.e., they are distinct and not on a line.

g. *collinear(x,y,u,v,w,z)* holds in **R** iff (x, y), (u, v), and (w, z) are three collinear points in the plane.

h. *parallel*$(x_1, y_1, x_2, y_2, u_1, v_1, u_2, v_2)$ holds in **R** iff the two points (x_1, y_1), (x_2, y_2) are distinct, the two points (u_1, v_1), (u_2, v_2) are distinct, and the line determined by (x_1, y_1), (x_2, y_2) is parallel to the line determined by (u_1, v_1), (u_2, v_2).

i. *perp*$(x_1, y_1, x_2, y_2, u_1, v_1, u_2, v_2)$ holds in **R** iff the points (x_1, y_1), (x_2, y_2) are distinct, the points (u_1, v_1), (u_2, v_2) are distinct, and the line determined by (x_1, y_1), (x_2, y_2) is perpendicular to the line determined by (u_1, v_1), (u_2, v_2).

j. *diag*(x,y,u,v) holds in **R** iff $\begin{bmatrix} x & y \\ u & v \end{bmatrix}$ is a diagonal matrix.

k. *upper_triangular*(x,y,u,v) holds in **R** iff $\begin{bmatrix} x & y \\ u & v \end{bmatrix}$ is an upper triangular matrix.

l. *invert*(x,y,u,v) holds in **R** iff $\begin{bmatrix} x & y \\ u & v \end{bmatrix}$ is an invertible matrix.

m. *eigenval*(x,y,u,v,w) holds in **R** iff w is an eigenvalue of the matrix $\begin{bmatrix} x & y \\ u & v \end{bmatrix}$.

n. *eigenvec*(x,y,u,v,w,z) holds in **R** iff (w, z) is an eigenvector of the matrix $\begin{bmatrix} x & y \\ u & v \end{bmatrix}$.

o. *distinct_eigenval*(x,y,u,v) holds in **R** iff $\begin{bmatrix} x & y \\ u & v \end{bmatrix}$ has two distinct eigenvalues.

p. *sim_diag*(x,y,u,v) holds in **R** iff $\begin{bmatrix} x & y \\ u & v \end{bmatrix}$ is similar to a diagonal matrix.

5.4.3 Using the preceding definitions, plus those in the text, express the following using first–order sentences:

a. The greatest common divisor of two integers can be expressed as a linear combination of the two integers.

b. If three points in the plane are not collinear, then there is a fourth point that is equidistant from all three.

c. If two lines in the plane are perpendicular to the same line, then they are parallel to each other.

d. If a two–by–two matrix with real entries has two distinct real eigenvalues, then it is invertible.

5.4.4 Find English statements that are true or false of the various number systems as indicated and that can be translated into first–order sentences. Give such a translation.

	N	**Z**	**Q**	**R**	
statement_1	true	true	true	true	*example in text*
statement_2	true	true	true	false	
statement_3	true	true	false	true	
statement_4	true	true	false	false	
statement_5	true	false	true	true	
statement_6	true	false	true	false	
statement_7	true	false	false	true	
statement_8	true	false	false	false	
statement_9	false	true	true	true	*example in text*
statement_10	false	true	true	false	
statement_11	false	true	false	true	
statement_12	false	true	false	false	
statement_13	false	false	true	true	*example in text*
statement_14	false	false	true	false	
statement_15	false	false	false	true	*example in text*
statement_16	false	false	false	false	*example in text*

5.5 FIRST–ORDER SYNTAX FOR (DIRECTED) GRAPHS

We chose the natural numbers to introduce important aspects of first–order logic because of the familiarity of the reader with the structure $\mathbf{N} = (N, +, \cdot, <, 0, 1)$.

Now, we will use directed graphs, and graphs, as introduced in Chapter 4, because of the simplicity of the language and the diversity of examples that are readily available. The first–order language of (directed) graphs is $\mathcal{L} = \{r\}$, where r is a binary relation symbol.

Since there are no function symbols nor constant symbols in our language, the only *terms* are the variables x, and *atomic* formulas look like $(x \approx y)$ or (rxy). Of course we are always eager to drop parentheses, where convenient, in formulas.

Example 5.5.1 The following are formulas:

$$\forall x\,(rxx), \;\; (rxy) \to (ryx), \;\; \forall x\,((rxy) \to \exists y\,(ryx)).$$

Example 5.5.2 The subformulas of $\forall x\,((rxy) \to \exists y\,(ryx))$ are

$$
\begin{array}{l}
\forall x\,((rxy) \to \exists y\,(ryx)) \\
(rxy) \to \exists y\,(ryx) \\
rxy \\
\exists y\,(ryx) \\
ryx.
\end{array}
$$

Exercises

The following problems are concerned with the language of graphs.

5.5.1 Which of the following are formulas? For the formulas find all subformulas.

a. $rx \vee y$

b. $rxy \vee z$

c. $x \vee (rxy)$

d. $rx\neg y$

e. $rxry$

f. $(y \approx x) \wedge \forall w(rvu)$

g. $\neg\,\exists z\big((ryz) \wedge (x \approx u)\big)$

h. $\big(\exists z(x \approx z)\big) \to \forall x(rwx)$

i. $\neg\,\Big(\forall u(\neg\,(u \approx w)) \to ((rxu) \wedge (rwz))\Big)$

j. $\neg\left((rvy) \to \Big(\exists u(\neg\,(y \approx z) \wedge (z \approx u))\Big)\right)$

k. $\neg\left(\Big(\neg\,((rux) \leftrightarrow (z \approx y)) \wedge (rwz)\Big) \to \neg\,(y \approx x)\right)$

5.5.2 Indicate the bound and free occurrences of variables in the following formulas.

a. $\forall w\big((v \approx w) \vee \exists v(rwv)\big)$

b. $\forall x \exists w\big((rux) \leftrightarrow \exists v(v \approx w)\big)$

c. $\neg\,\forall x\left(\Big(\exists v\big((rxw) \wedge (rwx)\big)\Big) \leftrightarrow (z \approx u)\right)$

d. $\Big(\forall y\Big(\neg\,(w \approx z) \vee \big(\forall v(v \approx y)\big)\Big)\Big) \vee (ryz)$

e. $\Big(\exists y\big((v \approx w) \to \exists v(ryv)\big)\Big) \wedge (x \approx y)$

f. $\forall y\Big((rxw) \rightarrow \Big((rvu) \vee \Big(\big(\forall z((ryz) \rightarrow (v \approx w))\big) \leftrightarrow (ruz)\Big)\Big)\Big)$

g. $\exists v\Big(\Big((v \approx y) \rightarrow \Big(\exists u\big(\neg \forall v(\neg (v \approx u))\big) \wedge \exists x(ruv)\Big)\Big) \rightarrow \neg (w \approx x)\Big)$.

5.6 THE SEMANTICS OF FIRST–ORDER SENTENCES IN (DIRECTED) GRAPHS

In this section we will look informally at how to interpret first–order sentences F in directed graphs, and in graphs; and how to translate from English to first–order. We will refer to the following two examples:

Example 5.6.1

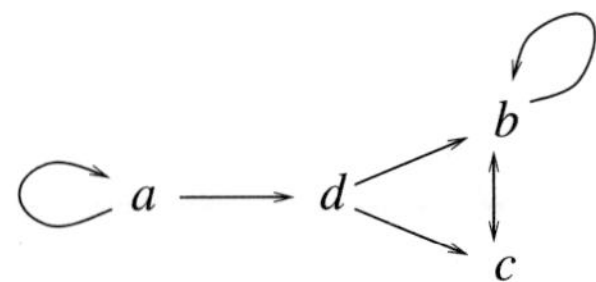

Figure 5.6.2 A directed graph

Example 5.6.3

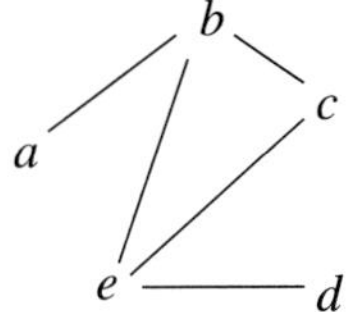

Figure 5.6.4 A graph

Example 5.6.5 The following first–order sentences about (directed) graphs

a. $\forall x\,(rxx)$
b. $\forall x \neg (rxx)$

c. $\forall x \forall y\, ((rxy) \rightarrow (ryx))$
d. $\exists x \exists y\, (rxy)$
e. $\forall x \forall y\, (rxy)$
f. $\forall x \exists y\, (rxy)$
g. $\exists y \forall x\, (rxy)$
h. $\exists x \exists y\, ((rxy) \wedge (ryx))$
i. $\exists x \exists y \exists z\, ((rxy) \wedge (rxz))$
j. $\exists x \exists y \exists z\, (\neg (y \approx z) \wedge (rxy) \wedge (rxz))$
k. $\exists x \exists y \exists z\, ((rxy) \wedge (ryz) \wedge (rzx))$

can be translated into English, and their truth for Examples 5.6.1 and 5.6.3 determined, as follows:

a. says "the (directed) graph is *reflexive*." False for both examples.

b. says "the (directed) graph is *irreflexive*." False for Example 5.6.1, true for Example 5.6.3.

c. says "the (directed) graph is *symmetric*." False for Example 5.6.1, true for Example 5.6.3.

d. says "there is an edge." True for both examples.

e. says "all possible edges are present." False for both examples.

f. says "for every vertex x there is an edge going from x to some vertex y." True for both examples.

g. says "there is a vertex y such that every possible edge to y is present." False for both examples.

h. says "there are vertices x and y with both directed edges between them present." True for both examples.

i. says "there are vertices x, y, z such that the edges (x, y) and (x, z) are present." True for both examples.

j. says "there are vertices x, y, z with $y \neq z$ such that the edges (x, y) and (x, z) are present." True for both examples.

k. says "there are vertices x, y, z such that the edges (x, y), (y, z) and (z, x) are present." True in both examples.

Now, let us practice translating English statements into first–order sentences about directed graphs.

Example 5.6.6 The following statements

a. The (directed) graph has at least two vertices.

b. Every vertex has an edge attached to it.

c. Every vertex has at most two edges directed from it to other vertices.

can be expressed by the first–order sentences

a. $\exists x \exists y \, (\neg \, (x \approx y))$.

b. $\forall x \exists y \, \big((rxy) \vee (ryx)\big)$.

c. $\forall x \forall y \forall z \forall w \, \Big(\big((rxy) \wedge (rxz) \wedge (rxw)\big) \rightarrow \big((y \approx z) \vee (y \approx w) \vee (w \approx z)\big)\Big)$.

Next, we turn to making statements about graphs. With graphs we know that the edge relation is irreflexive and symmetric, so we can make use of this information to shorten some translations. First, let us introduce two concepts from graph theory.

Definition 5.6.7 Let $\mathbf{G} = (G, r)$ be a graph.

- The *degree* of a vertex is the number of (undirected) edges attached to it.
- A *path of length* n from vertex x to vertex y is a sequence of vertices $a_1, \ldots, a_{n+1}$ with each (a_i, a_{i+1}) being an edge, and with $x = a_1$, $y = a_{n+1}$.
- Two vertices are *adjacent* if there is an edge connecting them.

Example 5.6.8 In Example 5.6.3 we see that the degree of a is 1, that of b is 3, and there is a path a, b, e, d of length 3 from a to d, as well as a path a, b, c, e, d of length 4.

Before expressing further statements about graphs we give some definable relations.

Proposition 5.6.9 The following relations are definable in graphs:

a. The degree of x is at least one.

b. The degree of x is at least two.

c. The degree of x is at least n (where n is fixed).

d. The degree of x is at most one.

e. The degree of x is at most two.

f. The degree of x is at most n (where n is fixed).

g. There is a path of length one from x to y.

h. There is a path of length two from x to y.

i. There is a path of length at most two from x to y.

j. There is a path of length n from x to y, where n is fixed.

Proof. We can use the following formulas:

a. $\exists y\,(rxy)$

b. $\exists y \exists z\,(\neg\,(y \approx z) \wedge (rxy) \wedge (rxz))$

c. $\exists y_1 \cdots \exists y_n \Big(\big(\bigwedge_{1 \leq i < j \leq n} \neg\,(y_i \approx y_j)\big) \wedge \big(\bigwedge_{1 \leq i \leq n} (rxy_i)\big) \Big)$

d. $\forall y \forall z \Big(\big((rxy) \wedge (rxz)\big) \rightarrow (y \approx z) \Big)$

e. $\forall y \forall z \forall w \Big(\big((rxy) \wedge (rxz) \wedge (rxw)\big) \rightarrow$
$\big((y \approx z) \vee (y \approx w) \vee (z \approx w)\big) \Big)$

f. $\forall y_1 \cdots \forall y_{n+1} \Big(\big(\bigwedge_{1 \leq i \leq n+1} (rxy_i)\big) \rightarrow \big(\bigvee_{1 \leq i < j \leq n+1} (y_i \approx y_j)\big) \Big)$

g. rxy

h. $\exists z\,\big((rxz) \wedge (rzy)\big)$

i. $(rxy) \vee \exists z\,\big((rxz) \wedge (rzy)\big)$

j. $\exists z_1 \cdots \exists z_{n+1} \Big((x \approx z_1) \wedge (y \approx z_{n+1}) \wedge \big(\bigwedge_{1 \leq i \leq n} (rz_iz_{i+1})\big) \Big).$

□

We could introduce the abbreviation $path_n(x, y)$ for the last formula in Proposition 5.6.9 (as well as any other abbreviations that we liked) and use this to express interesting statements about graphs. Of course, we would have to keep in mind the problems with substitution, to avoid binding new variables. As before, we could keep this problem in view by using the notation $path_n^\star$ when needed. However, in the following example, we prefer to simply write out the full first–order formulas.

Example 5.6.10 We can use the preceding formulas (or abbreviations for them) to express the following statements about graphs:

a. Some vertex has degree at least two.

b. Every vertex has degree at least two.

c. Every vertex has degree at least n.

d. Some vertex has degree at most two.

e. Every vertex has degree at most two.

f. Every vertex has degree at most n.

g. Some pair of distinct vertices is connected by a path of length at most two.

h. Every pair of distinct vertices is connected by a path of length two.

i. Every pair of distinct vertices is connected by a path of length at most two.

j. Every pair of distinct vertices is connected by a path of length n.

k. Every pair of distinct vertices is connected by a path of length at most n.

by the following first–order sentences:

a. $\exists x \exists y \exists z \left(\neg (y \approx z) \wedge (rxy) \wedge (rxz)\right)$

b. $\forall x \exists y \exists z \left(\neg (y \approx z) \wedge (rxy) \wedge (rxz)\right)$

c. $\forall x \exists y_1 \cdots \exists y_n \left(\left(\bigwedge_{1 \le i < j \le n} \neg (y_i \approx y_j)\right) \wedge \left(\bigwedge_{1 \le i \le n} (rxy_i)\right)\right)$

d. $\exists x \forall y \forall z \forall w \Big(\left((rxy) \wedge (rxz) \wedge (rxw)\right) \rightarrow$

$$\left((y \approx z) \vee (y \approx w) \vee (z \approx w)\right)\Big)$$

e. $\forall x \forall y \forall z \forall w \Big(\left((rxy) \wedge (rxz) \wedge (rxw)\right) \rightarrow$

$$\left((y \approx z) \vee (y \approx w) \vee (z \approx w)\right)\Big)$$

f. $\forall x \forall y_1 \cdots \forall y_{n+1} \left(\left(\bigwedge_{1 \le i \le n+1} (rxy_i)\right) \rightarrow \left(\bigvee_{1 \le i < j \le n+1} (y_i \approx y_j)\right)\right)$

g. $\exists x \exists y \exists z \left(\neg (x \approx y) \wedge (rxz) \wedge (rzy)\right)$

h. $\forall x \forall y \left(\left(\neg (x \approx y)\right) \rightarrow \exists z \left((rxz) \wedge (rzy)\right)\right)$

i. $\forall x \forall y \left(\left(\neg (x \approx y)\right) \rightarrow \left((rxy) \vee \exists z \left((rxz) \wedge (rzy)\right)\right)\right)$

j. $\forall x \forall y \exists z_1 \cdots \exists z_{n+1} \Big(\left(\neg (x \approx y)\right) \rightarrow$

$$\left((x \approx z_1) \;\wedge\; (y \approx z_{n+1}) \;\wedge\; \left(\bigwedge_{1 \le i \le n} (rz_i z_{i+1})\right)\right)\Big)$$

k. $\forall x \forall y \Bigg(\left(\neg (x \approx y)\right) \rightarrow \Big(\bigvee_{1 \le k \le n} \Big(\exists z_1 \cdots \exists z_{k+1} \left((x \approx z_1) \wedge\right.$

$$\left.(y \approx z_{k+1}) \;\wedge\; \left(\bigwedge_{1 \le i \le k} (rz_i z_{i+1})\right)\right)\Big)\Big)\Bigg).$$

5.7 SEMANTICS FOR FIRST–ORDER LOGIC

Given a first–order $\mathcal{L}$–structure $\mathbf{S} = (S, I)$, the interpretation I gives meaning to the symbols of the language $\mathcal{L}$, and we associate with each term $t(x_1, \cdots, x_n)$ the n–ary term function $t^{\mathbf{S}}$.

Now we will associate with each formula $\mathsf{F}(x_1, \cdots, x_n)$, with a specified *nonempty* sequence of variables $x_1, \ldots, x_n$, an n–ary relation $\mathsf{F}^{\mathbf{S}} \subseteq S^n$ as follows.

Definition 5.7.1 The relations $\mathsf{F}^{\mathbf{S}}$ are defined inductively as follows:

- $\mathsf{F}(\vec{x})$ is atomic:

  If F is $t_1(\vec{x}) \approx t_2(\vec{x})$, then $\mathsf{F}^{\mathbf{S}}(\vec{a})$ holds iff $t_1^{\mathbf{S}}(\vec{a}) = t_2^{\mathbf{S}}(\vec{a})$. If F is $rt_1(\vec{x}) \cdots t_n(\vec{x})$, then $\mathsf{F}^{\mathbf{S}}(\vec{a})$ holds iff $r^{\mathbf{S}}(t_1^{\mathbf{S}}(\vec{a}), \ldots, t_n^{\mathbf{S}}(\vec{a}))$ holds.
- F is $\neg\,\mathsf{G}$: Then $\mathsf{F}^{\mathbf{S}}(\vec{a})$ holds iff $\mathsf{G}^{\mathbf{S}}(\vec{a})$ does not hold.
- F is $\mathsf{G} \vee \mathsf{H}$: Then $\mathsf{F}^{\mathbf{S}}(\vec{a})$ holds iff $\mathsf{G}^{\mathbf{S}}(\vec{a})$ or $\mathsf{H}^{\mathbf{S}}(\vec{a})$ holds.
- F is $\mathsf{G} \wedge \mathsf{H}$: Then $\mathsf{F}^{\mathbf{S}}(\vec{a})$ holds iff $\mathsf{G}^{\mathbf{S}}(\vec{a})$ and $\mathsf{H}^{\mathbf{S}}(\vec{a})$ holds.
- F is $\mathsf{G} \to \mathsf{H}$: Then $\mathsf{F}^{\mathbf{S}}(\vec{a})$ holds iff $\mathsf{G}^{\mathbf{S}}(\vec{a})$ does not hold, or $\mathsf{H}^{\mathbf{S}}(\vec{a})$ holds.
- F is $\mathsf{G} \leftrightarrow \mathsf{H}$: Then $\mathsf{F}^{\mathbf{S}}(\vec{a})$ holds iff both or neither of $\mathsf{G}^{\mathbf{S}}(\vec{a})$ and $\mathsf{H}^{\mathbf{S}}(\vec{a})$ holds.
- $\mathsf{F}(\vec{x})$ is $\forall y\, \mathsf{G}(y, \vec{x})$: Then $\mathsf{F}^{\mathbf{S}}(\vec{a})$ holds iff $\mathsf{G}^{\mathbf{S}}(b, \vec{a})$ holds for *every* $b \in S$.
- $\mathsf{F}(\vec{x})$ is $\exists y\, \mathsf{G}(y, \vec{x})$: Then $\mathsf{F}^{\mathbf{S}}(\vec{a})$ holds iff $\mathsf{G}^{\mathbf{S}}(b, \vec{a})$ holds for *some* $b \in S$.

Now we need to take into consideration the fact that when we write $\mathsf{F}(x_1, \ldots, x_n)$ we require only that the variables with free occurrences in F be among the designated set of variables $x_1, \ldots, x_n$. If a variable x_i does not have a free occurrence in F, then the relation $\mathsf{F}^{\mathbf{S}}$ is completely free in the ith coordinate.

Lemma 5.7.2 If x_i does not have a free occurrence in the formula $\mathsf{F}(x_1, \ldots, x_i, \ldots, x_n)$ and $(a_1, \ldots, a_i, \ldots, a_n) \in \mathsf{F}^{\mathbf{S}}$ then, for any $a_i' \in S$, we have $(a_1, \ldots, a_i', \ldots, a_n) \in \mathsf{F}^{\mathbf{S}}$.

Proof. If we write out a proof that $(a_1, \ldots, a_i, \ldots, a_n) \in \mathsf{F}^{\mathbf{S}}$ using just the reasons permitted in Definition 5.7.1 of $\mathsf{F}^{\mathbf{S}}$, then we can replace a_i with a_i' throughout the proof and have a proof that $(a_1, \ldots, a_i', \ldots, a_n) \in \mathsf{F}^{\mathbf{S}}$. □

Definition 5.7.3 Given an $\mathcal{L}$–structure $\mathbf{S}$, we say that a relation $r \subseteq S^k$ is *definable in* $\mathbf{S}$ if there is a formula $\mathsf{F}(x_1, \ldots, x_k)$ such that $r = \mathsf{F}^{\mathbf{S}}$.

Example 5.7.4 Let us use the preceding definitions to determine the relation defined by a formula on a structure with a small number of elements. Suppose $\mathcal{L}$ consists of one *binary* relation symbol r and one *binary* function symbol f, and let $\mathbf{S}$ be the $\mathcal{L}$–structure given by

$\mathbf{S} = \{a, b\}$

r	a	b
a	0	1
b	1	0

f	a	b
a	a	a
b	b	a

a. First, let us find the unary relation determined by the formula

$$\mathsf{F}(x) = \forall y \exists z\, ((rfxyfxz) \vee (rfxzfzy)).$$

We will give a detailed development, based on our definition of $\mathsf{F}^{\mathbf{S}}$. To do this we introduce the following definitions:

$$\begin{aligned} \mathsf{H}_1(x,y,z) &= rfxyfxz \\ \mathsf{H}_2(x,y,z) &= rfxzfzy \\ \mathsf{H}(x,y,z) &= \mathsf{H}_1(x,y,z) \vee \mathsf{H}_2(x,y,z) \\ \mathsf{G}(x,y) &= \exists z\, \mathsf{H}(x,y,z) \\ \mathsf{F}(x) &= \forall y\, \mathsf{G}(x,y). \end{aligned}$$

Then we have

x	y	z	$\mathsf{H}_1(x,y,z)$	$\mathsf{H}_2(x,y,z)$	$\mathsf{H}(x,y,z)$
a	a	a	0	0	0
a	a	b	0	1	1
a	b	a	0	0	0
a	b	b	0	0	0
b	a	a	0	1	1
b	a	b	1	1	1
b	b	a	1	1	1
b	b	b	0	0	0

x	y	$\mathsf{G}(x,y)$
a	a	1
a	b	0
b	a	1
b	b	1

x	$\mathsf{F}(x)$
a	0
b	1

Thus $\mathsf{F}^{\mathsf{S}}(b)$ holds, and $\mathsf{F}^{\mathsf{S}}(a)$ does not hold, i.e., $\mathsf{F}^{\mathsf{S}} = \{b\}$.

b. Next, let us find the binary relation determined by the formula

$$\mathsf{F}(x, y) = \exists z\,((rxfyz) \wedge (fxy \approx z)) \to (fxy \approx fyx).$$

We introduce the following definitions:

$$\begin{aligned}
\mathsf{G}_1(x, y, z) &= rxfyz \\
\mathsf{G}_2(x, y, z) &= fxy \approx z \\
\mathsf{G}(x, y, z) &= \mathsf{G}_1(x, y, z) \wedge \mathsf{G}_2(x, y, z) \\
\mathsf{F}_1(x, y) &= \exists z\, \mathsf{G}(x, y, z) \\
\mathsf{F}_2(x, y) &= fxy \approx fyx \\
\mathsf{F}(x, y) &= \mathsf{F}_1(x, y) \to \mathsf{F}_2(x, y).
\end{aligned}$$

Then we have

x	y	z	$\mathsf{G}_1(x, y, z)$	$\mathsf{G}_2(x, y, z)$	$\mathsf{G}(x, y, z)$
a	a	a	0	1	0
a	a	b	0	0	0
a	b	a	1	1	1
a	b	b	0	0	0
b	a	a	1	0	0
b	a	b	1	1	1
b	b	a	0	1	0
b	b	b	1	0	0

x	y	$\mathsf{F}_1(x, y)$	$\mathsf{F}_2(x, y)$	$\mathsf{F}(x, y)$
a	a	0	1	1
a	b	1	0	0
b	a	1	0	0
b	b	0	1	1

Thus $\mathsf{F}^{\mathsf{S}}(a, a)$, $\mathsf{F}^{\mathsf{S}}(b, b)$ hold, and $\mathsf{F}^{\mathsf{S}}(a, b)$, $\mathsf{F}^{\mathsf{S}}(b, a)$ do not hold, i.e., $\mathsf{F}^{\mathsf{S}} = \{(a, a), (b, b)\}$.

Remark 5.7.5 We saw several examples of definable relations in the sections on natural numbers and graphs, but we did not give any examples of relations that are *not* definable. Powerful tools, e.g., the compactness theorem for first–order logic, are needed to show that certain natural relations are not definable. For example, the relation "there is a path from x to y" is not definable in graphs. We will prove this in Sec. 5.14.

Now we are ready to make one of our most important inductive definitions, namely, what it means for a sentence to be *true* in a structure.

Definition 5.7.6 Let F be a sentence and $\mathbf{S}$ a structure. Then F is *true* in $\mathbf{S}$ provided one of the following holds:

- F is $rt_1 \cdots t_n$ and $r^{\mathbf{S}}(t_1^{\mathbf{S}}, \ldots, t_n^{\mathbf{S}})$ holds;
- F is $t_1 \approx t_2$ and $t_1^{\mathbf{S}} = t_2^{\mathbf{S}}$;
- F is $\neg\,\mathsf{G}$ and G is not true in $\mathbf{S}$;
- F is $\mathsf{G} \vee \mathsf{H}$ and at least one of G, H is true in $\mathbf{S}$;
- F is $\mathsf{G} \wedge \mathsf{H}$ and both of G, H are true in $\mathbf{S}$;
- F is $\mathsf{G} \rightarrow \mathsf{H}$ and G is not true in $\mathbf{S}$ or H is true in $\mathbf{S}$;
- F is $\mathsf{G} \leftrightarrow \mathsf{H}$ and both or neither of G, H is true in $\mathbf{S}$;
- F is $\forall x\, \mathsf{G}(x)$ and $\mathsf{G}^{\mathbf{S}}(a)$ is true for every $a \in \mathbf{S}$;
- F is $\exists x\, \mathsf{G}(x)$ and $\mathsf{G}^{\mathbf{S}}(a)$ is true for some $a \in \mathbf{S}$.

If F is not true in $\mathbf{S}$, then we say F is *false* in $\mathbf{S}$.

Example 5.7.7 If we consider the language $\mathcal{L}$ in Example 5.7.4 and the structure $\mathbf{S}$ defined there, then we can use the results of the (a) part to show that $\forall x \forall y \exists z \left((rfxyfxz) \vee (rfxzfzy)\right)$ is false in $\mathbf{S}$, and $\exists x \forall y \exists z \left((rfxyfxz) \vee (rfxzfzy)\right)$ is true in $\mathbf{S}$.

The next result says that the relation $\mathsf{F}^{\mathbf{S}}$ defined by a sentence $\mathsf{F}(x_1, \ldots, x_n)$ with a designated sequence of variables $x_1, \ldots, x_n$ has all n–tuples of S if F is true in $\mathbf{S}$, and is the empty set of n–tuples when F is false in $\mathbf{S}$.

Lemma 5.7.8 If $\mathsf{F}(x_1, \ldots, x_n)$ is a *sentence*, then $\mathsf{F}^{\mathbf{S}} = S^n$ if $\mathbf{S} \models \mathsf{F}$, and otherwise $\mathsf{F}^{\mathbf{S}} = \emptyset$.

Proof. By Lemma 5.7.2 we see that $\mathsf{F}^{\mathbf{S}}$ must be either S^n or $\emptyset$. If F is true in $\mathbf{S}$, then a proof of this fact, based on Definition 5.7.6, will yield a proof that $(a_1, \ldots, a_n) \in \mathsf{F}^{\mathbf{S}}$ for any $(a_1, \ldots, a_n) \in S^n$. Similarly, if F is false in $\mathbf{S}$, then we can show that $(a_1, \ldots, a_n) \notin S^n$. $\square$

We have already looked at a number of examples of first–order sentences and, informally, determined if they were true or false in certain structures, in the sections on natural numbers and graphs. Of course, we would get the same answers, regarding their truth, if we were to work through Definition 5.7.6 with those examples.

Definition 5.7.9 Given a first–order language $\mathcal{L}$, let F be a sentence, $\mathcal{S}$ a set of sentences, and $\mathbf{S}$ a structure for this language.

- We write $\mathbf{S} \models \mathsf{F}$ if F is true in $\mathbf{S}$, and we also use the expressions $\mathbf{S}$ *satisfies* F, $\mathbf{S}$ *models* F, and $\mathbf{S}$ *is a model of* F.
- A sentence F that is true in all $\mathcal{L}$–structures is said to be *valid*.
- We write $\mathbf{S} \models \mathcal{S}$ if every sentence F in $\mathcal{S}$ is true in $\mathbf{S}$, and we also use the expressions $\mathbf{S}$ *satisfies* $\mathcal{S}$, $\mathbf{S}$ *models* $\mathcal{S}$, and $\mathbf{S}$ *is a model of* $\mathcal{S}$. We write $\mathrm{Sat}(\mathcal{S})$ to say $\mathcal{S}$ is satisfiable.
- We write $\mathcal{S} \models \mathsf{F}$ if every model of $\mathcal{S}$ is a model of F, and we use the expression F *is a consequence of* $\mathcal{S}$.

Example 5.7.10 We have, e.g., $\mathbf{N} \models \forall x \forall y\,(x \cdot y \approx y \cdot x)$. However, this sentence is not valid, i.e., it is not true in all structures for the language we chose for natural numbers. For example, consider the structure of 2-by-2 matrices over the reals, with usual matrix addition and multiplication, with 1 being the identity matrix, and 0 being the zero matrix, and let $<$ be the relation $det(A) < det(B)$. (Can you think of a simpler structure to use here?)

An example of a valid sentence would be $\forall x \forall y\,((x \approx y+1) \rightarrow (x+1 \approx (y+1)+1))$.

Example 5.7.11 Sometimes we want to try to find a small finite model $\mathbf{S}$ of a sentence F. Of course, the smallest possible model would have only one element in S. There is a fairly simple test for having a one–element model.

Let the *propositional skeleton*, $\mathrm{Skel}(\mathsf{F})$, of a formula be defined as follows:

- Delete all quantifiers and terms.
- Replace $\approx$ with 1.
- Replace the relation symbols r with propositional variables R.

Then F *has a one–element model iff Skel(*F*) is satisfiable.*

If the latter is satisfiable, then we choose an evaluation e that makes it true. Then we can construct a one–element model of F on $S = \{a\}$ by letting $r^{\mathbf{S}}(a, \ldots, a)$ hold iff $\mathsf{e}(R) = 1$. (Of course, there is exactly one choice for the interpretation of each function symbol and constant symbol in a one–element structure!)

To illustrate this, we consider the first–order sentence F given by $\forall x\,((0 < x) \rightarrow \exists y\,(x \approx y \cdot y))$. Then $\mathrm{Skel}(\mathsf{F})$ is the propositional formula $R \rightarrow 1$. As this is satisfiable, F has a one–element model, and such a model is given by letting $<$ be the relation $\{(a,a)\}$ on $S = \{a\}$.

Exercises

5.7.1 Determine if the following first–order sentences are true in the structure in Exercise 3.3.3, p. 147, where we also assume that we have a binary relation r in our language that is interpreted by rab iff a divides b.

a. $\forall x\,(rxfxxx)$

b. $\forall x \exists y\,(rfxyxfxxx)$

c. $\forall x \exists y\,((rfxyxy) \rightarrow (ryfxyx))$

d. $\forall x \exists y \forall z\,(rxfzyx)$

5.7.2 Determine if the following first–order sentences are true in the structure in Exercise 3.3.4, p. 147, where we also assume that we have a binary relation r in our language that is interpreted by rab iff a is less than b.

a. $\forall x\,(rxfxxx)$

b. $\forall x \exists y\,(rfxyxfxxx)$

c. $\forall x \exists y\,((rfxyxy) \rightarrow (ryfxyx))$

d. $\forall x \exists y \forall z\,(rxfzyx)$

5.8 EQUIVALENT FORMULAS

There are many ways to say the same thing in first–order logic. We assume we are working in a fixed language $\mathcal{L}$.

Definition 5.8.1 The sentences F and G are *equivalent*, written $\mathsf{F} \sim \mathsf{G}$, if they are true of the same $\mathcal{L}$–structures $\mathbf{S}$, i.e., for all structures $\mathbf{S}$ we have

$$\mathbf{S} \models \mathsf{F} \qquad \text{iff} \qquad \mathbf{S} \models \mathsf{G}.$$

Example 5.8.2 The sentences

$$\forall x\,(\neg\,(x \approx 0) \rightarrow \exists y\,(x \cdot y \approx 1))$$

and

$$\forall x \exists y\,(\neg\,(x \approx 0) \rightarrow (x \cdot y \approx 1))$$

are equivalent.

Proposition 5.8.3 The sentences F and G are equivalent iff $\mathsf{F} \leftrightarrow \mathsf{G}$ is a valid sentence.

Proof. This follows from the fact that **S** satisfies $\mathsf{F} \leftrightarrow \mathsf{G}$ iff **S** satisfies both or neither of F and G. □

We now define equivalence for formulas $\mathsf{F}(x_1, \ldots, x_n)$ with a specified *nonempty* sequence of variables $x_1, \ldots, x_n$.

Definition 5.8.4 The formulas $\mathsf{F}(x_1, \ldots, x_n)$ and $\mathsf{G}(x_1, \ldots, x_n)$ are *equivalent*, written $\mathsf{F}(x_1, \ldots, x_n) \sim \mathsf{G}(x_1, \ldots, x_n)$, iff F and G define the same relation on any $\mathcal{L}$–structure **S**, i.e., $\mathsf{F}^{\mathbf{S}} = \mathsf{G}^{\mathbf{S}}$.

Example 5.8.5 The formulas

$$\neg (x \approx 0) \rightarrow \exists y\, (x \cdot y \approx 1)$$

and

$$\exists y\, (\neg (x \approx 0) \rightarrow (x \cdot y \approx 1))$$

are equivalent.

Proposition 5.8.6 The formulas $\mathsf{F}(\vec{x})$ and $\mathsf{G}(\vec{x})$ are equivalent iff $\forall \vec{x}(\mathsf{F}(\vec{x}) \leftrightarrow \mathsf{G}(\vec{x}))$ is a valid sentence.

Proof. $\mathbf{S} \models \forall \vec{x}\,(\mathsf{F}(\vec{x}) \leftrightarrow \mathsf{G}(\vec{x}))$ iff for all $\vec{a}$ in S, either both or neither of $\mathsf{F}^{\mathbf{S}}(\vec{a})$ and $\mathsf{G}^{\mathbf{S}}(\vec{a})$ holds. But this is precisely the condition for $\mathsf{F}^{\mathbf{S}} = \mathsf{G}^{\mathbf{S}}$. □

Proposition 5.8.7 The relation $\sim$ is an equivalence relation on sentences as well as on formulas.

Proof. This is immediate from the definition of $\sim$ and the fact that ordinary equality $(=)$ is an equivalence relation. □

Next, we give a table of some fundamental equivalences involving quantifiers.

Proposition 5.8.8 [FUNDAMENTAL EQUIVALENCES]

$$\begin{array}{rcll}
\neg \exists x \, \mathsf{F} & \sim & \forall x \, (\neg \mathsf{F}) & \\
\neg \forall x \, \mathsf{F} & \sim & \exists x \, (\neg \mathsf{F}) & \\
(\forall x \, \mathsf{F}) \vee \mathsf{G} & \sim & \forall x \, (\mathsf{F} \vee \mathsf{G}) & \text{if } x \text{ is not free in G} \\
(\exists x \, \mathsf{F}) \vee \mathsf{G} & \sim & \exists x \, (\mathsf{F} \vee \mathsf{G}) & \text{if } x \text{ is not free in G} \\
(\forall x \, \mathsf{F}) \wedge \mathsf{G} & \sim & \forall x \, (\mathsf{F} \wedge \mathsf{G}) & \text{if } x \text{ is not free in G} \\
(\exists x \, \mathsf{F}) \wedge \mathsf{G} & \sim & \exists x \, (\mathsf{F} \wedge \mathsf{G}) & \text{if } x \text{ is not free in G} \\
(\forall x \, \mathsf{F}) \to \mathsf{G} & \sim & \exists x \, (\mathsf{F} \to \mathsf{G}) & \text{if } x \text{ is not free in G} \\
(\exists x \, \mathsf{F}) \to \mathsf{G} & \sim & \forall x \, (\mathsf{F} \to \mathsf{G}) & \text{if } x \text{ is not free in G} \\
\mathsf{F} \to (\forall x \, \mathsf{G}) & \sim & \forall x \, (\mathsf{F} \to \mathsf{G}) & \text{if } x \text{ is not free in F} \\
\mathsf{F} \to (\exists x \, \mathsf{G}) & \sim & \exists x \, (\mathsf{F} \to \mathsf{G}) & \text{if } x \text{ is not free in F} \\
\forall x \, (\mathsf{F} \wedge \mathsf{G}) & \sim & (\forall x \, \mathsf{F}) \wedge (\forall x \, \mathsf{G}) & \\
\exists x \, (\mathsf{F} \vee \mathsf{G}) & \sim & (\exists x \, \mathsf{F}) \vee (\exists x \, \mathsf{G}) &
\end{array}$$

Proof. These are obtained by a straightforward application of the definition of equivalence. □

Exercises

5.8.1 Determine if the following sentences are equivalent.

a. $\forall x \, \big((ffx \approx x) \wedge (fffx \approx x)\big)$ and $\forall x \, (fx \approx x)$

b. $\forall x \exists y \, (r_1 x \wedge r_2 y)$ and $\exists y \forall x \, (r_1 x \wedge r_2 y)$ [r_1, r_2 are unary.]

c. $\forall x \exists y \, (rxy)$ and $\exists y \forall x \, (ryx)$ [r is binary.]

d. $\forall x \forall y \, \big((fx \approx fy) \to (x \approx y)\big)$ and $\forall y \exists x \, (fx \approx y)$

5.8.2 Determine if the following formulas are equivalent.

a. $\forall y \, (rxy)$ and $\exists y \, (rxy)$ [r is binary.]

b. $\forall y \exists z \, ((rxy) \wedge (rxz))$ and $\exists z \forall y \, ((rxy) \wedge (rxz))$ [r is binary.]

c. $r_1 fx \wedge r_2 fx$ and $\exists y \, (r_1 y \wedge r_2 y \wedge (y \approx fx))$ [r_1, r_2 are unary.]

d. $\exists y \, (r_1 fy \wedge r_2 y \wedge (x \approx fy))$ and $\exists y \exists z \, (r_1 y \wedge r_2 z \wedge (x \approx fy) \wedge (x \approx fz))$

5.9 REPLACEMENT AND SUBSTITUTION

Equivalent propositional formulas lead to equivalent first–order formulas as follows.

Proposition 5.9.1 Suppose $\mathsf{F}(\vec{P})$ and $\mathsf{G}(\vec{P})$ are equivalent propositional formulas, and $\vec{\mathsf{H}}$ is a sequence of first–order formulas. Then $\mathsf{F}(\vec{\mathsf{H}}) \sim \mathsf{G}(\vec{\mathsf{H}})$.

Proof. First, we show this holds when $\mathsf{F}(\vec{P}) \sim \mathsf{G}(\vec{P})$ is a fundamental equivalence from Chapter 2. Then we use the fact that we can transform equivalent propositional formulas into one another using these fundamental propositional equivalences. □

Example 5.9.2 We can readily verify the following equivalence of propositional formulas:

$$P \to (Q \to R) \ \sim\ (P \wedge Q) \to R.$$

Thus we have, in first–order logic,

$$\begin{aligned}(\exists z\,(x \cdot z \approx y)) \to \Big((\exists u\,(y \cdot u \approx z)) \to \exists u\,(x \cdot u \approx z)\Big) \ \sim \\ \Big((\exists z\,(x \cdot z \approx y)) \wedge (\exists u\,(y \cdot u \approx z))\Big) \to \exists u\,(x \cdot u \approx z).\end{aligned}$$

Equivalence is compatible with our connectives and quantifiers in the obvious sense.

Lemma 5.9.3 Suppose $\mathsf{F}_i \sim \mathsf{G}_i$, $i = 1, 2$. Then

$$\begin{aligned} \neg\,\mathsf{F}_1 &\sim \neg\,\mathsf{G}_1 \\ \mathsf{F}_1 \vee \mathsf{F}_2 &\sim \mathsf{G}_1 \vee \mathsf{G}_2 \\ \mathsf{F}_1 \wedge \mathsf{F}_2 &\sim \mathsf{G}_1 \wedge \mathsf{G}_2 \\ \mathsf{F}_1 \to \mathsf{F}_2 &\sim \mathsf{G}_1 \to \mathsf{G}_2 \\ \mathsf{F}_1 \leftrightarrow \mathsf{F}_2 &\sim \mathsf{G}_1 \leftrightarrow \mathsf{G}_2 \\ \forall x\,\mathsf{F}_1 &\sim \forall x\,\mathsf{G}_1 \\ \exists x\,\mathsf{F}_1 &\sim \exists x\,\mathsf{G}_1. \end{aligned}$$

Proof. (Exercise.) □

This leads us to the basic replacement theorem for first–order logic.

Theorem 5.9.4 [REPLACEMENT THEOREM] Let F, G, G′ be formulas. If $\mathsf{G} \sim \mathsf{G}'$, and F′ is the result of replacing a certain occurrence of G in F with G′, then $\mathsf{F} \sim \mathsf{F}'$

Proof. The proof of this theorem is similar to the proof of Theorem 2.4.5 (p. 49), that is, by induction on F. □

As we saw in Remark 5.3.13 (p. 326), substitution requires something new—the need to rename variables. When using substitution we need to be careful to rename variables to avoid binding newly introduced occurrences of variables.

Definition 5.9.5 Given a first–order formula F, define a *conjugate* of F to be any formula $\overline{\mathsf{F}}$ obtained by renaming the occurrences of bound variables of F so that no free occurrences of variables in F become bound. This renaming must keep bound occurrences of distinct variables distinct.

Example 5.9.6 Let us consider the formula $\forall x\,\big((rxy) \vee \exists z\,(rxz)\big)$.

The replacement of bound x and bound z by w, which leads to $\forall w\,\big((rwy) \vee \exists w\,(rww)\big)$, does not give a conjugate, as x and z must be replaced by distinct variables.

The replacement of bound z by x, which leads to $\forall x\,\big((rxy) \vee \exists x\,(rxx)\big)$, does not give a conjugate, as the first x in rxx becomes bound by a different quantifier.

The replacement of bound x by y, which leads to $\forall y\,\big((ryy) \vee \exists z\,(ryz)\big)$, does not give a conjugate, as the second y in ryy becomes bound.

However, the replacement of bound x by w and bound z by y, which leads to the formula $\forall w\,\big((rwy) \vee \exists y\,(rwy)\big)$, does give a conjugate. Since there was a free occurrence of y in the formula, we do need to check that it did not become bound.

Still, the safest way to choose a conjugate is simply to change the names of the bound variables to entirely new variables, not mentioned in the formula, e.g., the replacement of bound x by u and bound z by v, which leads to $\forall u\,\big((ruy) \vee \exists v\,(ruv)\big)$, gives a conjugate.

The next result says that conjugates are equivalent.

Lemma 5.9.7 If $\overline{\mathsf{F}}$ is a conjugate of F, then $\overline{\mathsf{F}} \sim \mathsf{F}$.

Proof. By induction on F. □

Notational Convention 5.9.8 Suppose $\mathsf{F}(x_1, \ldots, x_n)$ is a formula and $t_1, \ldots, t_n$ are terms. By $\mathsf{F}(t_1, \ldots, t_n)$ we mean the formula that is obtained from F by simultaneously replacing each free occurrence of x_i with t_i, $i = 1, \ldots, n$.

The $\star$ in the next theorem reminds us to replace the formulas with conjugates to avoid binding newly introduced occurrences of variables. (Any such conjugate is acceptable.)

Theorem 5.9.9 [SUBSTITUTION THEOREM] If we have equivalent formulas, say $\mathsf{F}(x_1, \ldots, x_n) \sim \mathsf{G}(x_1, \ldots, x_n)$, and $t_1, \ldots, t_n$ are terms, then

$$\mathsf{F}^\star(t_1, \ldots, t_n) \sim \mathsf{G}^\star(t_1, \ldots, t_n).$$

This theorem is harder to prove than the others in this section. One way to prove it is to first show that if $\mathsf{H}(x_1, \ldots, x_n)$ is a valid formula, then so is $\mathsf{H}^\star(t_1, \ldots, t_n)$.

Exercises

5.9.1 Give the details of the proof of Lemma 5.9.3.

5.10 PRENEX FORM

Using the fundamental equivalences of Proposition 5.8.8 along with Lemma 5.9.7 (on conjugates) and replacement, we can transform any first–order formula into a formula with the quantifiers on the left.

Definition 5.10.1 A formula F is in *prenex* form if it looks like $Q_1x_1 \cdots Q_nx_n\, \mathsf{G}$, where the Q_i are quantifiers and G has no occurrences of quantifiers.

Example 5.10.2 The formula $\forall x\, \big((0 < x) \to \exists y\, (x \approx y \cdot y)\big)$ is not in prenex form, but it is indeed equivalent to the formula $\forall x \exists y\, \big((0 < x) \to (x \approx y \cdot y)\big)$, which is in prenex form.

Theorem 5.10.3 Every formula is equivalent to a formula in prenex form.

Proof. Carrying out the following steps puts F in prenex form:

- Rename the quantified variables so that distinct occurrences of quantifiers bind distinct variables, and no free variable is equal to a bound variable
- Eliminate all occurrences of $\to$ and $\leftrightarrow$ using
$$\begin{aligned} \mathsf{G} \leftrightarrow \mathsf{H} &\sim (\neg\, \mathsf{G} \vee \mathsf{H}) \wedge (\neg\, \mathsf{H} \vee \mathsf{F}) \\ \mathsf{G} \to \mathsf{H} &\sim \neg\, \mathsf{G} \vee \mathsf{H}. \end{aligned}$$

- Pull the quantifiers to the front by repeated use of the following (where $\Box$ can be $\vee$ or $\wedge$, and Q can be $\forall$ or $\exists$):

$$\begin{aligned}
\mathsf{G}\,\Box\,(Qx\,\mathsf{H}) &\sim Qx\,(\mathsf{G}\,\Box\,\mathsf{H})\\
(Qx\,\mathsf{G})\,\Box\,\mathsf{H} &\sim Qx\,(\mathsf{G}\,\Box\,\mathsf{H})\\
\neg\,(\mathsf{F}\vee\mathsf{G}) &\sim (\neg\,\mathsf{F}\wedge\neg\,\mathsf{G})\\
\neg\,(\mathsf{F}\wedge\mathsf{G}) &\sim (\neg\,\mathsf{F}\vee\neg\,\mathsf{G})\\
\neg\,\exists x\,\mathsf{G} &\sim \forall x\,\neg\,\mathsf{G}\\
\neg\,\forall x\,\mathsf{G} &\sim \exists x\,\neg\,\mathsf{G}.
\end{aligned}$$

Of course, we can use other equivalences to simplify the expression, e.g., eliminating double negations. Each of the steps preserves equivalence, and when the process terminates, the result is in prenex form.

$\Box$

Example 5.10.4 Let us put the formula

$$\forall x\left(\neg\left((\exists y\,(rxy)) \to \exists x\,(rxy)\right)\right)$$

in prenex form.

- Since the variable x is bound by two different occurrences of quantifiers, let us replace the subformula $\exists x(rxy)$ with its conjugate $\exists z(rzy)$ to obtain the equivalent formula

$$\forall x\left(\neg\left((\exists y\,(rxy)) \to \exists z\,(rzy)\right)\right).$$

- Since the variable y occurs both free and bound, let us replace the subformula $\exists y\,(rxy)$ with its conjugate $\exists w\,(rxw)$ to obtain the equivalent formula

$$\forall x\left(\neg\left((\exists w\,(rxw)) \to \exists z\,(rzy)\right)\right).$$

Now the variables have been properly renamed.

- The next step is to eliminate the $\to$. We do this by replacing $(\exists w\,(rxw)) \to \exists z\,(rzy)$ with $(\neg\,\exists w\,(rxw)) \vee \exists z\,(rzy)$ to obtain

$$\forall x\left(\neg\left((\neg\,\exists w\,(rxw)) \vee \exists z\,(rzy)\right)\right).$$

- Now we are ready to start pulling the quantifiers out, starting with an application of a DeMorgan law:

$$\forall x\left((\exists w\,(rxw)) \wedge (\neg\,\exists z\,(rzy))\right)$$

$$\forall x \exists w \left(rxw \wedge (\neg \exists z\, (rzy)) \right)$$

$$\forall x \exists w \left((rxw) \wedge \forall z\, (\neg\, (rzy)) \right)$$

$$\forall x \exists w \forall z\, ((rxw) \wedge \neg\, (rzy)).$$

Thus we have finally arrived at a prenex form for $\forall x \neg\, (\exists y\, (rxy) \rightarrow \exists z\, (rzy))$.

Exercises

5.10.1 Put the following formulas in the language of graphs into prenex form.

a. $\left(\forall y(\neg\, (y \approx z) \vee \forall y(y \approx z))\right) \vee (ryz)$

b. $(z \approx x) \vee \left(\left(\forall x(\neg \forall z(rxz))\right) \rightarrow (y \approx z)\right)$

c. $\left((rzy) \vee \neg \exists x(\neg \forall y(ryx))\right) \rightarrow (y \approx x)$

d. $\exists z\left(\left((rxz) \rightarrow ((ryx) \rightarrow \exists x(rxy))\right) \vee \neg\, (ryx)\right)$

e. $(x \approx y) \rightarrow \exists y\left(\left((y \approx z) \rightarrow (\exists z(y \approx z))\right) \wedge (y \approx z)\right)$

f. $\exists z\left(\neg\left(\left(((y \approx z) \wedge \exists y(z \approx y)) \wedge (rzz)\right) \vee \exists y(y \approx x)\right)\right)$

5.10.2 Put the following formulas in the language of the natural numbers into prenex form.

a. $(y \approx z) \vee \forall z(z \approx y)$

b. $((y + x) < x) \rightarrow \forall z(\neg\, (z < x))$

c. $\neg \forall x((y \cdot (z \cdot x)) \approx z)$

d. $\neg\left((\exists y(y \approx x)) \rightarrow \neg\, (y \approx x)\right)$

e. $\neg \exists x\left((\forall y \forall z(y \approx z)) \wedge \neg\, (z \approx x)\right)$

f. $\forall y\left((z < (x \cdot y)) \rightarrow ((z \approx z) \wedge \forall x(y < x))\right)$

g. $(\forall y(y < x)) \vee \left((y < (z \cdot x)) \rightarrow \left((\forall z(x < z)) \wedge (z \approx z)\right)\right)$

h. $\neg \forall z\left(((z \approx x) \wedge (x < z)) \vee \exists y \exists z(z \approx (y \cdot y))\right)$

5.11 VALID ARGUMENTS

We assume we are working with sentences in a fixed first–order language $\mathcal{L}$.

Definition 5.11.1 An argument $\mathsf{F}_1, \cdots, \mathsf{F}_n \therefore \mathsf{F}$ is *valid* (or *correct*) in first–order logic provided every structure $\mathbf{S}$ that makes $\mathsf{F}_1, \ldots, \mathsf{F}_n$ true also makes F true, i.e., for all $\mathbf{S}$

$$\mathbf{S} \models \{\mathsf{F}_1, \ldots, \mathsf{F}_n\} \qquad \text{implies} \qquad \mathbf{S} \models \mathsf{F}.$$

Proposition 5.11.2 An argument

$$\mathsf{F}_1, \cdots, \mathsf{F}_n \therefore \mathsf{F}$$

in first–order logic is valid iff

$$\mathsf{F}_1 \wedge \cdots \wedge \mathsf{F}_n \rightarrow \mathsf{F}$$

is valid; and this holds iff

$$\{\mathsf{F}_1, \cdots \mathsf{F}_n, \neg \mathsf{F}\}$$

is not satisfiable.

Proof. (Exercise.) □

Example 5.11.3 The argument

$$\forall x\, (fx \approx ffx)$$
$$\forall x\, (fffx \approx x)$$
$$\therefore \forall x\, (fx \approx x)$$

is valid.

Example 5.11.4 In the language of graphs choose two formulas and a sentence as follows:

- $deg_n(x)$ defining the relation "x has degree n";
- $path_2(x, y)$ defining the relation "there is a path of length 2 from x to y";
- $size_{\leq n}$ says "there are at most n vertices."

Then the argument

$$\forall x \neg (rxx)$$
$$\forall x \forall y((rxy) \rightarrow (ryx))$$
$$\forall x\, deg_2(x)$$
$$\forall x \forall y\, path_2(x, y)$$
$$\therefore size_{\leq 3}$$

is valid.

To show that an argument $\mathsf{F}_1, \cdots, \mathsf{F}_n \;\therefore\; \mathsf{F}$ is not valid it suffices to find a structure $\mathbf{S}$ such that each of the premisses $\mathsf{F}_1, \cdots, \mathsf{F}_n$ is true in $\mathbf{S}$, but F is false in $\mathbf{S}$. Such a structure is called a *counterexample* to the argument.

Example 5.11.5 The argument
$$\begin{array}{l} \forall x \exists y\,(rxy) \\ \therefore\; \exists y \forall x\,(rxy) \end{array}$$
is not valid; a two–element graph gives a counterexample.

Exercises

5.11.1 For the following arguments find a counterexample.

a. $\forall x\,\big(r_1 x \rightarrow (r_2 x \rightarrow r_3 x)\big)$
$\therefore\; \forall x\,\big((r_1 x \rightarrow r_2 x) \rightarrow r_3 x\big)$

b. $\exists x\,(r_1 x \wedge r_2 x)$
$\exists x\,(r_2 x \wedge r_3 x)$
$\therefore\; \exists x\,(r_1 x \wedge r_3 x)$

c. $\forall x \exists y\,(rxy)$
$\forall y \exists x\,(rxy)$
$\therefore\; \forall x \forall y\,\big(\neg\,(x \approx y) \rightarrow (rxy)\big)$

d. $\forall x\,(fffx \approx fffffx)$
$\forall y\,(ffy \approx ffffffy)$
$\therefore\; \forall z\,(ffz \approx fffz)$

e. $\exists z \forall x\,(fxz \approx x)$
$\therefore\; \exists z \forall x\,(fzx \approx x)$

5.11.2 Give details of the proof of Proposition 5.11.2.

5.12 SKOLEMIZING

Now we look at a famous technique for associating to a first–order sentence F a closely related (but usually not equivalent!) sentence F' in prenex form with all the quantifiers (if any) of F' being $\forall$. It is based on the following lemma.

Lemma 5.12.1

a. Given the sentence $\exists y\, \mathsf{G}(y)$, augment the language with a new constant c and form the sentence $\mathsf{G}(c)$. Then
$$\mathrm{Sat}\big(\exists y\, \mathsf{G}(y)\big) \qquad \text{iff} \qquad \mathrm{Sat}\big(\mathsf{G}(c)\big).$$

b. Given the sentence $\forall x_1 \cdots \forall x_n \exists y\, \mathsf{G}(\vec{x}, y)$, augment the language with a new n–ary function symbol f and form the sentence $\forall x_1 \cdots \forall x_n\, \mathsf{G}^\star(\vec{x}, f(\vec{x}))$. Then

$$\mathrm{Sat}\big(\forall x_1 \cdots \forall x_n \exists y\, \mathsf{G}(\vec{x}, y)\big) \quad \text{iff} \quad \mathrm{Sat}\big(\forall x_1 \cdots \forall x_n\, \mathsf{G}^\star(\vec{x}, f(\vec{x}))\big).$$

Proof. In each step we can expand a model of the sentence on the left–hand side to one on the right–hand side by adding the appropriately designated element for the constant, or by adding an appropriate function f. Conversely, we can take a model of the sentence on the right–hand side and remove the constant symbol, respectively function symbol, and get a model for the left–hand side. □

Definition 5.12.2 A first–order formula F is *universal* if it is in prenex form and all quantifiers are universal, i.e., F is of the form $\forall \vec{x}\, \mathsf{G}$, where G is quantifier–free. G is called the *matrix* of F.

Theorem 5.12.3 Given a first–order sentence F, there is an effective procedure for finding a universal sentence F' (usually in an extended language) such that

$$\mathrm{Sat}(\mathsf{F}) \qquad \text{iff} \qquad \mathrm{Sat}(\mathsf{F}').$$

Furthermore, we can choose F' such that every model of F can be expanded to a model of F', and every model of F' can be reducted to a model of F.

Proof. First, we put F in prenex form. Then we just apply Lemma 5.12.1 repeatedly until there are no existential quantifiers. □

Definition 5.12.4 The process of converting a sentence to such a universal sentence is called *skolemizing*. The new constants and functions are called *skolem constants* and *skolem functions*.

Example 5.12.5 Given the sentence $\mathsf{F} = \forall x \exists y\, (x < y)$, we note that it is already in prenex form. Applying Lemma 5.12.1(b), we introduce a new unary function symbol, say f, and arrive at the universal sentence $\mathsf{F}' = \forall x\, (x < fx)$.

The structure $\mathbf{R} = (R, <)$, consisting of the real numbers with the usual $<$, satisfies F. If we choose $f(r) = r + \frac{1}{2}$ for $r \in R$, we see that the expansion $(R, <, f)$ of $\mathbf{R}$ satisfies F'.

Example 5.12.6 Let $\mathsf{F} = \exists x \forall y\, (rxy)$. Applying Lemma 5.12.1(a), we introduce a new constant symbol, say c, and arrive at the universal sentence $\mathsf{F}' = \forall y\, (rcy)$.

The directed graph (S, r) described by the table

r	a	b
a	1	1
b	0	0

satisfies F. We choose the interpretation of c to be a, and this gives a model (S, r, c) of F'.

Example 5.12.7 Now let us do an example that brings all the pertinent ideas together. Let $\mathsf{F} = \exists x \forall y \big((y \not\approx 0) \rightarrow \exists z (y \cdot z \approx x)\big)$. First, we put this in prenex form:

$$\begin{aligned} &\exists x \forall y \big((y \not\approx 0) \rightarrow \exists z (y \cdot z \approx x)\big) \\ &\sim \ \exists x \forall y \exists z \big((y \not\approx 0) \rightarrow (y \cdot z \approx x)\big). \end{aligned}$$

Now, we take the prenex form, the last preceding step, and apply Lemma 5.12.1:

$$\begin{array}{ll} \exists x \forall y \exists z \big((y \not\approx 0) \rightarrow (y \cdot z \approx x)\big) & \text{prenex form} \\ \forall y \exists z \big((y \not\approx 0) \rightarrow (y \cdot z \approx c)\big) & \text{Use 5.12.1(a).} \\ \forall y \big((y \not\approx 0) \rightarrow (y \cdot fy \approx c)\big). & \text{Use 5.12.1(b).} \end{array}$$

Thus we arrive at the skolemized form of F, namely,

$$\mathsf{F}' = \forall y \big((y \not\approx 0) \rightarrow (y \cdot fy \approx c)\big).$$

Example 5.12.8 The steps for skolemizing the first–order sentence $\forall x \exists y \forall z \exists u \forall v \exists w \, (rxyzuvw)$ are as follows:

$$\begin{array}{l} \forall x \exists y \forall z \exists u \forall v \exists w \, (rxyzuvw) \\ \forall x \forall z \exists u \forall v \exists w \, (rxfxzuvw) \\ \forall x \forall z \forall v \exists w \, (rxfxzgxzvw) \\ \forall x \forall z \forall v \, (rxfxzgxzvhxzv), \end{array}$$

where f is unary, g is binary and h is ternary.

Now we extend these ideas to sets of sentences.

Theorem 5.12.9 Given a set of first–order sentences $\mathcal{S}$, there is a set $\mathcal{S}'$ of universal sentences (usually in an extended language) such that

$$\mathrm{Sat}(\mathcal{S}) \qquad \text{iff} \qquad \mathrm{Sat}(\mathcal{S}').$$

Furthermore, every model of $\mathcal{S}$ can be expanded to a model of $\mathcal{S}'$, and every model of $\mathcal{S}'$ can be reducted to a model of $\mathcal{S}$.

Proof. We skolemize each sentence in $\mathcal{S}$ as before, making sure that distinct sentences do not have any common skolem constants or functions. □

This theorem is at the heart of the translation from first–order logic to clause logic, as we shall see in the next section.

Example 5.12.10 We skolemize $\{\forall x \exists y\,(x < y),\ \forall x \exists y\,(x \nless y)\}$ and obtain $\{\forall x\,(x < fx),\ \forall x\,(x \nless gx)\}$.

Exercises

5.12.1 Skolemize the following sentences:

a. $\forall x \exists y\,(x < y)$

b. $\exists x \forall y\,(x < y)$

c. $\exists x \forall y \exists z\,\Big(\big((rxy) \wedge (ryz)\big) \rightarrow (rxz)\Big)$

d. $\forall x\,\big((0 < x) \rightarrow \exists y\,(y \cdot y \approx x)\big)$

e. $\forall x \forall y\,\Big((rxy) \rightarrow \exists z\,\big((rxz) \wedge (rzy)\big)\Big)$

f. $\forall x\,\big(\exists y\,(x \approx y + y) \vee \exists y\,(x \approx y + y + 1)\big)$.

5.13 THE REDUCTION OF FIRST–ORDER LOGIC TO PREDICATE CLAUSE LOGIC

In Sec. 4.5 we saw that the satisfiability of a set of clauses in the predicate clause logic could be reduced to propositional clause logic by using ground clauses over the Herbrand universe. Now we show that first–order formulas can be reduced to the predicate clause logic, and thus to the propositional clause logic.

First we want to look at some basic translations between clauses and first–order formulas.

Definition 5.13.1 Let $\mathsf{C} = \{\mathsf{L}_1(\vec{x}), \ldots, \mathsf{L}_k(\vec{x})\}$ be a clause. Define the universal sentence F_C by

$$\mathsf{F}_\mathsf{C} = \forall \vec{x}\ (\mathsf{L}_1(\vec{x}) \vee \cdots \vee \mathsf{L}_k(\vec{x})).$$

Lemma 5.13.2 Given a clause C, for any structure **S**,

$$\mathbf{S} \models \mathsf{C} \qquad \text{iff} \qquad \mathbf{S} \models \mathsf{F}_\mathsf{C}.$$

Proof. This is a simple consequence of the definition of $\models$ for clauses and for universal sentences. □

Example 5.13.3 For the clause $\mathsf{C} = \{\neg rxy,\ \neg ryz,\ rxz\}$ we have $\mathsf{F_C} = \forall x \forall y \forall z\,(\neg\,(rxy) \vee \neg\,(ryz) \vee (rxz))$. Now, $\mathsf{F_C} \sim \forall x \forall y \forall z\,\Big(((rxy) \wedge (ryz)) \rightarrow (rxz)\Big)$. Thus for $\mathbf{S}$ a structure,

$$\mathbf{S} \models \mathsf{C} \qquad \text{iff} \qquad \mathbf{S} \models \forall x \forall y \forall z\,\Big(((rxy) \wedge (ryz)) \rightarrow (rxz)\Big).$$

There is also a translation from universal first–order sentences to clauses.

Lemma 5.13.4 One can effectively construct, for any given universal sentence F, a finite set $\mathcal{C}_\mathsf{F}$ of clauses such that

$$\mathbf{S} \models \mathsf{F} \qquad \text{iff} \qquad \mathbf{S} \models \mathcal{C}_\mathsf{F},$$

for any structure $\mathbf{S}$.

Proof. Let F be $\forall \vec{x}\, \mathsf{G}(\vec{x}\,)$ with $\mathsf{G}(\vec{x}\,)$ quantifier–free. By propositional logic reasoning we can put $\mathsf{G}(\vec{x}\,)$ in a conjunctive form, i.e.,

$$\mathsf{G}(\vec{x}\,) \ \sim\ \mathsf{G}_1(\vec{x}\,) \wedge \cdots \wedge \mathsf{G}_k(\vec{x}\,),$$

where each $\mathsf{G}_i(\vec{x}\,)$ is a disjunction of literals, i.e.,

$$\mathsf{G}_i(\vec{x}\,) \ \sim\ \mathsf{G}_{i1}(\vec{x}\,) \vee \cdots \vee \mathsf{G}_{im_i}(\vec{x}\,),$$

with each G_{ij} being an atomic formula or a negated atomic formula. Then

$$\begin{aligned} \forall \vec{x}\ \mathsf{G}(\vec{x}\,) &\sim \forall \vec{x}\ (\mathsf{G}_1(\vec{x}\,) \wedge \cdots \wedge \mathsf{G}_k(\vec{x}\,)) \\ &\sim (\forall \vec{x}\ \mathsf{G}_1(\vec{x}\,)) \wedge \cdots \wedge (\forall \vec{x}\ \mathsf{G}_k(\vec{x}\,)) \\ &\sim \mathsf{F}_1 \wedge \cdots \wedge \mathsf{F}_k, \end{aligned}$$

where $\mathsf{F}_i = \forall \vec{x}\ \mathsf{G}_i(\vec{x}\,)$. Thus

$$\mathbf{S} \models \mathsf{F} \qquad \text{iff} \qquad \mathbf{S} \models \{\mathsf{F}_1, \cdots, \mathsf{F}_k\}.$$

Let C_i be the clause $\mathsf{C}_{\mathsf{F}_i}$ associated with F_i, i.e., $\mathsf{C}_i = \{\mathsf{G}_{i1}(\vec{x}\,), \cdots, \mathsf{G}_{im_i}(\vec{x}\,)\}$, and let $\mathcal{C}_\mathsf{F} = \{\mathsf{C}_1, \ldots, \mathsf{C}_k\}$. Then

$$\mathbf{S} \models \mathsf{F} \qquad \text{iff} \qquad \mathbf{S} \models \mathcal{C}_\mathsf{F}.$$

□

Example 5.13.5 Let $\mathsf{F} = \forall x \forall y \forall z\,((rxy) \rightarrow ((rxz) \wedge (rzy)))$. First, we put the matrix of F in conjunctive form:

$$\begin{aligned} (rxy) \rightarrow ((rxz) \wedge (rzy)) &\sim \neg\,(rxy) \vee ((rxz) \wedge (rzy)) \\ &\sim (\neg\,(rxy) \vee (rxz)) \wedge (\neg\,(rxy) \vee (rzy)). \end{aligned}$$

Now we can read off the clauses from the conjuncts, namely,

$$\mathcal{C}_{\mathsf{F}} = \{\{\neg\, rxy,\ rxz\},\ \{\neg\, rxy,\ rzy\}\}.$$

Definition 5.13.6 For $\mathcal{S}$ a set of universal sentences let

$$\mathcal{C}_{\mathcal{S}} = \bigcup_{\mathsf{F}\in\mathcal{S}} \mathcal{C}_{\mathsf{F}},$$

the union of the collection of all sets $\mathcal{C}_{\mathsf{F}}$ of clauses obtained from the universal sentences F in $\mathcal{S}$.

Lemma 5.13.7 Given a set $\mathcal{S}$ of universal sentences and $\mathbf{S}$ a structure,

$$\mathbf{S} \models \mathcal{S} \qquad \text{iff} \qquad \mathbf{S} \models \mathcal{C}_{\mathcal{S}}.$$

Thus

$$\text{Sat}(\mathcal{S}) \qquad \text{iff} \qquad \text{Sat}(\mathcal{C}_{\mathcal{S}}).$$

Proof. This follows immediately from Lemma 5.13.4. □

Now we can formulate a connection between satisfiability of a set of first–order sentences (not necessarily universal) and a set of clauses.

Lemma 5.13.8 Given a set $\mathcal{S}$ of first–order sentences, let $\mathcal{S}'$ be a skolemization of $\mathcal{S}$. Then

$$\text{Sat}(\mathcal{S}) \qquad \text{iff} \qquad \text{Sat}(\mathcal{C}_{\mathcal{S}'}).$$

Proof. From the basic result on skolemization, Theorem 5.12.9 (p. 355), we have

$$\text{Sat}(\mathcal{S}) \qquad \text{iff} \qquad \text{Sat}(\mathcal{S}').$$

And by Lemma 5.13.7

$$\text{Sat}(\mathcal{S}') \qquad \text{iff} \qquad \text{Sat}(\mathcal{C}_{\mathcal{S}'}).$$

Combining these two facts gives the lemma. □

Example 5.13.9 For $\mathcal{S} = \{\exists x \forall y\,(rxy),\ \neg\,\forall y \exists x\,(rxy)\}$ we have

$$\begin{aligned} \mathcal{S}' &= \{\forall y\,(ray),\ \forall x\,\neg\,(rxb)\} \\ \mathcal{C}'_{\mathcal{S}} &= \{\{ray\},\ \{\neg\, rxb\}\}. \end{aligned}$$

We can easily derive the empty clause from $\mathcal{C}'_{\mathcal{S}}$ (in one resolution step), so $\mathcal{C}'_{\mathcal{S}}$ is not satisfiable. Thus $\mathcal{S}$ is also not satisfiable. Consequently, the sentence

$$\mathsf{F} = (\exists x \forall y\,(rxy)) \wedge (\neg\,\forall y \exists x\,(rxy)),$$

obtained by conjuncting the members of $\mathcal{S}$, is not satisfiable. So $\neg\,\mathsf{F}$ is a valid sentence, as every structure $\mathbf{S}$ must satisfy it. Now,

$$\neg\,\mathsf{F} \;\sim\; (\exists x \forall y\,(rxy)) \;\rightarrow\; (\forall y \exists x\,(rxy)),$$

so

$$(\exists x \forall y\,(rxy)) \;\rightarrow\; (\forall y \exists x\,(rxy))$$

is a valid sentence, a fact that we observed earlier in the chapter.

Now we can state the fundamental transformation from first–order logic to clause logic.

Theorem 5.13.10 Given $\mathcal{S}$, a set of sentences, and F, a sentence,

$$\mathcal{S} \models \mathsf{F} \qquad \text{iff} \qquad \neg\,\mathrm{Sat}(\mathsf{C}_{\mathcal{S}(\neg\,\mathsf{F})'}),$$

where $\mathcal{S}(\neg\,\mathsf{F})$ is the set $\mathcal{S} \cup \{\neg\,\mathsf{F}\}$, and the prime symbol refers to taking the skolemization, as before.

Thus the argument $\mathcal{S} \;\; \therefore \; \mathsf{F}$ is valid iff one cannot satisfy the set $\mathsf{C}_{\mathcal{S}(\neg\,\mathsf{F})'}$ of clauses.

Proof. By Proposition 5.11.2 (p. 352) and Lemma 5.13.8,

$$\begin{array}{lll} \mathcal{S} \models \mathsf{F} & \text{iff} & \neg\,\mathrm{Sat}(\mathcal{S} \cup \{\neg\,\mathsf{F}\}) \\ & \text{iff} & \neg\,\mathrm{Sat}(\mathsf{C}_{\mathcal{S}(\neg\,\mathsf{F})'}). \end{array}$$

□

Example 5.13.11 Let us look at the argument which says if we have an equivalence relation on a set and two elements are not related, then their cosets are disjoint. We can express this in first–order logic as follows:

$$\begin{array}{l} \forall x\,(rxx) \\ \forall x \forall y\,((rxy) \rightarrow (ryx)) \\ \forall x \forall y \forall z\,\Big(((rxy) \wedge (ryz)) \rightarrow (rxz)\Big) \\ \therefore\;\; \forall x \forall y\,\Big((\neg\,(rxy)) \rightarrow \forall u\,((rxu) \rightarrow \neg\,(ryu))\Big). \end{array}$$

The premisses say that r is an equivalence relation, and the conclusion says that if x and y are not equivalent (by r), then any u in the coset of x is not in the coset of y. Let us translate this into the clause logic. Let $\mathcal{S}$ be the set of premisses, and F the conclusion. Now,

$$\neg\,\mathsf{F} \;\sim\; \exists x \exists y\,\Big((\neg\,(rxy)) \wedge \exists u\,((rxu) \wedge (ryu))\Big).$$

So putting $\neg\,\mathsf{F}$ in prenex form gives

$$\neg\,\mathsf{F} \;\sim\; \exists x \exists y \exists u \Big((\neg\,(rxy)) \wedge (rxu) \wedge (ryu) \Big).$$

The premisses $\mathcal{S}$ are already skolemized, as they are universal sentences. For $\neg\,\mathsf{F}$ we obtain the skolemized form

$$(\neg\, r(ab)) \wedge (rac) \wedge (rbc).$$

Thus the set of clauses corresponding to the argument is

$$\begin{array}{l} \{rxx\} \\ \{\neg\, rxy,\ ryx\} \\ \{\neg\, rxy,\ \neg\, ryz,\ rxz\} \\ \{\neg\, rab\},\ \{rac\},\ \{rbc\}. \end{array}$$

We can also handle sentences with equality.

Example 5.13.12 Consider the argument

$$\begin{array}{l} \forall x\,(x \cdot 1 \approx x) \\ \forall x \exists y\,(x \cdot y \approx 1) \\ \forall x \forall y \forall z\,\big((x \cdot y) \cdot z \approx x \cdot (y \cdot z)\big) \\ \therefore \forall x\,(1 \cdot x \approx x). \end{array}$$

The negation of the conclusion is $\exists x\,(1 \cdot x \not\approx x)$. Thus skolemizing the premisses with the negated conclusion gives

$$\begin{array}{l} \forall x\,(x \cdot 1 \approx x) \\ \forall x\,(x \cdot fx \approx 1) \\ \forall x \forall y \forall z\,\big((x \cdot y) \cdot z \approx x \cdot (y \cdot z)\big) \\ 1 \cdot a \not\approx a. \end{array}$$

These are ready to be converted to clauses, giving

$$\begin{array}{l} \{x \cdot 1 \approx x\} \\ \{x \cdot fx \approx 1\} \\ \{(x \cdot y) \cdot z \approx x \cdot (y \cdot z)\} \\ \{1 \cdot a \not\approx a\}. \end{array}$$

We can convert this into a set of clauses without equality, as in Sec. 4.9:

$$\begin{array}{l}\{x \equiv x\}\\ \{x \not\equiv y,\ y \equiv x\}\\ \{x \not\equiv y,\ y \not\equiv z,\ x \equiv z\}\\ \{x \not\equiv y,\ fx \equiv fy\}\\ \{x_1 \not\equiv y_1,\ x_2 \not\equiv y_2,\ x_1 \cdot x_2 \equiv y_1 \cdot y_2\}\\ \{x \cdot 1 \equiv x\}\\ \{x \cdot fx \equiv 1\}\\ \{(x \cdot y) \cdot z \equiv x \cdot (y \cdot z)\}\\ \{1 \cdot a \not\equiv a\}.\end{array}$$

Exercises

5.13.1 For each of the following first–order sentences F find a finite set of clauses $\mathcal{C}$ such that F is satisfiable iff $\mathcal{C}$ is satisfiable.

a. $\forall x \exists y \left((rxy) \rightarrow (ryx)\right)$

b. $\exists y \forall x \,(x < y)$

c. $\forall x \exists y \left((0 < x) \rightarrow (x \approx y \cdot y)\right)$

d. $\forall x \exists y \left((y \cdot y) \cdot y \approx x\right)$

e. $\forall x \forall y \Big((x < y) \rightarrow \exists z \left((x < z) \wedge (z < y)\right)\Big)$

f. $\forall x \Big(\left(\exists y \,(x \approx y + y)\right) \vee \exists y \left(x \approx (y + y) + 1\right)\Big)$

5.13.2 For each of the following arguments in first–order logic find a finite set $\mathcal{C}$ of clauses such that the argument is valid iff $\mathcal{C}$ is not satisfiable. r_k denotes a k–ary relation symbol, and f_k a k–ary function symbol.

a. $\forall x \exists y \left((r_1 x) \rightarrow \neg (r_2 xy)\right)$
$\exists x \forall y \left((r_2 xy) \rightarrow \neg (r_1 y)\right)$
$\therefore \exists x \forall y \left((r_1 x) \wedge (r_1 y)\right)$

b. $\forall x \exists y \left((r_2 xy) \rightarrow (r_2 yx)\right)$
$\exists x \forall y \left(\neg (r_2 xy)\right)$
$\therefore \forall x \forall y \,(r_2 xy)$

c. $\forall x \exists y \left((r_2 x f_1 y) \rightarrow (r_2 y f_1 x)\right)$
$\exists x \forall y \exists z \left(\neg (r_2 x f_1 y) \rightarrow \neg (r_2 y f_1 z)\right)$
$\therefore \forall x \exists y \forall z \left((r_2 f_1 xy) \vee \neg (r_2 y f_1 z)\right)$

d. $\forall x \forall y\,((f_1 x \approx f_1 y) \rightarrow (x \approx y))$
$\therefore \forall y \exists x\,(f_1 x \approx y)$

e. $\forall x\,(f_1 x \approx f_1 f_1 x)$
$\forall x\,(f_1 f_1 f_1 x \approx x)$
$\therefore \forall x\,(f_1 x \approx x)$

f. $\forall x \forall y\,((f_1 x \approx f_1 y) \rightarrow (x \approx y))$
$\therefore \forall y \exists x\,(f_1 x \approx y)$

g. $\forall x \exists y\,((f_2 xy \approx f_2 yx) \rightarrow (f_1 x \approx f_1 y))$
$\forall x \exists y\,((f_1 x \approx f_1 y) \rightarrow (x \approx y))$
$\therefore \forall x \exists y\,((f_2 xy \approx f_2 yx) \rightarrow (x \approx y))$

5.14 THE COMPACTNESS THEOREM

The following theorem is one of the most powerful for first–order logic. With it we can often show that from the existence of rather mundane models follows the existence of some more interesting and/or rather nonintuitive ones.

Theorem 5.14.1 Let $\mathcal{S}$ be a set of first–order formulas. Then

$$\mathrm{Sat}(\mathcal{S}) \qquad \text{iff} \qquad \mathrm{Sat}(\mathcal{S}_0) \text{ for all finite } \mathcal{S}_0 \subseteq \mathcal{S}.$$

Proof. This follows from Lemma 5.13.8 (p. 358) on the reduction of satisfiability in first–order logic to satisfiability in clause logic, since we have a compactness theorem for clause logic with equality (Theorem 4.10.8, p. 312). □

Corollary 5.14.2 Let $\mathcal{S}$ be a set of sentences, and F a sentence. Then

$$\mathcal{S} \models \mathsf{F} \qquad \text{iff} \qquad \mathcal{S}_0 \models \mathsf{F}, \text{ for some finite } \mathcal{S}_0 \subseteq \mathcal{S}.$$

Proof. From Proposition 5.11.2 (p. 352) we have

$$\mathcal{S} \models \mathsf{F} \qquad \text{iff} \qquad \neg\,\mathrm{Sat}(\mathcal{S} \cup \{\neg\,\mathsf{F}\}).$$

By compactness, the right–hand side holds iff for some finite $\mathcal{S}_0 \subseteq \mathcal{S}$ we have $\neg\,\mathrm{Sat}(\mathcal{S}_0 \cup \{\neg\,\mathsf{F}\})$. When we apply Proposition 5.11.2 again, the latter holds for a given $\mathcal{S}_0$ iff $\mathcal{S}_0 \models \mathsf{F}$. □

The following three examples indicate fairly typical applications of the compactness theorem for first–order logic.

Example 5.14.3 A first application of the compactness theorem is to look at models of *first–order arithmetic*, i.e., all the first–order sentences that are true of $\mathbf{N} = (N, +, \cdot, <, 0, 1)$.

We have the *standard* model, namely, $\mathbf{N}$. However, we claim there are other models that start off like the natural numbers but continue on with numbers that are greater than all the natural numbers.

To show this, we take a new constant symbol c and let $\mathcal{S}$ be the set of sentences true of $\mathbf{N}$ along with the infinitely many axioms $c \not\approx n$, for $n = 0, 1, \cdots$. Given any finite subset $\mathcal{S}_0$ of $\mathcal{S}$ we can easily satisfy them in the standard model. So by compactness there must be a model of $\mathcal{S}$.

We can define the natural order $\leq$ on such a model by $x \leq y$ iff $\exists z(x + z \approx y)$ and prove (from $\mathcal{S}$) that this order is linear. Then we can continue to show that the usual integers (i.e., $0, 1, \cdots$) that are sums of 1's come first, since $\mathbf{N}$ satisfies

$$\forall x \Big((x \leq n) \rightarrow ((x \approx 0) \vee \cdots \vee (x \approx n))\Big).$$

Then there are the strange new big numbers that someone suggested be called the *supernatural* numbers.

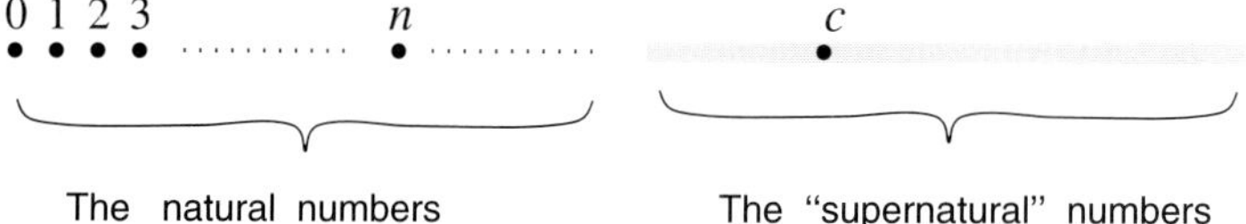

Figure 5.14.4 A nonstandard model of first–order arithmetic

Example 5.14.5 Let our language be $\mathcal{L} = \{r\}$, the language of graphs. Can we find a formula $path(x, y)$ in the language of graphs that, in any graph $\mathbf{G}$, defines the relation "there is a path from x to y"? Suppose we have such a formula.

Let $path_n(x, y)$ define the relation "there is a path from x to y of length n." Then in the language of graphs plus two constants a, b, we have

$$\{\neg\, path_n(a, b) : n \geq 1\} \models \neg\, path(a, b),$$

since we have a path from a to b iff for some positive integer n there is a path of length n from a to b. By Corollary 5.14.2 there must be a positive integer k such that

$$\{\neg\, path_n(a, b) : 1 \leq n \leq k\} \models \neg\, path(a, b).$$

But this is false, for consider the graph

Figure 5.14.6 A counterexample to a definability claim

There is no path of length at most k connecting the endpoints, but there is a path connecting them. Thus we have proved that the relation "there is a path from x to y" is not definable in graphs.

Definition 5.14.7 A class of structures K is said to be an *elementary* class if there is a set of first–order sentences $\mathcal{S}$ such that K is the class of all models of $\mathcal{S}$. If this is the case, then $\mathcal{S}$ is said to *axiomatize* K. An elementary class is *strictly* elementary if we can choose $\mathcal{S}$ to be finite.

We have looked at several classes that are defined by first–order sentences, e.g., semigroups, groups, Boolean algebras, and graphs, and hence they are all elementary classes. Now we will use the compactness theorem to show a naturally occurring class that is not an elementary class.

Example 5.14.8 The class of finite sets is not an elementary class, for otherwise we could add the statements $\{c_m \not\approx c_n : m < n\}$, where the indices m, n run over the natural numbers, to the axioms and obtain a satisfiable set of sentences (by compactness), and any model would be infinite.

Exercises

5.14.1 Suppose $\mathcal{S}$ is a set of first–order sentences with arbitrarily large finite models, i.e., for any given positive integer n we can find a structure $\mathbf{S}$ that satisfies $\mathcal{S}$ and has at least n elements. Prove that $\mathcal{S}$ must have an infinite model.

Thus the collection of finite $\mathcal{L}$–structures, for any first–order language $\mathcal{L}$, is not an elementary class.

5.14.2 Suppose that $\mathcal{S}$ is a set of first–order sentences and $\mathsf{F}(x)$ is a formula such that for any positive integer n we can find a model $\mathbf{S}_n$ of $\mathcal{S}$ such that there are at least n elements of $\mathbf{S}_n$ that satisfy $\mathsf{F}(x)$. Prove that there is a model of $\mathcal{S}$ with infinitely many elements satisfying $\mathsf{F}(x)$.

5.15 HISTORICAL REMARKS

First–order formulas appeared toward the end of Vol. III of Schröder's *Algebra der Logik*, but there was no focused development of this area. The idea that first–order formulas could lead to an interesting logic seems to have originated with Hilbert. In the previous work of Frege, and of Whitehead and Russell, first–order notions were a part of their systems, but first–order proof systems were not isolated and given special treatment. Hilbert and Ackermann were the first to do this in their classic book of 1928.

The first significant results on first–order logic were published in 1915 by Löwenheim. He showed that if a first–order sentence F had a model, then it had a countable model, i.e., one could choose the universe to be either the natural numbers or finitely many natural numbers. This had remarkable consequences for the study of foundations, but for our purposes the important contribution was his method of proof.

In a somewhat cumbersome notation based on "fleeing–variables" Löwenheim essentially skolemized the sentence F, so that he was working with a universal sentence $\forall\vec{x}\ \mathsf{G}(\vec{x}\,)$. Then he looked for truth assignments to ground instances of the matrix $\mathsf{G}(\vec{x}\,)$ of the universal sentence. By making a list of the ground instances $\mathsf{A}_1, \mathsf{A}_2, \ldots$ of the atomic subformulas of G, one can think of the possible truth assignments to the A_i as forming a finitely branching tree. By invoking what was essentially König's Lemma, Löwenheim realized that if one could find arbitrarily long finite branches of truth assignments that satisfied the corresponding ground instances of G, then one could find a satisfying assignment along some infinite branch. Such an infinite branch would yield a countable model. Otherwise, there was no model for F.

In 1920 Skolem gave an elegant proof of Löwenheim's theorem, but his new method did not lead to the reduction of first–order logic to quantifier–free logic that was implicit in Löwenheim's proof.

The more interesting direction of Löwenheim's proof, for our purposes, is that if there is no model of F, then there cannot be arbitrarily long finite branches of satisfying truth assignments, and thus at some finite stage the ground instances, say $\mathsf{G}_1, \ldots, \mathsf{G}_k$, are not satisfiable. Löwenheim was not concerned that the G_i be clauses. The focus on clause logic did not come until the 1960s.

In 1928 Skolem first presented a "new" method for proving first–order theorems, a method that he thought was superior to traditional axiomatic approaches with rules of inference. It was in a rather simple paper in 1929, based on a talk he gave in 1928, that what are now called *skolem functions* appeared. To prove that a first–order sentence F was valid he skolemized $\neg$ F and proceeded to show (in a simple example) that some finite collection of ground instances was not satisfiable. This demonstration was clearly a successor to the ideas in Löwenheim's proof.

What Hilbert and Ackermann did was to give a system of axioms and rules of inference for first–order logic. Thus they had the notion of derivation, i.e., of $\mathcal{S} \vdash$ F. Löwenheim and Skolem worked with the semantic notions, i.e., of $\mathcal{S} \models$ F.

Herbrand wanted to develop first–order logic without reference to infinite models, to tie together the first–order proof system of Hilbert and Ackermann with the study of ground instances by Löwenheim and Skolem. His work was in his Ph.D. thesis of 1930. The following triangle shows the basic connections, where (G_n) denotes the list of ground instances of the skolemization of $\neg$ F:

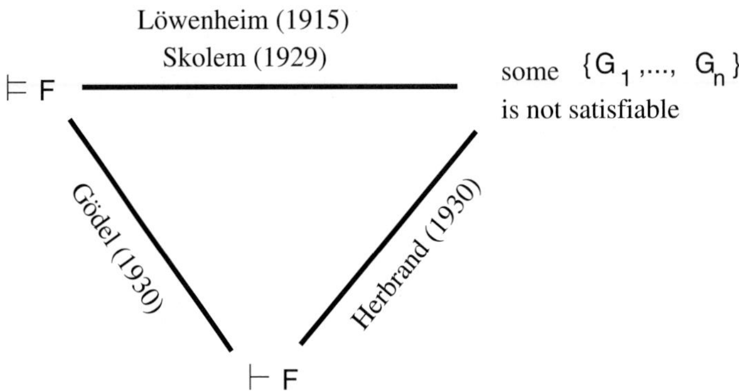

Figure 5.15.1 Approaches to first–order logic

The notion of $\vdash$, and the connection between $\vdash$ and $\models$, i.e., the completeness theorem of Gödel, is the subject of the next, and last, chapter.

Chapter 6

A Proof System for First–Order Logic, and Gödel's Completeness Theorem

A set of axioms, with rules of inference, just for first–order logic was first proposed by Hilbert and Ackermann in 1928. Such a first–order proof system has two main requirements:

- (*Soundness*) Only valid sentences can be derived.
- (*Completeness*) All valid sentences can be derived.

The first is essential—we don't want to derive sentences that are not valid. The second is icing on the cake.

For the first–order proof system of Hilbert and Ackermann the axioms were readily seen to be valid, and the rules sound. In 1930 K. Gödel published his proof that this system was also complete (his Ph.D. thesis at the University of Vienna). As a corollary he gave the first proof of the compactness theorem for first–order logic.

6.1 A PROOF SYSTEM

SYMBOLS	
connectives $\rightarrow, \neg$	quantifier $\exists$
function symbols	equality $\approx$
relation symbols	constant symbols
variables	

There are now many sound and complete proof systems for first–order logic in the literature. We have selected one that uses the (adequate) propositional connectives $\neg$ and $\rightarrow$, and the single quantifier $\exists$. Since we have the equivalence $\forall x\, \mathsf{F} \sim \neg \exists x \neg \mathsf{F}$, it follows that any formula in our previous development of first–order logic is equivalent to a formula using just these connectives and the one quantifier.

AXIOMS FROM TAUTOLOGIES

If $\mathsf{F}(P_1, \ldots, P_k)$ is a propositional tautology and the G_i are first–order formulas, then we take $\mathsf{F}(\mathsf{G}_1, \ldots, \mathsf{G}_k)$ as an axiom

AXIOMS FOR EQUALITY

$s \approx s$

$(s \approx t) \rightarrow (\mathsf{F}(\cdots s \cdots) \rightarrow \mathsf{F}(\cdots t \cdots))$ for F quantifier–free
(replacing a single occurrence of the term s with the term t)

MODUS PONENS

$$\frac{\mathsf{F}, \mathsf{F} \rightarrow \mathsf{G}}{\mathsf{G}}$$

RULES FOR QUANTIFICATION

$$\frac{\mathsf{F} \rightarrow \mathsf{G}}{(\exists x\, \mathsf{F}) \rightarrow \mathsf{G}}$$

provided G does not have a free occurrence of x

$$\frac{\mathsf{F} \rightarrow \mathsf{G}^\star(t, x_1, \ldots, x_n)}{\mathsf{F} \rightarrow \exists x\, \mathsf{G}(x, x_1, \ldots, x_n)}$$

Recall that $\mathsf{G}^\star(t, x_1, \ldots, x_n)$ means any conjugate of G that does not bind the variables of the indicated occurrences of t.

We will use the name *tautologies* for the axioms $\mathsf{F}(\mathsf{G}_1, \ldots, \mathsf{G}_k)$ that are derived from (propositional) tautologies.

Thus we have taken a generalized sound and complete system from the propositional logic, namely, tautologies with modus ponens, and added two axiom schemata to handle equality, and two rules to handle the existential quantifier.

Remark 6.1.1 We can replace the tautology axioms with axiom schemata that, with modus ponens, provide a propositional proof system with generalized soundness and completeness, e.g., we could use the axiom schemata from the Frege-Łukasiewicz propositional proof system presented in Appendix D:

AXIOMS FROM PROPOSITIONAL LOGIC
$\mathsf{F} \to (\mathsf{G} \to \mathsf{F})$
$(\mathsf{F} \to (\mathsf{G} \to \mathsf{H})) \to ((\mathsf{F} \to \mathsf{G}) \to (\mathsf{F} \to \mathsf{H}))$
$(\neg\mathsf{F} \to \neg\mathsf{G}) \to (\mathsf{G} \to \mathsf{F})$

Doing so would require modifications in a handful of the propositional logic steps in this chapter. For those steps see the proof of the generalized soundness and completeness of the Frege-Łukasiewicz proof system given in Appendix D.

Definition 6.1.2 Given a set $\mathcal{S}$ of *formulas* and a *formula* F we say that $\mathcal{S} \vdash \mathsf{F}$ if there is a finite sequence $\mathsf{F}_1, \ldots, \mathsf{F}_n$ of formulas such that each is either an axiom or a member of $\mathcal{S}$ or results from applying one of the rules of inference to previous F_j's, and F_n is F. We call $\mathsf{F}_1, \ldots, \mathsf{F}_n$ a *derivation* of F from $\mathcal{S}$ and say F *is derivable from* $\mathcal{S}$.

Application of the first quantifier rule

$$\left[\frac{\mathsf{F} \to \mathsf{G}}{(\exists x\, \mathsf{F}) \to \mathsf{G}} \quad \text{provided } \mathsf{G} \text{ does not have a free occurrence of } x\right]$$

requires that there be no free occurrence of x *in any member of* $\mathcal{S} \cup \{\mathsf{G}\}$.

Remark 6.1.3 It would seem natural to want a first–order proof system for deriving sentences from sentences, with all the steps in derivations also being sentences. However, the use of formulas has long been accepted as the simplest way to develop a proof system in first–order logic.

6.2 FIRST FACTS ABOUT DERIVATIONS

Now, let us look at some simple derivations using this system.

Lemma 6.2.1 If $\mathcal{S} \vdash \mathsf{F}$ and $\mathcal{S} \vdash \mathsf{F} \to \mathsf{G}$, then $\mathcal{S} \vdash \mathsf{G}$.

Proof. (Exercise.) □

Lemma 6.2.2 [PROPOSITIONAL IMPLICATION] Given a propositional tautology $\mathsf{F}(P_1, \ldots, P_k) \to \mathsf{G}(P_1, \ldots, P_k)$ and a set $\mathcal{S}$ of formulas, and first–order formulas $\mathsf{H}_1, \ldots, \mathsf{H}_k$, if $\mathcal{S} \vdash \mathsf{F}(\mathsf{H}_1, \ldots, \mathsf{H}_k)$, then $\mathcal{S} \vdash \mathsf{G}(\mathsf{H}_1, \ldots, \mathsf{H}_k)$.

Proof. As $\mathsf{F}(\vec{\mathsf{H}}) \to \mathsf{G}(\vec{\mathsf{H}})$ is a tautology, $\mathcal{S} \vdash \mathsf{G}(\vec{\mathsf{H}})$ follows by Lemma 6.2.1. □

We will refer to this lemma by its initials, PI.

Lemma 6.2.3 If $\mathcal{S} \vdash \mathsf{G}$, then $\mathcal{S} \vdash \mathsf{F} \to \mathsf{G}$.

Proof. Let $\mathsf{G}_1, \ldots, \mathsf{G}$ be a derivation of G from $\mathcal{S}$. Then

$$\mathsf{G}_1, \ldots, \mathsf{G},\ \mathsf{G} \to (\mathsf{F} \to \mathsf{G}),\ \mathsf{F} \to \mathsf{G}$$

is a derivation of $\mathsf{F} \to \mathsf{G}$ from $\mathcal{S}$. □

The next lemma shows that equality has the usual reflexive, symmetric, and transitive properties.

Lemma 6.2.4

a. $\mathcal{S} \vdash s \approx s$
b. $\mathcal{S} \vdash s \approx t$ implies $\mathcal{S} \vdash t \approx s$
c. If $\mathcal{S} \vdash s \approx t$ and $\mathcal{S} \vdash t \approx t'$, then $\mathcal{S} \vdash s \approx t'$.

Proof.

a. We use the first scheme of equality axioms.
b. We use the second scheme of equality axioms to obtain

$$\vdash (s \approx t) \to ((s \approx s) \to (t \approx s)).$$

Then we apply Lemma 6.2.1 twice.
c. We use the second scheme of equality axioms to obtain

$$\vdash (s \approx t) \to (\neg (s \approx t') \to \neg (t \approx t')).$$

Then applying PI yields

$$\vdash (s \approx t) \to ((t \approx t') \to (s \approx t')).$$

Now we apply Lemma 6.2.1 twice.

□

The next lemma gives the important *replacement property* for atomic formulas.

Lemma 6.2.5 If $\mathcal{S} \vdash s_1 \approx s_2$, then for t a term and r a relation symbol, we have

a. $\mathcal{S} \vdash t(\cdots s_1 \cdots) \approx t(\cdots s_2 \cdots)$, and
b. $\mathcal{S} \vdash r(\cdots s_1 \cdots) \rightarrow r(\cdots s_2 \cdots)$.

Proof. Both of these claims follow from the equality axioms, using Lemma 6.2.1. □

Now we make the first connections between semantics and syntax.

Definition 6.2.6 A formula $\mathsf{F}(\vec{x})$ is *valid* if $\mathbf{S} \models \mathsf{F}(\vec{a})$ for every structure $\mathbf{S}$ belonging to the language and every $\vec{a} \in S$. A sentence F is *valid* if $\mathbf{S} \models \mathsf{F}$ for every structure $\mathbf{S}$.

Lemma 6.2.7 If $\vdash \mathsf{F}$, then F is a valid formula.

Proof. This is an argument by induction on the length of a derivation of F. The axioms are valid formulas, and the rules produce only valid formulas from valid formulas. □

Exercises

6.2.1 Provide a proof for Lemma 6.2.1.

6.2.2 Fill in the induction proof of Lemma 6.2.7.

6.2.3 If $\mathcal{S} \vdash \mathsf{F}^\star(t, x_1, \ldots, x_n)$, show that $\mathcal{S} \vdash \exists x\, \mathsf{F}(x, x_1, \ldots, x_n)$.

Let $\forall x\, \mathsf{F}$ be an abbreviation for $\neg \exists x \neg \mathsf{F}$.

6.2.4 If $\mathcal{S} \vdash \mathsf{F} \rightarrow \mathsf{G}$, show that $\mathcal{S} \vdash \mathsf{F} \rightarrow \forall x\, \mathsf{G}$, provided x does not occur free in F or in $\mathcal{S}$.

6.3 HERBRAND'S DEDUCTION LEMMA

The following looks just like the deduction lemma for propositional logic, except now it applies to first–order formulas.

Lemma 6.3.1 [Deduction Lemma]
Let $\mathcal{S}$ be a set of first–order formulas, and F, G first–order formulas. Then

$$\mathcal{S} \cup \{\mathsf{F}\} \vdash \mathsf{G} \qquad \text{iff} \qquad \mathcal{S} \vdash \mathsf{F} \rightarrow \mathsf{G}.$$

Proof. The direction ($\Leftarrow$) follows immediately from Lemma 6.2.1. For ($\Rightarrow$) we proceed by induction on the number n of steps in a derivation $\mathsf{F}_1, \ldots, \mathsf{F}_n$ of G from $\mathcal{S} \cup \{\mathsf{F}\}$.

Ground case: $(n = 1)$

1. If F_1 $(= \mathsf{G})$ is an axiom or $\mathsf{F}_1 \in \mathcal{S}$, then we have $\mathcal{S} \vdash \mathsf{G}$. Then $\mathcal{S} \vdash \mathsf{F} \rightarrow \mathsf{G}$ by Lemma 6.2.3.
2. If F_1 is F, then $\mathcal{S} \vdash \mathsf{F} \rightarrow \mathsf{F}$, as the latter formula is a tautology.

Induction hypothesis: The deduction lemma is true provided the derivation has less than n steps.

Induction step: The derivation has n steps, $\mathsf{F}_1, \ldots, \mathsf{F}_n$.
We consider the possibilities for F_n:

1. F_n is an axiom, or $\mathsf{F}_n \in \mathcal{S}$, or F_n is F. Then we proceed as in the ground case.
2. F_n is the result of applying modus ponens. Then we have F_j and F_k with $j, k < n$ such that $\mathsf{F}_k = \mathsf{F}_j \rightarrow \mathsf{F}_n$. By the induction hypothesis we see that $\mathcal{S} \vdash \mathsf{F} \rightarrow \mathsf{F}_j$ as well as $\mathcal{S} \vdash \mathsf{F} \rightarrow \mathsf{F}_k$, i.e., $\mathcal{S} \vdash \mathsf{F} \rightarrow (\mathsf{F}_j \rightarrow \mathsf{F}_n)$. Now we have

$$\mathcal{S} \vdash (\mathsf{F} \rightarrow (\mathsf{F}_j \rightarrow \mathsf{F}_n)) \rightarrow ((\mathsf{F} \rightarrow \mathsf{F}_j) \rightarrow (\mathsf{F} \rightarrow \mathsf{F}_n)),$$

as the latter formula is a tautology. Applying Lemma 6.2.1 twice gives $\mathcal{S} \vdash \mathsf{F} \rightarrow \mathsf{F}_n$, the desired induction step.
3. F_n is the result of applying the first quantifier rule, say F_n is $(\exists x\, \mathsf{G}) \rightarrow \mathsf{H}$. Then x does not have a free occurrence in $\mathcal{S} \cup \{\mathsf{F}, \mathsf{H}\}$, and we have $\mathsf{G} \rightarrow \mathsf{H}$ at an earlier step, so by the induction hypothesis

$$\mathcal{S} \vdash \mathsf{F} \rightarrow (\mathsf{G} \rightarrow \mathsf{H}).$$

Then applying PI gives

$$\mathcal{S} \vdash \mathsf{G} \rightarrow (\mathsf{F} \rightarrow \mathsf{H}).$$

Now we apply the first quantifier rule to this result (as x is not free in $\mathcal{S} \cup \{\mathsf{F} \rightarrow \mathsf{H}\}$) to obtain

$$\mathcal{S} \vdash (\exists x\, \mathsf{G}) \rightarrow (\mathsf{F} \rightarrow \mathsf{H})\,;$$

and another application of PI gives

$$\mathcal{S} \vdash \mathsf{F} \rightarrow ((\exists x\, \mathsf{G}) \rightarrow \mathsf{H}).$$
4. F_n is the result of applying the second quantifier rule, say F_n is $\mathsf{G} \rightarrow \exists x\, \mathsf{H}$. We have $\mathsf{G} \rightarrow \mathsf{H}^\star(t, x_1, \ldots, x_n)$ at an earlier step, so by the induction hypothesis

$$\mathcal{S} \vdash \mathsf{F} \rightarrow (\mathsf{G} \rightarrow \mathsf{H}^\star(t, x_1, \ldots, x_n))\,;$$

and then by applying PI we have

$$\mathcal{S} \vdash \neg(\mathsf{F} \rightarrow \neg\mathsf{G}) \rightarrow \mathsf{H}^{\star}(t, x_1, \dots, x_n).$$

We can apply the second quantifier rule to this result to obtain

$$\mathcal{S} \vdash \neg(\mathsf{F} \rightarrow \neg\mathsf{G}) \rightarrow \exists x\, \mathsf{H}(x, x_1, \dots, x_n);$$

and another application of PI gives

$$\mathcal{S} \vdash \mathsf{F} \rightarrow (\mathsf{G} \rightarrow \exists x\, \mathsf{H}).$$

This finishes the proof of the deduction lemma. □

From the deduction lemma we have "proof by cases."

Lemma 6.3.2 If $\mathcal{S} \cup \{\mathsf{F}\} \vdash \mathsf{G}$ and $\mathcal{S} \cup \{\neg\mathsf{F}\} \vdash \mathsf{G}$, then $\mathcal{S} \vdash \mathsf{G}$.

Proof. Using the deduction lemma we have $\mathcal{S} \vdash \mathsf{F} \rightarrow \mathsf{G}$ and $\mathcal{S} \vdash \neg\mathsf{F} \rightarrow \mathsf{G}$. Now $\mathcal{S} \vdash (\mathsf{F} \rightarrow \mathsf{G}) \rightarrow ((\neg\mathsf{F} \rightarrow \mathsf{G}) \rightarrow \mathsf{G})$, as the formula is a tautology. Applying Lemma 6.2.1 twice gives the desired conclusion. □

Theorem 6.3.3 [SOUNDNESS] The first–order proof system is sound, i.e., given a set of sentences $\mathcal{S}$ and a sentence F we have

$$\mathcal{S} \vdash \mathsf{F} \quad \Longrightarrow \quad \mathcal{S} \models \mathsf{F}.$$

Proof. If $\mathcal{S} \vdash \mathsf{F}$, then for some finite subset $\{\mathsf{F}_1, \dots, \mathsf{F}_n\}$ of $\mathcal{S}$ we have $\{\mathsf{F}_1, \dots, \mathsf{F}_n\} \vdash \mathsf{F}$, as derivations are finite. But then $\vdash \mathsf{F}_1 \rightarrow (\mathsf{F}_2 \rightarrow (\cdots \rightarrow \mathsf{F}_n)\cdots)$ by the deduction lemma. By Lemma 6.2.7 we see that $\mathsf{F}_1 \rightarrow (\mathsf{F}_2 \rightarrow (\cdots \rightarrow \mathsf{F}_n)\cdots)$ must be valid. But then $\{\mathsf{F}_1, \dots, \mathsf{F}_n\} \models \mathsf{F}$, and consequently $\mathcal{S} \models \mathsf{F}$. □

Exercises

6.3.1

a. If $\mathsf{F}_1, \dots, \mathsf{F}_k$ is a derivation to witness $\vdash \mathsf{L} \rightarrow \mathsf{F}$, and $\mathsf{G}_1, \dots, \mathsf{G}_m$ is a derivation to witness $\vdash \mathsf{L} \rightarrow (\mathsf{F} \rightarrow \mathsf{G})$, then show how to find a derivation to witness $\vdash \mathsf{L} \rightarrow \mathsf{G}$.

b. If $\mathsf{F}_1, \dots, \mathsf{F}_k$ is a derivation to witness $\vdash \mathsf{L} \rightarrow (\mathsf{F} \rightarrow \mathsf{G})$, and $\mathsf{G}_1, \dots, \mathsf{G}_m$ is a derivation to witness $\vdash \mathsf{L} \rightarrow \neg\mathsf{G}$, then show how to find a derivation to witness $\vdash \mathsf{L} \rightarrow \neg\mathsf{F}$.

c. If $\mathsf{F}_1, \dots, \mathsf{F}_k$ is a derivation to witness $\vdash \mathsf{L} \rightarrow (\mathsf{F} \rightarrow \mathsf{G})$, and $\mathsf{G}_1, \dots, \mathsf{G}_m$ is a derivation to witness $\vdash \mathsf{L} \rightarrow (\mathsf{G} \rightarrow \mathsf{H})$, then show how to find a derivation to witness $\vdash \mathsf{L} \rightarrow (\mathsf{F} \rightarrow \mathsf{H})$.

6.4 CONSISTENT SETS OF FORMULAS

Definition 6.4.1 A set of formulas $\mathcal{S}$ is *consistent* if there is some formula that cannot be derived from $\mathcal{S}$. We use the notation $\mathsf{Consis}(\mathcal{S})$.

Lemma 6.4.2 Let F be a formula, and let $\mathcal{S}$ be a set of formulas. Then $\mathcal{S}$ is consistent iff $\neg(\mathsf{F} \to \mathsf{F})$ cannot be derived from $\mathcal{S}$.

Proof. For any formula G the formula $\neg(\mathsf{F} \to \mathsf{F}) \to \mathsf{G}$ is a tautology. Thus if $\mathcal{S} \vdash \neg(\mathsf{F} \to \mathsf{F})$, then it follows by Lemma 6.2.1 that $\mathcal{S} \vdash \mathsf{G}$ for all formulas G. But then $\mathcal{S}$ is not consistent.

The converse, that if $\mathcal{S}$ is inconsistent then $\neg(\mathsf{F} \to \mathsf{F})$ is derivable, is clear. □

Lemma 6.4.3 A set $\mathcal{S}$ of formulas is consistent iff every finite subset is consistent.

Proof. A derivation from $\mathcal{S}$ can involve only finitely many members from $\mathcal{S}$. Thus $\mathcal{S} \vdash \neg(\mathsf{F} \to \mathsf{F})$ iff $\mathcal{S}_0 \vdash \neg(\mathsf{F} \to \mathsf{F})$, for some finite subset $\mathcal{S}_0$ of $\mathcal{S}$. Now we apply Lemma 6.4.2. □

Lemma 6.4.4 If a set $\mathcal{S}$ of sentences is satisfiable, then $\mathcal{S}$ is consistent, i.e.,

$$\mathsf{Sat}(\mathcal{S}) \quad \Longrightarrow \quad \mathsf{Consis}(\mathcal{S}).$$

Proof. Suppose $\mathcal{S}$ is *not* consistent. We will prove that $\mathcal{S}$ is not satisfiable. Let F be a first–order sentence. We have $\mathcal{S} \vdash \neg(\mathsf{F} \to \mathsf{F})$, and hence $\mathcal{S} \models \neg(\mathsf{F} \to \mathsf{F})$, by Theorem 6.3.3. Since $\neg(\mathsf{F} \to \mathsf{F})$ is false in every structure but true in every model of $\mathcal{S}$, it follows that $\mathcal{S}$ has no models, so $\mathcal{S}$ is not satisfiable. □

6.5 MAXIMAL CONSISTENT SETS OF FORMULAS

The concept that we need to prove completeness is that of a maximal consistent set of formulas.

Definition 6.5.1 A set of formulas $\mathcal{S}$ is *maximal consistent* if it is consistent, but adding any formula not in $\mathcal{S}$ gives an inconsistent set of formulas.

Now we examine the key properties of a maximal consistent set of formulas. The convenient notation $x \notin \mathsf{F}$ means *x does not occur free in* F.

Lemma 6.5.2 Let $\mathcal{M}$ be a maximal consistent set of formulas. Then for any formulas F and G, we have

a. $F \notin \mathcal{M}$ implies $\mathcal{M} \cup \{F\} \vdash G$.
b. $F \notin \mathcal{M}$ implies $\mathcal{M} \nvdash \neg F$.
c. $\mathcal{M} \vdash F$ implies $F \in \mathcal{M}$.
d. $F \in \mathcal{M}$ or $\neg F \in \mathcal{M}$.
e. $F \notin \mathcal{M}$ or $\neg F \notin \mathcal{M}$.
f. $F \to G \notin \mathcal{M}$ holds iff $F \in \mathcal{M}$ and $G \notin \mathcal{M}$.
g. $F \in \mathcal{M}$ and $F \to G \in \mathcal{M}$ implies $G \in \mathcal{M}$.

Proof. In the following W is $\neg(F \to F)$.

a. This is from the definition of maximal consistent.

b. From $F \notin \mathcal{M}$ we have $\mathcal{M} \cup \{F\} \vdash W$, so by the deduction lemma $\mathcal{M} \vdash F \to W$. If $\mathcal{M} \vdash F$, then $\mathcal{M} \vdash W$ by Lemma 6.2.1. But this contradicts the assumption that $\mathcal{M}$ is consistent. Thus $\mathcal{M} \nvdash F$.

c. Suppose $\mathcal{M} \vdash F$ but $F \notin \mathcal{M}$. Then $\mathcal{M} \cup \{F\} \vdash W$ by (a). Then $\mathcal{M} \vdash F \to W$ by the deduction lemma, and so $\mathcal{M} \vdash W$ by Lemma 6.2.1. But this contradicts consistency.

d. Otherwise, $\mathcal{M} \cup \{F\} \vdash W$ and $\mathcal{M} \cup \{\neg F\} \vdash W$ by (b), so by Lemma 6.3.2 $\mathcal{M} \vdash W$, which contradicts consistency.

e. Otherwise, $\mathcal{M} \vdash F$, $\mathcal{M} \vdash \neg F$, and $\mathcal{M} \vdash (\neg F) \to (F \to W)$, as the last formula is a tautology. Thus $\mathcal{M} \vdash W$ follows by using Lemma 6.2.1 twice. This contradicts consistency.

f. If $F \notin \mathcal{M}$, then $\mathcal{M} \cup \{F\} \vdash G$. Applying the deduction lemma, we have $\mathcal{M} \vdash F \to G$, and then (c) implies $F \to G \in \mathcal{M}$.

If $G \in \mathcal{M}$, then observing that $G \to (F \to G)$ is a tautology, we have $\mathcal{M} \vdash F \to G$ by Lemma 6.2.1. Then by (c) we have $F \to G \in \mathcal{M}$.

Finally, if $F \in \mathcal{M}$ and $G \notin \mathcal{M}$, then if we also had $F \to G \in \mathcal{M}$, it would follow that $\mathcal{M} \vdash F$ and $\mathcal{M} \vdash F \to G$. Applying Lemma 6.2.1 would give $\mathcal{M} \vdash G$, so $G \in \mathcal{M}$ by (c). But this contradicts our assumption on G. Consequently, $F \to G \notin \mathcal{M}$.

g. From $F \in \mathcal{M}$ and $F \to G \in \mathcal{M}$ we have $\mathcal{M} \vdash F$ and $\mathcal{M} \vdash F \to G$. Then Lemma 6.2.1 gives $\mathcal{M} \vdash G$, so $G \in \mathcal{M}$ by (c).

□

In the proof of the completeness theorem we will start out with a consistent set of formulas. The next lemmas describe how we arrive at a maximal consistent set of formulas.

Lemma 6.5.3 If $\mathcal{S}$ is a consistent set of formulas and F is a formula, then at least one of $\mathcal{S} \cup \{F\}$ and $\mathcal{S} \cup \{\neg F\}$ is consistent.

Proof. If both of the extensions of $\mathcal{S}$ were inconsistent, then by Lemma 6.4.3, there would be a finite subset $\{\mathsf{F}_1, \cdots, \mathsf{F}_k\}$ of $\mathcal{S}$ such that both $\{\mathsf{F}_1, \ldots, \mathsf{F}_k, \mathsf{F}\}$ and $\{\mathsf{F}_1, \ldots, \mathsf{F}_k, \neg\,\mathsf{F}\}$ are inconsistent. Then we would have

$$\mathsf{F}_1, \ldots, \mathsf{F}_k, \mathsf{F} \vdash \neg\,(\mathsf{F} \to \mathsf{F})$$
$$\mathsf{F}_1, \ldots, \mathsf{F}_k, \neg\,\mathsf{F} \vdash \neg\,(\mathsf{F} \to \mathsf{F}),$$

so by Lemma 6.3.2 we would have $\mathsf{F}_1, \ldots, \mathsf{F}_k \vdash \neg\,(\mathsf{F} \to \mathsf{F})$. Then by Lemma 6.4.2, we see that $\mathsf{F}_1, \ldots, \mathsf{F}_k$, and hence $\mathcal{S}$, is not consistent, contradicting our assumption. □

Lemma 6.5.4 Any consistent set of formulas $\mathcal{S}$ can be extended to a maximal consistent set of formulas.

Proof. We can do this by listing *all* the formulas $\mathsf{G}_1, \mathsf{G}_2, \ldots$ and adding, in turn, one of each pair G_i, $\neg\,\mathsf{G}_i$ to the set, so that the set remains consistent. This is possible by Lemma 6.5.3. The resulting set $\mathcal{S}'$ will be consistent (as every finite subset is consistent), and it will be maximal, since (1) for every formula F either $\mathsf{F} \in \mathcal{S}'$ or $\neg\,\mathsf{F} \in \mathcal{S}'$, and (2) a consistent set of formulas cannot have both F and $\neg\,\mathsf{F}$ as members, for any formula F. □

6.6 ADDING WITNESS FORMULAS TO A CONSISTENT SENTENCE

In this section we show there is a consistent collection of formulas that give witnesses for existential quantifiers with the property that if any consistent *sentence* F is added to the collection, then the collection is still consistent.

Definition 6.6.1 Let $\mathsf{F}_1, \mathsf{F}_2, \ldots$ be an enumeration of all formulas in the language that start with an existential quantifier. If F_n is $\exists x \mathsf{G}_n(x, \vec{z}\,)$, let

$$\widehat{\mathsf{F}}_n \;=\; \mathsf{F}_n \to \mathsf{G}_n(y_n, \vec{z}\,),$$

where y_n does not occur in $\widehat{\mathsf{F}_1}, \ldots, \widehat{\mathsf{F}}_{n-1}, \mathsf{F}_n$. The formulas $\widehat{\mathsf{F}}_n$ are called *witness* formulas, as the free variables y_n witness the existence of an x.

Lemma 6.6.2 If F is a consistent sentence, then adding the witness formulas to F gives a consistent set of formulas, i.e.,

$$\mathsf{Consis}(\mathsf{F}) \qquad \Longrightarrow \qquad \mathsf{Consis}(\{\mathsf{F}, \widehat{\mathsf{F}}_1, \ldots, \widehat{\mathsf{F}}_n, \ldots\}).$$

Proof. Otherwise, there is a positive n such that

- $\text{Consis}(\{\mathsf{F}, \widehat{\mathsf{F}}_1, \ldots, \widehat{\mathsf{F}}_{n-1}\})$.
- $\neg\,\text{Consis}(\{\mathsf{F}, \widehat{\mathsf{F}}_1, \ldots, \widehat{\mathsf{F}}_n\})$.

Then for any *sentence* H we have $\mathsf{F}, \widehat{\mathsf{F}}_1, \ldots, \widehat{\mathsf{F}}_n \vdash \mathsf{H}$, and since either of the formulas $\neg\,\exists x\, \mathsf{G}_n(x, \vec{z}\,)$ and $\mathsf{G}_n(y_n, \vec{z}\,)$ implies $\widehat{\mathsf{F}}_n$, we have

$$\begin{aligned} \mathsf{F}, \widehat{\mathsf{F}}_1, \ldots, \widehat{\mathsf{F}}_{n-1}, \neg\,\exists x\, \mathsf{G}_n(x, \vec{z}\,) &\ \vdash\ \mathsf{H} \\ \mathsf{F}, \widehat{\mathsf{F}}_1, \ldots, \widehat{\mathsf{F}}_{n-1}, \mathsf{G}_n(y_n, \vec{z}\,) &\ \vdash\ \mathsf{H}. \end{aligned}$$

Applying the deduction lemma to the last statement, we have

$$\vdash\ \mathsf{G}_n(y_n, \vec{z}\,) \to (\mathsf{F} \to (\widehat{\mathsf{F}}_1 \to \cdots \to (\widehat{\mathsf{F}}_{n-1} \to \mathsf{H}) \cdots)).$$

We can apply the first rule for quantifiers to this result, since y_n does not occur as a free variable in any of the formulas $\mathsf{F}, \widehat{\mathsf{F}}_1, \ldots, \widehat{\mathsf{F}}_{n-1}, \mathsf{H}$. This gives

$$\vdash\ \exists x\, \mathsf{G}_n(x, \vec{z}\,) \to (\mathsf{F} \to (\widehat{\mathsf{F}}_1 \to \cdots \to (\widehat{\mathsf{F}}_{n-1} \to \mathsf{H}) \cdots)).$$

Now we reverse the applications of the deduction lemma to obtain

$$\mathsf{F}, \widehat{\mathsf{F}}_1, \ldots, \widehat{\mathsf{F}}_{n-1}, \exists x\, \mathsf{G}_n(x, \vec{z}\,) \vdash \mathsf{H}.$$

But then, applying Lemma 6.3.2, we have

$$\mathsf{F}, \widehat{\mathsf{F}}_1, \ldots, \widehat{\mathsf{F}}_{n-1} \vdash \mathsf{H},$$

which contradicts the consistency of the left–hand side. □

6.7 CONSTRUCTING A MODEL USING A MAXIMAL CONSISTENT SET OF FORMULAS WITH WITNESS FORMULAS

Now we are ready to begin the construction of a structure **S** that models the sentences in a maximal consistent set $\mathcal{M}$ of formulas that has the witness formulas in it. Such maximal consistent sets of formulas are quite amazing—the atomic formulas tell us how to construct the desired structure **S** from the set $T(X)$ of terms. There are parallels with the construction used in the proof of the completeness theorem for equational logic, Theorem 3.10.1 (p. 184).

First, we need to know which terms are to be identified (or "collapsed").

Definition 6.7.1 Let $\equiv$ be the relation defined on the set of terms $T(X)$ by $s \equiv t$ iff $s \approx t \in \mathcal{M}$.

We will need the following version of the replacement property.

Lemma 6.7.2 If $s_1 \equiv s_2$, then for t a term and r a relation symbol, we have

a. $t(\cdots s_1 \cdots) \equiv t(\cdots s_2 \cdots) \in \mathcal{M}$, and
b. $r(\cdots s_1 \cdots) \in \mathcal{M}$ implies $r(\cdots s_2 \cdots) \in \mathcal{M}$.

Proof. These claims follow from Lemma 6.2.5 (p. 370) and Lemma 6.5.2 (p. 375). □

The equivalence classes $[t]$ of the equivalence relation $\equiv$ will give the elements of the set that we use to create the model of $\mathcal{M}$.

Definition 6.7.3 Let S be the set of *equivalence classes* $[t]$ of terms with respect to $\equiv$, i.e., $S = \{[t] : t \in T(X)\}$. We define an interpretation I on S as follows:

- $I(c) = [c]$ for c a constant symbol;
- $I(f)([t_1], \ldots, [t_n]) = [ft_1 \cdots t_n]$;
- $I(r)([t_1], \ldots, [t_n])$ holds iff $rt_1 \cdots t_n \in \mathcal{M}$.

I is well defined by Lemma 6.7.2. Let $\mathbf{S}$ be the structure (S, I).

The notation $[\vec{x}\,]$ means $[x_1], [x_2], \ldots$.

Lemma 6.7.4 For any formula $\mathsf{F}(\vec{x}\,)$ we have

$$\mathbf{S} \models \mathsf{F}([\vec{x}\,]) \qquad \text{iff} \qquad \mathsf{F}(\vec{x}\,) \in \mathcal{M}.$$

Thus, for F a sentence, we have

$$\mathbf{S} \models \mathsf{F} \qquad \text{iff} \qquad \mathsf{F} \in \mathcal{M}.$$

Proof. We proceed by induction.

Ground case: For atomic formulas we have

$$\begin{aligned} \mathbf{S} \models s([\vec{x}\,]) \approx t([\vec{x}\,]) &\iff s^{\mathbf{S}}([\vec{x}\,]) = t^{\mathbf{S}}([\vec{x}\,]) \\ &\iff [s(\vec{x}\,)] = [t(\vec{x}\,)] \\ &\iff s(\vec{x}\,) \equiv t(\vec{x}\,) \\ &\iff s(\vec{x}\,) \approx t(\vec{x}\,) \in \mathcal{M}; \end{aligned}$$

and

$$\begin{aligned} \mathbf{S} \models r([\vec{t}\,]) &\iff r^{\mathbf{S}}([\vec{t}\,]) \text{ holds} \\ &\iff r(\vec{t}\,) \in \mathcal{M}. \end{aligned}$$

Induction hypothesis: The claim holds if the total number of connective and quantifier occurrences in F is less than n.

Induction step: F has n occurrences of connectives and quantifiers. This step breaks down into three subcases:

1. F is $\neg\,\mathsf{G}$. Then

$\mathbf{S} \models \neg\,\mathsf{G}([\vec{x}\,])$	$\iff \mathbf{S} \not\models \mathsf{G}([\vec{x}\,])$	definition of $\models$
	$\iff \mathsf{G}(\vec{x}\,) \notin \mathcal{M}$	induction hypothesis
	$\iff \neg\,\mathsf{G}(\vec{x}\,) \in \mathcal{M}.$	Lemma 6.5.2.

2. F is $\mathsf{G} \rightarrow \mathsf{H}$. Then

$\mathbf{S} \not\models \mathsf{G}([\vec{x}\,]) \rightarrow \mathsf{H}([\vec{x}\,])$

$\iff \mathbf{S} \models \mathsf{G}([\vec{x}\,])$ and $\mathbf{S} \not\models \mathsf{H}([\vec{x}\,])$	definition of $\models$
$\iff \mathsf{G}(\vec{x}\,) \in \mathcal{M}$ and $\mathsf{H}(\vec{x}\,) \notin \mathcal{M}$	induction hypothesis
$\iff \mathsf{G}(\vec{x}\,) \rightarrow \mathsf{H}(\vec{x}\,) \notin \mathcal{M}.$	Lemma 6.5.2.

3. F is $\exists x\,\mathsf{G}(x, \vec{x}\,)$. Then

$\mathbf{S} \models \exists x\,\mathsf{G}(x, [\vec{x}\,])$

$\Longrightarrow \mathbf{S} \models \mathsf{G}([t], [\vec{x}\,])$	for some term t
$\Longrightarrow \mathbf{S} \models \overline{\mathsf{G}}([t], [\vec{x}\,])$	where we choose $\overline{\mathsf{G}}$ to be a conjugate of G such that the bound variables of G are disjoint from the variables of t
$\Longrightarrow \overline{\mathsf{G}}(t, \vec{x}\,) \in \mathcal{M}$	induction hypothesis
$\Longrightarrow \exists x\,\mathsf{G}(x, \vec{x}\,) \in \mathcal{M}$	Lemma 6.5.2, Exercise 6.2.3
$\Longrightarrow \mathsf{F}(\vec{x}\,) \in \mathcal{M}.$	

For the converse

$\mathsf{F}(\vec{x}\,) \in \mathcal{M}$	$\Longrightarrow \exists x\,\mathsf{G}(x, \vec{x}\,) \in \mathcal{M}$	
	$\Longrightarrow \mathsf{G}(y_n, \vec{x}\,) \in \mathcal{M}$	for some y_n (as the witness formulas are in $\mathcal{M}$)
	$\Longrightarrow \mathbf{S} \models \mathsf{G}([y_n], [\vec{x}\,])$	induction hypothesis
	$\Longrightarrow \mathbf{S} \models \exists x\,\mathsf{G}(x, [\vec{x}\,])$	definition of $\models$
	$\Longrightarrow \mathbf{S} \models \mathsf{F}([\vec{x}\,]).$	

□

6.8 CONSISTENT SETS OF SENTENCES ARE SATISFIABLE

Lemma 6.8.1 If F is a consistent sentence, then F is satisfiable, i.e.,

$$\mathsf{Consis}(\mathsf{F}) \qquad \Longrightarrow \qquad \mathsf{Sat}(\mathsf{F}).$$

Proof. By Lemma 6.6.2 we see that we can add the witness formulas to F and still have a consistent set $\mathcal{S}$ of formulas. We extend $\mathcal{S}$ to a maximal consistent set $\mathcal{M}$ of formulas by Lemma 6.5.4. By Lemma 6.7.4 there is a structure **S** satisfying the sentences of $\mathcal{M}$. Then **S** is a model of the sentence F. □

Lemma 6.8.2 If $\mathcal{S}$ is a consistent set of sentences, then $\mathcal{S}$ is satisfiable, i.e.,

$$\mathsf{Consis}(\mathcal{S}) \qquad \Longrightarrow \qquad \mathsf{Sat}(\mathcal{S}).$$

Proof. For $\mathcal{S}_0$ a finite subset of $\mathcal{S}$ let F be the consistent sentence obtained by taking the conjunction of the sentences in $\mathcal{S}_0$. By Lemma 6.6.2 the set of formulas $\{\mathsf{F}, \widehat{\mathsf{F}_1}, \widehat{\mathsf{F}_2}, \ldots\}$ is consistent. Consequently, $\mathcal{S} \cup \{\widehat{\mathsf{F}_1}, \widehat{\mathsf{F}_2}, \ldots\}$ is consistent by Lemma 6.4.3, so we take a maximal consistent extension $\mathcal{M}$ and proceed exactly as in the proof of Lemma 6.8.1 to obtain a model **S** of $\mathcal{S}$. □

6.9 GÖDEL'S COMPLETENESS THEOREM

Theorem 6.9.1 The axioms and rules presented at the beginning of the section are complete for first–order logic, i.e., given a set $\mathcal{S}$ of sentences, and a first–order sentence F,

$$\mathcal{S} \models \mathsf{F} \qquad \Longrightarrow \qquad \mathcal{S} \vdash \mathsf{F}.$$

Proof. Suppose $\mathcal{S} \models \mathsf{F}$. Then we argue as follows, using Proposition 5.11.2 (p. 352) and Lemma 6.8.2:

$$\begin{aligned} \mathcal{S} \models \mathsf{F} \quad &\Longleftrightarrow \quad \neg\,\mathrm{Sat}(\mathcal{S} \cup \{\neg\,\mathsf{F}\}) \\ &\Longrightarrow \quad \neg\,\mathrm{Consis}(\mathcal{S} \cup \{\neg\,\mathsf{F}\}) \\ &\Longrightarrow \quad \mathcal{S} \cup \{\neg\,\mathsf{F}\} \vdash \mathsf{F}. \end{aligned}$$

But clearly, $\mathcal{S} \cup \{\mathsf{F}\} \vdash \mathsf{F}$, which leads to $\mathcal{S} \vdash \mathsf{F}$ by Lemma 6.3.2. □

Although we have defined $\mathcal{S} \vdash \mathsf{F}$ for *formulas*, $\mathcal{S} \models \mathsf{F}$ has been defined only for *sentences*. We can remedy this asymmetry by modifying the semantics. We require that an interpretation I in S also map each variable x to a member of S, i.e., $I(x) \in S$. Then let $\mathbf{S} = (S, I) \models \mathsf{F}(x_1, \ldots, x_n)$ mean $(Ix_1, \ldots, Ix_n) \in \mathsf{F}^{\mathbf{S}}$. With this modification the soundness and completeness theorems can be extended to *formulas*, i.e.,

$$\mathcal{S} \models \mathsf{F} \qquad \Longleftrightarrow \qquad \mathcal{S} \vdash \mathsf{F},$$

for $\mathcal{S}$ a set of formulas, and for F a formula.

6.10 COMPACTNESS

From our previous lemmas we also have Gödel's method of proof of the compactness theorem for first–order logic.

Theorem 6.10.1 For $\mathcal{S}$ a set of first–order sentences we have $\mathsf{Sat}(\mathcal{S})$ iff for every finite $\mathcal{S}_0 \subseteq \mathcal{S}$ we have $\mathsf{Sat}(\mathcal{S}_0)$.

Proof. By Lemmas 6.4.4 and 6.8.2 we see that $\text{Sat}(\mathcal{S})$ holds iff $\text{Consis}(\mathcal{S})$ holds, and by Lemma 6.4.3 the latter holds iff every finite subset of $\mathcal{S}$ is consistent. And again by the just mentioned lemmas, each finite subset of $\mathcal{S}$ is consistent iff it is satisfiable. □

For some applications of first–order compactness see Sec. 5.14.

6.11 HISTORICAL REMARKS

Frege's pioneering work of 1879 did not look at first–order logic but at a *higher–order logic*, where one could quantify over the relation symbols as well, e.g., in his system one could write $\forall r \neg (\forall x \forall y\, (rxy))$, and so forth. Frege claimed that not only had he found a nice set of axioms but he could go farther than Euclid and also specify the rules of inference that were to be used.

His system was very simple and powerful—too powerful it turns out. In 1901 Bertrand Russell found that one could derive a contradiction in Frege's system, and then one could derive any sentence. This caused a considerable commotion in the "foundations of mathematics" circles.

It seems, as we said earlier, that David Hilbert was the first to recognize that first–order proof systems were worthy of study. This was not an obvious step, as many of the most important mathematical concepts seemed to be naturally formulated in a higher–order logic. It is all the more surprising that Hilbert provided this strong focus on first–order logic given that he had not yet realized the power of first–order logic to express "real" mathematics, for in the 1928 book of Hilbert and Ackermann we have the following:

> But as soon as one makes the foundations of theories, especially of mathematical theories, as the object of investigation ... the higher–order calculus is indispensable. ([**26**], p. 86).

In this book they posed the question of whether or not their proof system for first–order logic was complete:

> It is still an unsolved problem as to whether the axiom system is complete in the sense that all logical formulas which are valid in every domain can be derived. It can only be stated on empirical grounds that this axiom system has always been adequate in the applications. The independence of the axioms has not been investigated. ([**26**], p. 68).

Actually, the 1928 book was remarkable for the fact that very little was actually proved in it, not even Löwenheim's theorem, which was the outstanding result prior to 1930.

Kurt Gödel, a student in Vienna, proved the completeness theorem in his doctoral dissertation. He submitted his dissertation in 1929 and published a paper with the completeness theorem in 1930, with an added corollary, the compactness theorem. The proof we gave is based on an idea of Leon Henkin's, to add formulas that give witnesses for the existential quantifiers. Gödel's proof was based on transforming the sentences into skolem normal form (not to be confused with the process of skolemization). He first proved it for first–order logic without equality, and then with equality. And he did not include function symbols. Before the 1950s n–ary functions were treated as $n+1$–ary relations in mathematical logic. This simplified certain matters, since the only terms were the variables.

The use of complete proof systems is, at present, not a crucial factor in automated theorem proving. The systems currently used tend to run out of time and memory long before desired theorems are proved—the crucial question is how to find efficient methods of proving theorems.

Also, as the famous logician Mostowski noted, such formal systems are not actually used by mathematicians to help prove theorems in mathematics; indeed, the one that we have presented is not particularly easy to work with. He noted that working through a volume of *Principia* you would prove statements like $2+2 = 4$, but not much more.

However, the study of such formal systems can shed light on the power of methods being used in ordinary mathematics. Also, they play a role in the design of automated theorem checkers and theorem provers.

Various attempts have been made to develop first–order logics that more closely mimic the way we reason in everyday mathematics, in particular, natural deduction systems and Gentzen systems.

Now that we have examined the concepts of truth and validity for first–order logic we come to two crucial questions:

- How can one determine if a first–order sentence F is true in a particular structure, e.g., in $\mathbf{N}$?
- How can one determine if a first–order sentence or argument is valid?

It has been known since the mid 1930s that in both cases the answer is that no algorithm exists, nor will it ever exist, to determine the truth of first–order sentences in the natural numbers or to determine the valid statements and arguments for a simple language such as the language of graphs.

Appendix A

A Simple Timetable of Mathematical Logic and Computing

The formalization of correct reasoning has been motivated by three major goals: (1) to provide a foundation for mathematics, (2) to eliminate errors from reasoning, and (3) to search for efficient means to find justification for a conclusion from given premisses. The following timetable highlights some of the most important developments.

1. 1850– The formalization of mathematics
2. 1850– The search for good algorithms
3. 1920– The discovery of completeness theorems
4. 1930–1935 The discovery of undecidability
5. 1940–1950 The invention of electronic computers
6. 1950–1960 Early implementations of theorem provers
7. 1960– The search for better algorithms
8. 1965– The study of complexity
9. 1980– Microcomputers

Let us say a few words about each of the preceding items.

1. There are two aspects to the formalization of mathematics:

- the reduction of mathematical systems to simpler systems, and
- the formulation of a logic suitable to carrying out traditional mathematics.

The reduction of mathematics to simple systems was well under way by 1850. In particular, Dirichlet said (according to Dedekind) that all statements in analysis or higher algebra could be reformulated as assertions about the natural numbers.

Mathematics applied to logic started with Boole's *algebra of logic*, a translation of certain kinds of (everyday) logical arguments into the language of classes, in which one could apply algebraic computations.

The development of *logic for mathematics* started in 1879 when Frege undertook to define the concept of natural number in a formal logic system that had quantifiers, axioms, and rules of inference. Frege's work was the main logical influence on Whitehead and Russell's *Principia Mathematica* (1910–1913).

Both inspired Hilbert's study of mathematical logic, a synthesis of foundations and algorithms, with a special fondness for formal number theory.

By 1931 Gödel could say (see van Heijenoort [**41**], p. 596):

> The development of mathematics toward greater precision has led, as is well known, to the formalization of large tracts of it, so that one can prove any theorem using nothing but a few mechanical rules. The most comprehensive formal systems that have been set up hitherto are the system of Principia Mathematica on the one hand, and the Zermelo–Fraenkel axiom system of set theory (further developed by J. von Neumann) on the other. These two systems are so comprehensive that in them all methods of proof used in mathematics today are formalized, that is, reduced to a few axioms and rules of inference.

2. When Boole introduced his algebra for the logic of classes he gave a general procedure for determining the validity of arguments. His procedure was slow, and several mathematicians (including Boole) experimented with trying to speed up the process of determining the validity of an argument, with no clear success. Boole's main focus was solving the *Elimination Problem* for the logic of classes, a procedure to determine the most general conclusion involving some of the classes in the premisses, eliminating the others.

Schröder pointed out that the use of equations in the logic of classes was inadequate for the Aristotelian syllogisms that involve the *particular* statements, that one needed to use negated equations as well. Subsequently, Schröder put considerable effort into the *Elimination Problem* for sets of equations and negated equations. In modern terminology this is the problem of eliminating quantifiers in the logic of classes. A clear solution to this problem was not available until Skolem's work of 1919, which showed that the addition of *numerical* predicates made elimination possible. (Schröder had indicated that such was possible.) Again the algorithm was slow.

However, by this time it was generally recognized that the logic of classes was too weak to handle typical arguments in mathematics. For this one needed a logic with predicates of several variables, or an axiomatic set theory (with a binary predicate). In this more general setting Hilbert and Ackermann (1928) said that the *Decision Problem,* i.e., finding an algorithm to determine which statements were true, was the most important problem of mathematical logic.

In 1929 Presburger found such an algorithm for the first–order statements about $(\mathrm{Z}, +)$. The set of true first–order statements about the integers with $+$ is now called *Presburger arithmetic.* In the 1940s Tarski published his proof that the first–order theory of the real line $(\mathrm{R}, +, \times)$ is decidable. Both of these algorithms were based on quantifier elimination.

In 1928 Skolem proposed a reduction of first–order logic to propositional logic as a more efficient approach to theorem proving. His reduction method was essentially Löwenheim's method of proof to show that a sentence is true iff one cannot construct a countable countermodel to its negation. One tries to construct a countable countermodel, and if at some finite stage all avenues for continuing the construction are blocked off, then the sentence is true. A closely related theorem, obtained by replacing "true" by "derivable," was proved by Herbrand in his thesis (1930).

3. A *completeness theorem* for a proof system says that the axioms and rules of inference one is using are adequate to prove all true statements in the logic. The first completeness theorem was for propositional logic in 1921. Emil Post showed that the propositional logic part of Whitehead and Russell's *Principia Mathematica* was complete, i.e., one could prove all tautologies with it.

Hilbert and Ackermann asked, in their 1928 book, if the axioms and rules they had presented for first–order logic were complete. They said their investigations had certainly supported this possibility. In 1929 the young student Gödel proved that indeed their first–order system was complete.

From this proof it follows that one can "generate" all first–order theorems by simply generating all finite sequences of formulas and keeping those that check out to be proofs. In this sense a completeness theorem can guarantee that one at least has the theoretical possibility of generating all theorems by a computer.

In 1935 Garrett Birkhoff proved that the usual rules for manipulating equations that one learns in high school are complete for equational logic.

4. In 1931 Gödel showed that in spite of the fact that at least two proof systems were strong enough to prove all known true statements of mathematics, no known formalization of mathematics was actually strong enough to prove *all* the first–order sentences that are true in $\mathbf{N} = (N, +, \cdot)$, the natural numbers with addition and multiplication. This is his famous *incompleteness theorem.* His proof made use of primitive recursive functions.

In the mid 1930s Alan Turing introduced the Turing machine as a model of computation that captured the notion of an algorithm, and he showed that the halting problem could not be solved by a Turing machine.

At about this same time Alonzo Church proposed that functions definable in his lambda calculus were precisely the functions computable by an algorithm. Church's class of functions agree with the partial recursive functions, a generalization of Gödel's primitive recursive functions, and functions that are computable by Turing machines.

Under his hypothesis Church showed that first–order logic did not have a decision procedure. Rosser extended Church's work on the undecidability of the theorems of Peano arithmetic to show that there is no algorithm to decide which first–order sentences are true in $\mathbf{N}$.

In particular, Church's work showed that the *Decision Problem* of Hilbert and Ackermann for first–order logic could not be solved in the affirmative.

5. The need for electronic computers grew out of efforts to break coded radio messages during World War II and to calculate tables of the trajectories of shells fired from heavy artillery.

6. M. Davis was one of the first to use computers for automatic theorem proving. In 1954 he worked on implementing the algorithm for Presburger arithmetic. In the years 1958–1960 H. Wang wrote a program that could automatically prove all the approximately 350 first–order theorems in *Principia Mathematica* in about 10 minutes. Davis and Putnam, as well as Wang, wrote general theorem proving programs based on the ideas of Löwenheim, Skolem, and Herbrand.

7. The work in the 1950s on general automated theorem proving revealed that the algorithms used were too slow to be of much use. In the 1960s we saw an effort made to find better algorithms. The major successes were *resolution*, introduced by John A. Robinson (1965) for clause logic; *paramodulation*, introduced by Larry Wos and George Robinson (1970) to deal with clause logic with equality; and the *Knuth–Bendix algorithm* (1967/1970) for equational logic. These methods are currently widely used. Resolution is used in the popular language Prolog.

8. Still, there were no great breakthroughs in automated theorem proving, and by 1970 there was considerable suspicion that some problems might be computationally intractable. In the 1960s Cobham and Edmonds recognized the importance of *polynomial time* (P) and *nondeterministic polynomial time* (NP) for the evaluation of the efficiency of algorithms. By the early 1970s the work of Cook and Karp showed that several problems of finite mathematics were polynomial time equivalent to the problem of determining if a propositional formula is satisfiable, and among the problems that could be solved in nondeterministic polynomial time, they were the most difficult. Thus an efficient (= polynomial time) algorithm for the propositional calculus would have wide–ranging ramifications in discrete mathematics. The current view is that no such algorithm exists. Some mathematicians consider the settling of this open question to be among the most important problems in mathematics.

Also, in the early 1970s the work of Meyer and Stockmeyer led to the discovery that some simple decidable first–order theories are such that any decision algorithm eventually requires an exponential amount of time. Fischer and Rabin then showed that this is also true of Presburger arithmetic and the first–order theory of the reals.

9. The decade of the 1980s saw the introduction of microcomputers, that led to the involvement of a greater number of mathematicians with computers. The increases in memory and speed have led to theorem provers that have been able to show (the already known facts) that the ring axioms plus $x^n \approx x$ imply $xy \approx yx$ for $n = 2, 3, 4, 6$, and that strongly regular rings are regular.

And, as we mentioned at the end of Chapter 3, these developments led to the remarkable solution in 1996 of the *Robbins Conjecture* on Boolean algebras by the automated theorem proving program EQP written by William McCune.

Looking at various ways of formalizing mathematics, we sense the hope of finding a symbolic notation that will guide us toward the truth.

Appendix B

The Dedekind–Peano Number System

Let P be the positive natural numbers, and let $'$ be the successor function, i.e., n' is the next natural number after n. We assume the following:

P1: 1 is not the successor of any number.
P2: If $m' = n'$, then $m = n$.
P3: (*Induction*) If $X \subseteq P$ is closed under successor, and if $1 \in X$, then $X = P$.

From just these simple assumptions we will be able to define the basic operations on the natural numbers and prove their basic properties. The presentation is reasonably close to the original work of Dedekind (1888) but starts from Peano's axioms (1889). The justifications for most of the steps have been left to the reader.

Definition B.0.1 [ADDITION] Let addition be defined as follows:

i. $n + 1 = n'$
ii. $m + n' = (m + n)'$.

Lemma B.0.2 $m' + n = m + n'$.

PROOF. By induction on n.
For $n = 1$:

$m' + 1$	$=$ $(m')'$	by	
	$=$ $(m + 1)'$	by	
	$=$ $m + 1'$.	by	

Induction Hypothesis: $m' + n = m + n'$.

$m' + n'$	$=$ $(m' + n)'$	by	
	$=$ $(m + n')'$	by	
	$=$ $m + n''$.	by	

Lemma B.0.3 $m' + n = (m + n)'$.

PROOF.

$$\begin{aligned} m' + n &= m + n' && \text{by } \boxed{} \\ &= (m+n)'. && \text{by } \boxed{} \end{aligned}$$

Lemma B.0.4 $1 + n = n'$.

PROOF. By induction on n.

For $n = 1$:

$$1 + 1 = 1'. \quad \text{by } \boxed{}$$

Induction Hypothesis: $1 + n = n'$.

$$\begin{aligned} 1 + n' &= (1+n)' && \text{by } \boxed{} \\ &= n''. && \text{by } \boxed{} \end{aligned}$$

Lemma B.0.5 $1 + n = n + 1$.

PROOF.

$$\begin{aligned} 1 + n &= n' && \text{by } \boxed{} \\ &= n + 1. && \text{by } \boxed{} \end{aligned}$$

Lemma B.0.6 $m + n = n + m$.

PROOF. By induction on n.

For $n = 1$:

$$m + 1 = 1 + m. \quad \text{by } \boxed{}$$

Induction Hypothesis: $m + n = n + m$.

$$\begin{aligned} m + n' &= (m+n)' && \text{by } \boxed{} \\ &= (n+m)' && \text{by } \boxed{} \\ &= n + m' && \text{by } \boxed{} \\ &= n' + m. && \text{by } \boxed{} \end{aligned}$$

Lemma B.0.7 $(l + m) + n = l + (m + n)$.

PROOF. By induction on n.

For $n = 1$:

$$\begin{aligned} (l+m) + 1 &= (l+m)' && \text{by } \boxed{} \\ &= l + m' && \text{by } \boxed{} \\ &= l + (m+1). && \text{by } \boxed{} \end{aligned}$$

Induction Hypothesis: $(l+m)+n = l+(m+n)$.

$$\begin{aligned}(l+m)+n' &= ((l+m)+n)' &&\text{by } \boxed{}\\ &= (l+(m+n))' &&\text{by } \boxed{}\\ &= l+(m+n)' &&\text{by } \boxed{}\\ &= l+(m+n'). &&\text{by } \boxed{}\end{aligned}$$

Definition B.0.8 [MULTIPLICATION] Let multiplication be defined as follows:

i. $n \cdot 1 = n$
ii. $m \cdot n' = (m \cdot n) + m$.

Lemma B.0.9 $m' \cdot n = (m \cdot n) + n$.

PROOF. By induction on n.
For $n = 1$:

$$\begin{aligned}m' \cdot 1 &= m' &&\text{by } \boxed{}\\ &= m+1 &&\text{by } \boxed{}\\ &= (m \cdot 1)+1. &&\text{by } \boxed{}\end{aligned}$$

Induction Hypothesis: $m' \cdot n = (m \cdot n) + n$.

$$\begin{aligned}m' \cdot n' &= (m' \cdot n) + m' &&\text{by } \boxed{}\\ &= ((m \cdot n)+n)+m' &&\text{by } \boxed{}\\ &= (m \cdot n)+(n+m') &&\text{by } \boxed{}\\ &= (m \cdot n)+(m'+n) &&\text{by } \boxed{}\\ &= (m \cdot n)+(m+n') &&\text{by } \boxed{}\\ &= ((m \cdot n)+m)+n' &&\text{by } \boxed{}\\ &= (m \cdot n')+n'. &&\text{by } \boxed{}\end{aligned}$$

Lemma B.0.10 $1 \cdot n = n$.

PROOF. By induction on n.
For $n = 1$:

$$1 \cdot 1 = 1. \quad \text{by } \boxed{}$$

Induction Hypothesis: $1 \cdot n = n$.

$$\begin{aligned}1 \cdot n' &= (1 \cdot n)+1 &&\text{by } \boxed{}\\ &= n+1 &&\text{by } \boxed{}\\ &= n'. &&\text{by } \boxed{}\end{aligned}$$

Lemma B.0.11 $m \cdot n = n \cdot m$.

PROOF. By induction on n.
For $n = 1$:

$$\begin{aligned} m \cdot 1 &= m && \text{by } \boxed{} \\ &= 1 \cdot m. && \text{by } \boxed{} \end{aligned}$$

Induction Hypothesis: $m \cdot n = n \cdot m$.

$$\begin{aligned} m \cdot n' &= (m \cdot n) + m && \text{by } \boxed{} \\ &= (n \cdot m) + m && \text{by } \boxed{} \\ &= n' \cdot m. && \text{by } \boxed{} \end{aligned}$$

Lemma B.0.12 $l \cdot (m + n) = (l \cdot m) + (l \cdot n)$.

PROOF. By induction on n.
For $n = 1$:

$$\begin{aligned} l \cdot (m + 1) &= l \cdot m' && \text{by } \boxed{} \\ &= (l \cdot m) + l && \text{by } \boxed{} \\ &= (l \cdot m) + (l \cdot 1). && \text{by } \boxed{} \end{aligned}$$

Induction Hypothesis: $l \cdot (m + n) = (l \cdot m) + (l \cdot n)$.

$$\begin{aligned} l \cdot (m + n') &= l \cdot (m + n)' && \text{by } \boxed{} \\ &= (l \cdot (m + n)) + l && \text{by } \boxed{} \\ &= ((l \cdot m) + (l \cdot n)) + l && \text{by } \boxed{} \\ &= (l \cdot m) + ((l \cdot n) + l) && \text{by } \boxed{} \\ &= (l \cdot m) + (l \cdot n'). && \text{by } \boxed{} \end{aligned}$$

Lemma B.0.13 $(m + n) \cdot l = (m \cdot l) + (n \cdot l)$.

PROOF.

$$\begin{aligned} (m + n) \cdot l &= l \cdot (m + n) && \text{by } \boxed{} \\ &= (l \cdot m) + (l \cdot n) && \text{by } \boxed{} \\ &= (m \cdot l) + (n \cdot l). && \text{by } \boxed{} \end{aligned}$$

Lemma B.0.14 $(l \cdot m) \cdot n = l \cdot (m \cdot n)$.

PROOF. By induction on n.
For $n = 1$:

$$\begin{aligned} (l \cdot m) \cdot 1 &= l \cdot m && \text{by } \boxed{} \\ &= l \cdot (m \cdot 1). && \text{by } \boxed{} \end{aligned}$$

Induction Hypothesis:$(l \cdot m) \cdot n = l \cdot (m \cdot n)$.

$$\begin{aligned}(l \cdot m) \cdot n' &= ((l \cdot m) \cdot n) + (l \cdot m) && \text{by } \boxed{}\\ &= (l \cdot (m \cdot n)) + (l \cdot m) && \text{by } \boxed{}\\ &= l \cdot ((m \cdot n) + m) && \text{by } \boxed{}\\ &= l \cdot (m \cdot n'). && \text{by } \boxed{}\end{aligned}$$

Definition B.0.15 [EXPONENTIATION] Let exponentiation be defined as follows:

i. $a^1 = a$
ii. $a^{n'} = a^n \cdot a$.

Lemma B.0.16 $a^{m+n} = a^m \cdot a^n$.

PROOF. By induction on n. For $n = 1$:

$$\begin{aligned}a^{m+1} &= a^{m'} && \text{by } \boxed{}\\ &= a^m \cdot a && \text{by } \boxed{}\\ &= a^m \cdot a^1. && \text{by } \boxed{}\end{aligned}$$

Induction Hypothesis: $a^{m+n} = a^m \cdot a^n$.

$$\begin{aligned}a^{m+n'} &= a^{(m+n)'} && \text{by } \boxed{}\\ &= a^{m+n} \cdot a && \text{by } \boxed{}\\ &= (a^m \cdot a^n) \cdot a && \text{by } \boxed{}\\ &= a^m \cdot (a^n \cdot a) && \text{by } \boxed{}\\ &= a^m \cdot a^{n'}. && \text{by } \boxed{}\end{aligned}$$

Lemma B.0.17 $(a^m)^n = a^{m \cdot n}$.

PROOF. By induction on n.

For $n = 1$:

$$\begin{aligned}(a^m)^1 &= a^m && \text{by } \boxed{}\\ &= a^{m \cdot 1}. && \text{by } \boxed{}\end{aligned}$$

Induction Hypothesis: $(a^m)^n = a^{m \cdot n}$.

$$\begin{aligned}(a^m)^{n'} &= (a^m)^n \cdot a^m && \text{by } \boxed{}\\ &= a^{m \cdot n} a^m && \text{by } \boxed{}\\ &= a^{(m \cdot n)+m} && \text{by } \boxed{}\\ &= a^{m \cdot n'}. && \text{by } \boxed{}\end{aligned}$$

Lemma B.0.18 $(a \cdot b)^n = a^n \cdot b^n$.

PROOF. By induction on n.

For $n = 1$:

$$\begin{aligned} (a \cdot b)^1 &= a \cdot b && \text{by } \boxed{} \\ &= a^1 \cdot b^1. && \text{by } \boxed{} \end{aligned}$$

Induction Hypothesis: $(a \cdot b)^n = a^n \cdot b^n$.

$$\begin{aligned} \text{Thus } (a \cdot b)^{n'} &= (a \cdot b)^n \cdot (a \cdot b) && \text{by } \boxed{} \\ &= ((a \cdot b)^n \cdot a) \cdot b && \text{by } \boxed{} \\ &= (a \cdot (a \cdot b)^n) \cdot b && \text{by } \boxed{} \\ &= (a \cdot (a^n \cdot b^n)) \cdot b && \text{by } \boxed{} \\ &= ((a \cdot a^n) \cdot b^n) \cdot b && \text{by } \boxed{} \\ &= ((a^n \cdot a) \cdot b^n) \cdot b && \text{by } \boxed{} \\ &= (a^{n'} \cdot b^n) \cdot b && \text{by } \boxed{} \\ &= a^{n'} \cdot (b^n \cdot b) && \text{by } \boxed{} \\ &= a^{n'} \cdot b^{n'}. && \text{by } \boxed{} \end{aligned}$$

Appendix C

Writing Up an Inductive Definition or Proof

C.1 INDUCTIVE DEFINITIONS

The net result of a definition by induction over propositional formulas is that something has been assigned to each propositional formula, e.g., its length, or a truth value, or its subformulas. Suppose one wants to give a definition by induction of something new, say $\mathsf{SweetHeart}(\mathsf{F})$. Here is one way to go about it.

TEMPLATE FOR DEFINITION BY INDUCTION
OVER PROPOSITIONAL FORMULAS

GROUND CASES

- $\mathsf{SweetHeart}\,(0)$ is ... .
- $\mathsf{SweetHeart}\,(1)$ is ... .
- For P a propositional variable, $\mathsf{SweetHeart}\,(P)$ is ... .

Now, assuming that Sweetheart(F) and Sweetheart(G) have been defined, proceed with the INDUCTIVE STEPS

- $\mathsf{SweetHeart}\,(\neg\,\mathsf{F})$ is ... .
- $\mathsf{SweetHeart}\,(\mathsf{F} \vee \mathsf{G})$ is ... .
- $\mathsf{SweetHeart}\,(\mathsf{F} \wedge \mathsf{G})$ is ... .
- $\mathsf{SweetHeart}\,(\mathsf{F} \rightarrow \mathsf{G})$ is ... .
- $\mathsf{SweetHeart}\,(\mathsf{F} \leftrightarrow \mathsf{G})$ is ... .

Remark C.1.1 Sometimes we use the same definition for each of the binary connectives. In this case, rather than writing out all four cases, it suffices to write out the single case:

- $\mathsf{SweetHeart}\,(\mathsf{F} \,\square\, \mathsf{G})$ is ... .

C.2 INDUCTIVE PROOFS

The following is the general result that we use for induction proofs over formulas.

Proposition C.2.1 [PROOF BY INDUCTION OVER PROPOSITIONAL FORMULAS]
Let Assert be a property of propositional formulas. Suppose

- Assert(0) and Assert(1) hold.
- Assert(P) holds for propositional variables P.
- Assert(F) holds implies Assert($\neg$ F) holds.
- Assert(F) holds and Assert(G) holds implies Assert(F $\square$ G) holds, where $\square$ is any of the four binary connectives $\vee$, $\wedge$, $\rightarrow$, $\leftrightarrow$.

Then Assert(F) holds for any propositional formula F.

Here is one format for an induction proof that some assertion Assert(F) holds for any propositional formula F.

TEMPLATE FOR PROOF BY INDUCTION
OVER PROPOSITIONAL FORMULAS

GROUND CASES

- *Give a proof that* Assert(0) *holds.*
- *Give a proof that* Assert(1) *holds.*
- *For P a propositional variable, give a proof that* Assert(P) *holds.*

INDUCTIVE STEPS

- **[Induction Hypothesis]**
  Suppose Assert(F) and Assert(G) hold.

- *Give a proof that* Assert($\neg$ F) *holds.*
- *Give a proof that* Assert(F $\vee$ G) *holds.*
- *Give a proof that* Assert(F $\wedge$ G) *holds.*
- *Give a proof that* Assert(F $\rightarrow$ G) *holds.*
- *Give a proof that* Assert(F $\leftrightarrow$ G) *holds.*

Remark C.2.2 Sometimes we can give the same proof in the inductive steps for the binary connectives. In this case, rather than writing out all four cases, it suffices to write out the single case:

- *Give a proof that* Assert(F $\square$ G) *holds.*

Remark C.2.3 If we want to do a definition or proof by induction over formulas that involve only some of the connectives, say just $\neg$ and $\rightarrow$, then we include only the steps for those connectives, and the propositional variables. Definitions and proofs by induction for other logics follow a similar pattern.

Appendix D

The FŁ Propositional Logic

The propositional logic PC has a generous set of axioms and rules of inference. A good deal of attention has been devoted to finding "smallest possible" propositional proof systems. We will discuss one such in this section—it has three axiom schemata and one rule of inference. The price of this economy is that derivations quickly become quite long.

In Frege's work of 1879 we find a set of six axiom schemata for propositional logic using the connectives $\neg$ and $\rightarrow$, and the single rule of modus ponens. This was simplified in 1928 by Łukasiewicz to just three axiom schemata. We will now look at this propositional proof system, which we call simply FŁ. Most of the reasons for the steps have been omitted. The $\star$'s to the right–hand side of lines of proof indicate omissions for the reader to fill.

Connectives: $\neg, \rightarrow$
Rule of inference: modus ponens
Axiom schemata:
A1: $\mathsf{F} \rightarrow (\mathsf{G} \rightarrow \mathsf{F})$
A2: $(\mathsf{F} \rightarrow (\mathsf{G} \rightarrow \mathsf{H})) \rightarrow ((\mathsf{F} \rightarrow \mathsf{G}) \rightarrow (\mathsf{F} \rightarrow \mathsf{H}))$
A3: $(\neg\mathsf{F} \rightarrow \neg\mathsf{G}) \rightarrow (\mathsf{G} \rightarrow \mathsf{F})$

Definition D.0.4 $\mathcal{S} \vdash \mathsf{F}$ [read: F *can be derived from* $\mathcal{S}$] means there is a sequence of propositional formulas $\mathsf{F}_1, \ldots, \mathsf{F}_n$, with $\mathsf{F} = \mathsf{F}_n$, such that for each i

i. either F_i is an axiom;
ii. or F_i is in $\mathcal{S}$;
iii. or F_i is obtained from two previous F_j's by an application of modus ponens.

Such a sequence is an $\mathcal{S}$–*derivation* (or $\mathcal{S}$–*proof*) of F. A $\emptyset$–derivation is simply called a *derivation.*

The preceding axioms are tautologies and the rule is sound. Now we start to develop the completeness theorem.

Lemma D.0.5 $\vdash \mathsf{F} \rightarrow \mathsf{F}$

Proof. The following sequence of five formulas is a proof of $\mathsf{F} \rightarrow \mathsf{F}$:

1.	$\mathsf{F} \rightarrow ((\mathsf{F} \rightarrow \mathsf{F}) \rightarrow \mathsf{F})$	$\star$
2.	$(\mathsf{F} \rightarrow ((\mathsf{F} \rightarrow \mathsf{F}) \rightarrow \mathsf{F})) \rightarrow ((\mathsf{F} \rightarrow (\mathsf{F} \rightarrow \mathsf{F})) \rightarrow (\mathsf{F} \rightarrow \mathsf{F}))$	$\star$
3.	$(\mathsf{F} \rightarrow (\mathsf{F} \rightarrow \mathsf{F})) \rightarrow (\mathsf{F} \rightarrow \mathsf{F})$	$\star$
4.	$\mathsf{F} \rightarrow (\mathsf{F} \rightarrow \mathsf{F})$	$\star$
5.	$\mathsf{F} \rightarrow \mathsf{F}$.	$\star$

□

Lemma D.0.6 If $\mathcal{S} \vdash \mathsf{F}$ and $\mathcal{S} \vdash \mathsf{F} \rightarrow \mathsf{G}$, then $\mathcal{S} \vdash \mathsf{G}$.

Proof. Let $\mathsf{F}_1, \ldots, \mathsf{F}_m$ be an $\mathcal{S}$–derivation of F, and let $\mathsf{F}_{m+1}, \ldots, \mathsf{F}_n$ be an $\mathcal{S}$–derivation of $\mathsf{F} \rightarrow \mathsf{G}$. Then $\mathsf{F}_1, \ldots, \mathsf{F}_n, \mathsf{G}$ is an $\mathcal{S}$–derivation of G. □

Lemma D.0.7 If $\mathcal{S} \vdash \mathsf{F}$ and $\mathcal{S} \subseteq \mathcal{S}'$, then $\mathcal{S}' \vdash \mathsf{F}$.

Proof. Let $\mathsf{F}_1, \ldots, \mathsf{F}_n$ be an $\mathcal{S}$–derivation of F. Then it is also an $\mathcal{S}'$–derivation of F. □

Lemma D.0.8 [DEDUCTION LEMMA / HERBRAND 1930]

$$\mathcal{S} \cup \{\mathsf{F}\} \vdash \mathsf{G} \qquad \text{iff} \qquad \mathcal{S} \vdash \mathsf{F} \rightarrow \mathsf{G}.$$

Proof. ($\Longleftarrow$) Let $\mathsf{F}_1, \ldots, \mathsf{F}_n$ be an $\mathcal{S}$–derivation of $\mathsf{F} \rightarrow \mathsf{G}$. Then $\mathsf{F}_1, \ldots, \mathsf{F}_n, \mathsf{F}, \mathsf{G}$ is a proof of G from $\mathcal{S} \cup \{\mathsf{F}\}$ because

i. F is in $\mathcal{S} \cup \{\mathsf{F}\}$;

ii. G follows from F_n, F by modus ponens.

Thus $\mathcal{S} \cup \{\mathsf{F}\} \vdash \mathsf{G}$.

($\Longrightarrow$) Let $\mathsf{F}_1, \ldots, \mathsf{F}_n$ be an $\mathcal{S} \cup \{\mathsf{F}\}$–derivation of G. We proceed by induction on n.

For $n = 1$: We must have $\mathsf{F}_1 = \mathsf{G}$, so $\mathsf{G} \in$ Axioms $\cup\, \mathcal{S} \cup \{\mathsf{F}\}$. If $\mathsf{G} = \mathsf{F}$, then $\mathcal{S} \vdash \mathsf{F} \rightarrow \mathsf{G}$ by Lemma D.0.5. If $\mathsf{G} \neq \mathsf{F}$, then $\mathsf{G} \in \mathcal{S} \cup$ Axioms and then

1.	G	$\star$
2.	$\mathsf{G} \rightarrow (\mathsf{F} \rightarrow \mathsf{G})$	$\star$
3.	$\mathsf{F} \rightarrow \mathsf{G}$	$\star$

is an $\mathcal{S}$–derivation of $\mathsf{F} \rightarrow \mathsf{G}$.

For $n > 1$: Suppose that $\mathsf{F}_1, \ldots, \mathsf{F}_n$ is an $\mathcal{S} \cup \{\mathsf{F}\}$–derivation of G. We proceed by induction on n, so the induction hypothesis is as follows: If $\mathsf{F}_1, \ldots, \mathsf{F}_m$ is an $\mathcal{S} \cup \{\mathsf{F}\}$–derivation and $m < n$, then $\mathcal{S} \vdash \mathsf{F} \rightarrow \mathsf{F}_m$.

If $\mathsf{G} \in$ Axioms $\cup\, \mathcal{S} \cup \{\mathsf{F}\}$, then we have (as for $n = 1$) the two cases: If $\mathsf{G} = \mathsf{F}$, then $\mathcal{S} \vdash \mathsf{F} \to \mathsf{G}$ by Lemma D.0.5; and if $\mathsf{G} \neq \mathsf{F}$, then

1.	G	$\star$
2.	$\mathsf{G} \to (\mathsf{F} \to \mathsf{G})$	$\star$
3.	$\mathsf{F} \to \mathsf{G}$	$\star$

is an $\mathcal{S}$–derivation of $\mathsf{F} \to \mathsf{G}$.

If $\mathsf{G} \notin$ Axioms $\cup\, \mathcal{S} \cup \{\mathsf{F}\}$, then G follows from two members of $\{\mathsf{F}_1, \ldots, \mathsf{F}_n\}$ by modus ponens, so for some $i, j < n$ we have F_i is $\mathsf{F}_j \to \mathsf{G}$. Then

1.	$\mathcal{S}$	$\vdash$	$\mathsf{F} \to \mathsf{F}_j$	$\star$
2.	$\mathcal{S}$	$\vdash$	$\mathsf{F} \to (\mathsf{F}_j \to \mathsf{G})$	$\star$
3.	$\mathcal{S}$	$\vdash$	$(\mathsf{F} \to (\mathsf{F}_j \to \mathsf{G})) \to ((\mathsf{F} \to \mathsf{F}_j) \to (\mathsf{F} \to \mathsf{G}))$	$\star$
4.	$\mathcal{S}$	$\vdash$	$(\mathsf{F} \to \mathsf{F}_j) \to (\mathsf{F} \to \mathsf{G})$	$\star$
5.	$\mathcal{S}$	$\vdash$	$\mathsf{F} \to \mathsf{G}.$	$\star$

$\square$

Lemma D.0.9 If $\mathcal{S} \vdash \mathsf{F} \to \mathsf{G}$ and $\mathcal{S} \vdash \mathsf{G} \to \mathsf{H}$, then $\mathcal{S} \vdash \mathsf{F} \to \mathsf{H}$.

Proof.

1.	$\mathcal{S}$	$\vdash$	$\mathsf{F} \to \mathsf{G}$	$\star$
2.	$\mathcal{S}$	$\vdash$	$\mathsf{G} \to \mathsf{H}$	$\star$
3.	$\mathcal{S} \cup \{\mathsf{F}\}$	$\vdash$	G	$\star$
4.	$\mathcal{S} \cup \{\mathsf{F}\}$	$\vdash$	$\mathsf{G} \to \mathsf{H}$	$\star$
5.	$\mathcal{S} \cup \{\mathsf{F}\}$	$\vdash$	H	$\star$
6.	$\mathcal{S}$	$\vdash$	$\mathsf{F} \to \mathsf{H}.$	$\star$

$\square$

Lemma D.0.10 If $\mathcal{S} \vdash \mathsf{F} \to (\mathsf{G} \to \mathsf{H})$ and $\mathcal{S} \vdash \mathsf{G}$, then $\mathcal{S} \vdash \mathsf{F} \to \mathsf{H}$.

Proof.

1.	$\mathcal{S}$	$\vdash$	$\mathsf{F} \to (\mathsf{G} \to \mathsf{H})$	$\star$
2.	$\mathcal{S}$	$\vdash$	G	$\star$
3.	$\mathcal{S} \cup \{\mathsf{F}\}$	$\vdash$	G	$\star$
4.	$\mathcal{S} \cup \{\mathsf{F}\}$	$\vdash$	$\mathsf{G} \to \mathsf{H}$	$\star$
5.	$\mathcal{S} \cup \{\mathsf{F}\}$	$\vdash$	H	$\star$
6.	$\mathcal{S}$	$\vdash$	$\mathsf{F} \to \mathsf{H}.$	$\star$

$\square$

Lemma D.0.11 $\vdash \neg F \to (F \to G)$.

Proof.

1.	$\vdash$	$\neg F \to (\neg G \to \neg F)$	$\star$
2.	$\vdash$	$(\neg G \to \neg F) \to (F \to G)$	$\star$
3.	$\vdash$	$\neg F \to (F \to G)$.	$\star$

□

Lemma D.0.12 $\vdash \neg\neg F \to F$.

Proof.

1.		$\vdash$	$\neg\neg F \to (\neg F \to \neg\neg\neg F)$	$\star$
2.		$\vdash$	$(\neg F \to \neg\neg\neg F) \to (\neg\neg F \to F)$	$\star$
3.		$\vdash$	$\neg\neg F \to (\neg\neg F \to F)$	$\star$
4.	$\neg\neg F$	$\vdash$	$\neg\neg F \to F$	$\star$
5.	$\neg\neg F$	$\vdash$	$\neg\neg F$	$\star$
6.	$\neg\neg F$	$\vdash$	F	$\star$
7.		$\vdash$	$\neg\neg F \to F$.	$\star$

□

Lemma D.0.13 $\vdash F \to \neg\neg F$.

Proof.

1.	$\vdash$	$\neg\neg\neg F \to \neg F$	$\star$
2.	$\vdash$	$(\neg\neg\neg F \to \neg F) \to (F \to \neg\neg F)$	$\star$
3.	$\vdash$	$F \to \neg\neg F$.	$\star$

□

Lemma D.0.14 $\vdash (F \to G) \to (\neg G \to \neg F)$.

Proof.

1.	$F \to G$	$\vdash$	$\neg\neg F \to F$	$\star$
2.	$F \to G$	$\vdash$	$F \to G$	$\star$
3.	$F \to G$	$\vdash$	$\neg\neg F \to G$	$\star$
4.	$F \to G$	$\vdash$	$G \to \neg\neg G$	$\star$
5.	$F \to G$	$\vdash$	$\neg\neg F \to \neg\neg G$	$\star$
6.	$F \to G$	$\vdash$	$(\neg\neg F \to \neg\neg G) \to (\neg G \to \neg F)$	$\star$
7.	$F \to G$	$\vdash$	$\neg G \to \neg F$	$\star$
8.		$\vdash$	$(F \to G) \to (\neg G \to \neg F)$.	$\star$

□

Lemma D.0.15 $\vdash \mathsf{F} \to (\neg\,\mathsf{G} \to \neg\,(\mathsf{F} \to \mathsf{G}))$.

Proof.

1.	$\mathsf{F}, \mathsf{F} \to \mathsf{G}$	$\vdash$	F	$\star$
2.	$\mathsf{F}, \mathsf{F} \to \mathsf{G}$	$\vdash$	$\mathsf{F} \to \mathsf{G}$	$\star$
3.	$\mathsf{F}, \mathsf{F} \to \mathsf{G}$	$\vdash$	G	$\star$
4.	F	$\vdash$	$(\mathsf{F} \to \mathsf{G}) \to \mathsf{G}$	$\star$
5.		$\vdash$	$\mathsf{F} \to ((\mathsf{F} \to \mathsf{G}) \to \mathsf{G})$	$\star$
6.		$\vdash$	$((\mathsf{F} \to \mathsf{G}) \to \mathsf{G}) \to (\neg\,\mathsf{G} \to \neg\,(\mathsf{F} \to \mathsf{G}))$	$\star$
7.		$\vdash$	$\mathsf{F} \to (\neg\,\mathsf{G} \to \neg\,(\mathsf{F} \to \mathsf{G}))$.	$\star$

□

Lemma D.0.16 If $\mathcal{S} \cup \{\mathsf{F}\} \vdash \mathsf{G}$ and $\mathcal{S} \cup \{\neg\,\mathsf{F}\} \vdash \mathsf{G}$, then $\mathcal{S} \vdash \mathsf{G}$.

Proof.

1.	$\mathcal{S} \cup \{\mathsf{F}\}$	$\vdash$	G	$\star$
2.	$\mathcal{S}$	$\vdash$	$\mathsf{F} \to \mathsf{G}$	$\star$
3.	$\mathcal{S}$	$\vdash$	$(\mathsf{F} \to \mathsf{G}) \to (\neg\,\mathsf{G} \to \neg\,\mathsf{F})$	$\star$
4.	$\mathcal{S}$	$\vdash$	$\neg\,\mathsf{G} \to \neg\,\mathsf{F}$	$\star$
5.	$\mathcal{S} \cup \{\neg\,\mathsf{F}\}$	$\vdash$	G	$\star$
6.	$\mathcal{S}$	$\vdash$	$\neg\,\mathsf{F} \to \mathsf{G}$	$\star$
7.	$\mathcal{S}$	$\vdash$	$(\neg\,\mathsf{F} \to \mathsf{G}) \to (\neg\,\mathsf{G} \to \neg\,\neg\,\mathsf{F})$	$\star$
8.	$\mathcal{S}$	$\vdash$	$\neg\,\mathsf{G} \to \neg\,\neg\,\mathsf{F}$	$\star$
9.	$\mathcal{S}$	$\vdash$	$\neg\,\mathsf{G} \to (\neg\,\neg\,\mathsf{F} \to \neg\,(\neg\,\mathsf{G} \to \neg\,\mathsf{F}))$	$\star$
10.	$\mathcal{S}$	$\vdash$	$(\neg\,\mathsf{G} \to (\neg\,\neg\,\mathsf{F} \to \neg\,(\neg\,\mathsf{G} \to \neg\,\mathsf{F})))$ $\to ((\neg\,\mathsf{G} \to \neg\,\neg\,\mathsf{F}) \to (\neg\,\mathsf{G} \to \neg\,(\neg\,\mathsf{G} \to \neg\,\mathsf{F})))$	$\star$
11.	$\mathcal{S}$	$\vdash$	$(\neg\,\mathsf{G} \to \neg\,\neg\,\mathsf{F}) \to (\neg\,\mathsf{G} \to \neg\,(\neg\,\mathsf{G} \to \neg\,\mathsf{F}))$	$\star$
12.	$\mathcal{S}$	$\vdash$	$\neg\,\mathsf{G} \to \neg\,(\neg\,\mathsf{G} \to \neg\,\mathsf{F})$	$\star$
13.	$\mathcal{S}$	$\vdash$	$(\neg\,\mathsf{G} \to \neg\,(\neg\,\mathsf{G} \to \neg\,\mathsf{F})) \to ((\neg\,\mathsf{G} \to \neg\,\mathsf{F}) \to \mathsf{G})$	$\star$
14.	$\mathcal{S}$	$\vdash$	$(\neg\,\mathsf{G} \to \neg\,\mathsf{F}) \to \mathsf{G}$	$\star$
15.	$\mathcal{S}$	$\vdash$	G.	$\star$

□

In the following material we use the notion of a truth evaluation e that is slightly differently from that in Chapter 2. Namely, e denotes a mapping from the set of variables X to the truth values $\{0, 1\}$ that is extended to all propositional formulas by the inductive definition $\mathsf{e}(\neg\,\mathsf{F}) = \neg\,\mathsf{e}(\mathsf{F})$ and $\mathsf{e}(\mathsf{F} \to \mathsf{G}) = \mathsf{e}(\mathsf{F}) \to \mathsf{e}(\mathsf{G})$.

Definition D.0.17 Let $\mathsf{F}(P_1, \ldots, P_n)$ be a propositional formula and let $\widetilde{P_1}, \ldots, \widetilde{P_n}$ be a sequence where $\widetilde{P_i} \in \{P_i, \neg P_i\}$. Let e be a truth evaluation such that $\mathsf{e}(\widetilde{P_i}) = 1$, for $i \leq n$. Then let

$$\widetilde{\mathsf{F}} = \begin{cases} \mathsf{F} & \text{if } \mathsf{e}(\mathsf{F}) = 1 \\ \neg\,\mathsf{F} & \text{if } \mathsf{e}(\mathsf{F}) = 0. \end{cases}$$

Lemma D.0.18 [KALMAR]) Let $\mathsf{F}, \widetilde{P_1}, \ldots, \widetilde{P_n}, \widetilde{\mathsf{F}}$ and e be as in Definition D.0.17. Then $\widetilde{P_1}, \ldots, \widetilde{P_n} \vdash \widetilde{\mathsf{F}}$.

Proof. The proof is by induction on the number k of connectives in the formula F.

If $k = 0$: Then F is P_i for some i. If $\widetilde{P_i} = P_i$, then $\widetilde{\mathsf{F}} = \mathsf{F}$. If $\widetilde{P_i} = \neg\,P_i$, then $\widetilde{\mathsf{F}} = \neg\,\mathsf{F}$. Thus $\widetilde{P_i} = \widetilde{\mathsf{F}}$, so $\widetilde{P_i} \vdash \widetilde{\mathsf{F}}$.

If $k > 0$: Then our induction hypothesis is that the lemma holds for any propositional formula with fewer than k connectives.

Case i: Suppose F is $\neg\,\mathsf{G}$. Then G has fewer connectives than F, so $\widetilde{P_1}, \ldots, \widetilde{P_n} \vdash \widetilde{\mathsf{G}}$.

i$_a$: Suppose $\mathsf{e}(\mathsf{G}) = 1$. Then $\widetilde{\mathsf{G}} = \mathsf{G}$ and

$$\begin{array}{llll|l} 1. & \widetilde{P_1}, \ldots, \widetilde{P_n} & \vdash & \mathsf{G} & \star \\ 2. & \widetilde{P_1}, \ldots, \widetilde{P_n} & \vdash & \mathsf{G} \to \neg\neg\,\mathsf{G} & \star \\ 3. & \widetilde{P_1}, \ldots, \widetilde{P_n} & \vdash & \neg\neg\,\mathsf{G}. & \star \end{array}$$

Since $\mathsf{e}(\mathsf{F}) = 0$, we have $\widetilde{\mathsf{F}} = \neg\,\mathsf{F} = \neg\neg\,\mathsf{G}$, so $\widetilde{P_1}, \ldots, \widetilde{P_n} \vdash \widetilde{\mathsf{F}}$.

i$_b$: Suppose $\mathsf{e}(\mathsf{G}) = 0$. Then $\widetilde{\mathsf{G}} = \neg\,\mathsf{G}$ and $\widetilde{P_1}, \ldots, \widetilde{P_n} \vdash \neg\,\mathsf{G}$. Since $\mathsf{e}(\mathsf{F}) = 1$, we have $\widetilde{\mathsf{F}} = \mathsf{F} = \neg\,\mathsf{G}$, so $\widetilde{P_1}, \ldots, \widetilde{P_n} \vdash \widetilde{\mathsf{F}}$.

Case ii. F is $\mathsf{G} \to \mathsf{H}$. Then G and H have fewer connectives than F so

$$\begin{array}{lll} \widetilde{P_1}, \ldots, \widetilde{P_n} & \vdash & \widetilde{\mathsf{G}} \\ \widetilde{P_1}, \ldots, \widetilde{P_n} & \vdash & \widetilde{\mathsf{H}}. \end{array}$$

ii$_a$: Suppose $\mathsf{e}(\mathsf{G}) = 0$. Then $\widetilde{\mathsf{G}} = \neg\,\mathsf{G}$, and as $\mathsf{e}(\mathsf{F}) = 1$, $\widetilde{\mathsf{F}} = \mathsf{F}$. Thus we have

1.	$\widetilde{P}_1, \ldots, \widetilde{P}_n$	$\vdash$	$\neg\,\mathsf{G}$	$\star$
2.	$\widetilde{P}_1, \ldots, \widetilde{P}_n$	$\vdash$	$\neg\,\mathsf{G} \to (\mathsf{G} \to \mathsf{H})$	$\star$
3.	$\widetilde{P}_1, \ldots, \widetilde{P}_n$	$\vdash$	$\mathsf{G} \to \mathsf{H}$.	$\star$

As $\widetilde{\mathsf{F}} = \mathsf{G} \to \mathsf{H}$, we have $\widetilde{P}_1, \ldots, \widetilde{P}_n \vdash \widetilde{\mathsf{F}}$.

ii$_b$: Suppose $\mathsf{e}(\mathsf{H}) = 1$. Then $\widetilde{\mathsf{H}} = \mathsf{H}$, and as $\mathsf{e}(\mathsf{F}) = 1$, $\widetilde{\mathsf{F}} = \mathsf{F}$. Then we have

1.	$\widetilde{P}_1, \ldots, \widetilde{P}_n$	$\vdash$	H	$\star$
2.	$\widetilde{P}_1, \ldots, \widetilde{P}_n$	$\vdash$	$\mathsf{H} \to (\mathsf{G} \to \mathsf{H})$	$\star$
3.	$\widetilde{P}_1, \ldots, \widetilde{P}_n$	$\vdash$	$\mathsf{G} \to \mathsf{H}$.	$\star$

As $\widetilde{\mathsf{F}} = \mathsf{G} \to \mathsf{H}$, $\widetilde{P}_1, \ldots, \widetilde{P}_n \vdash \widetilde{\mathsf{F}}$.

ii$_c$: Suppose $\mathsf{e}(\mathsf{G}) = 1$, $\mathsf{e}(\mathsf{H}) = 0$. Then $\mathsf{e}(\mathsf{F}) = 0$, so $\widetilde{\mathsf{G}} = \mathsf{G}$, $\widetilde{\mathsf{H}} = \neg\,\mathsf{H}$, $\widetilde{\mathsf{F}} = \neg\,\mathsf{F}$. Then we have

1.	$\widetilde{P}_1, \ldots, \widetilde{P}_n$	$\vdash$	G	$\star$
2.	$\widetilde{P}_1, \ldots, \widetilde{P}_n$	$\vdash$	$\neg\,\mathsf{H}$	$\star$
3.	$\widetilde{P}_1, \ldots, \widetilde{P}_n$	$\vdash$	$\mathsf{G} \to (\neg\,\mathsf{H} \to \neg\,(\mathsf{G} \to \mathsf{H}))$	$\star$
4.	$\widetilde{P}_1, \ldots, \widetilde{P}_n$	$\vdash$	$\neg\,\mathsf{H} \to \neg\,(\mathsf{G} \to \mathsf{H})$	$\star$
5.	$\widetilde{P}_1, \ldots, \widetilde{P}_n$	$\vdash$	$\neg\,(\mathsf{G} \to \mathsf{H})$.	$\star$

As $\widetilde{\mathsf{F}} = \neg\,(\mathsf{G} \to \mathsf{H})$, we have $\widetilde{P}_1, \ldots, \widetilde{P}_n \vdash \widetilde{\mathsf{F}}$.

□

Theorem D.0.19 [COMPLETENESS: FŁ]

$$\vdash \mathsf{F} \quad \text{if} \quad \mathsf{F} \text{ is a tautology.}$$

Proof. Let $\mathsf{F}(P_1, \ldots, P_n)$ be a tautology. Then for any $\widetilde{P}_1, \ldots, \widetilde{P}_n$ and e we have $\widetilde{\mathsf{F}} = \mathsf{F}$. Thus

1.	$\widetilde{P}_1, \ldots, \widetilde{P}_{n-1}, P_n$	$\vdash$	F	$\star$
2.	$\widetilde{P}_1, \ldots, \widetilde{P}_{n-1}, \neg\,P_n$	$\vdash$	F	$\star$
3.	$\widetilde{P}_1, \ldots, \widetilde{P}_{n-1}$	$\vdash$	F.	$\star$

Continuing, we have $\vdash \mathsf{F}$. □

Remark D.0.20 Although the proof of the completeness of FŁ looks theoretical, we emphasize that this (like the situation for PC) is a *constructive proof*; it gives an *algorithm* such that when presented with a tautology F one can find a derivation of F. This algorithm can be [and has been!] implemented on a computer. Unfortunately, these derivations tend to be *extremely* long.

Theorem D.0.21 [GENERALIZED SOUNDNESS AND COMPLETENESS] In the FŁ propositional proof system we have $\mathcal{S} \vdash \mathsf{F}$ iff $\mathcal{S} \models \mathsf{F}$.

Proof. ($\Longrightarrow$: Generalized soundness) Suppose $\mathcal{S} \vdash \mathsf{F}$. Then since derivations are finite, there are formulas $\mathsf{F}_1, \ldots, \mathsf{F}_n \in \mathcal{S}$ such that $\mathsf{F}_1, \ldots, \mathsf{F}_n \vdash \mathsf{F}$. Thus by repeated application of Lemma D.0.8 we have $\vdash \mathsf{F}_1 \to (\mathsf{F}_2 \to \cdots (\mathsf{F}_n \to \mathsf{F}) \cdots)$, and thus by the completeness of FŁ (see Theorem D.0.19) $\mathsf{F}_1 \to (\mathsf{F}_2 \to \cdots (\mathsf{F}_n \to \mathsf{F}) \cdots)$ is a tautology. Thus any assignment that satisfies $\mathcal{S}$ will make each of $\mathsf{F}_1, \ldots, \mathsf{F}_n$ true and thus will also make F true. Thus $\mathcal{S} \models \mathsf{F}$.
($\Longleftarrow$: Generalized completeness) Suppose $\mathcal{S} \nvdash \mathsf{F}$. Then for $\mathsf{F}_1, \ldots, \mathsf{F}_n \in \mathcal{S}$, certainly $\mathsf{F}_1, \ldots, \mathsf{F}_n \nvdash \mathsf{F}$. Using Lemma D.0.8 again, we have $\nvdash \mathsf{F}_1 \to (\mathsf{F}_2 \to \cdots (\mathsf{F}_n \to \mathsf{F}) \cdots)$, so by the completeness of FŁ, $\mathsf{F}_1 \to (\mathsf{F}_2 \to \cdots (\mathsf{F}_n \to \mathsf{F}) \cdots)$ is not a tautology. Thus it is possible to find an assignment of truth values that satisfies $\mathsf{F}_1, \ldots, \mathsf{F}_n$ but makes F false. Thus $\mathsf{F}_1, \ldots, \mathsf{F}_n \not\models \mathsf{F}$. In other words, every finite subset of $\mathcal{S} \cup \{\neg \mathsf{F}\}$ is satisfiable. So by the compactness theorem (Theorem 2.8.1), $\mathcal{S} \cup \{\neg \mathsf{F}\}$ itself is satisfiable. But then $\mathcal{S} \not\models \mathsf{F}$. $\square$

Exercises

D.0.1 Fill in the missing details (noted by $\star$'s) in the preceding proof of completeness.

Bibliography

[1] F. Baader and J. Siekmann, Unification theory. In *Handbook of Logic in Artificial Intelligence and Logic Programming*, vol. 2. Edited by D. M. Gabbay, C. J. Hogger, and J. A. Robinson, 41–125, Oxford Univ. Press, 1994.

[2] L. Bachmair and N. Dershowitz, Equational inference, canonical proofs, and proof orderings. *J. Assoc. Comput. Mach.* **41** (1994), 236–276.

[3] W. W. Bartley, III, ed., *Lewis Carroll's Symbolic Logic.* New York: Clarkson N. Potters, 1977.

[4] G. Birkhoff, On the structure of abstract algebras. *Proc. of the Camb. Phil. Soc.* **29** (1935), 433–454.

[5] G. Boole, *An Investigation of the Laws of Thought.* Toronto: A Dover reprint of the 1854 edition.

[6] S. Burris and S. Lee, Tarski's high school identities. *Amer. Math. Monthly* **100**, no. 3 (1993), 231–236.

[7] L. Carroll, *Symbolic Logic and The Game of Logic.* 1897/87. New York: Two books bound as one by Dover, 1958.

[8] A. Church, *Introduction to Mathematical Logic.* vol. 1. Princeton Univ. Press, 1956.

[9] V. Chvátal and E. Szemerédi, Many hard examples for resolution. *J. Assoc. Comput. Mach.* **35** (1988), 759–768.

[10] S. A. Cook, The complexity of theorem proving procedures. *Proc. 3d Annual ACM Symposium on Theory of Computing* (1971), 151–158.

[11] I. M. Copi, *Symbolic Logic.* New York: Macmillan, 1954.

[12] R. Dedekind, *Was sind und was sollen die Zahlen?* Braunschweig, 1888.

[13] H. DeLong, *A Profile of Mathematical Logic.* Reading: Addison–Wesley, 1971.

[14] H. D. Ebbinghaus, J. Flum, and W. Thomas, *Mathematical Logic.* Undergraduate Texts in Math., New York: Springer–Verlag, 1981.

[15] T. Evans, On multiplicative systems defined by generators and relations, I. *Proc. Cambr. Philos. Soc.* **47** (1951), 637–649.

[16] J. E. Fenstad, ed., *Th. Skolem, Selected Works in Logic.* Oslo: Scand. Univ. Books, Universitetsforlaget, 1970.

[17] G. Frege, *Begriffsschrift. Eine der arithmetischen nachgebildeten Formelsprache des reinen Denkens.* Halle, 1879.

[18] G. Frege, *Grundlagen der Arithmetik, eine logisch–mathematische Untersuchung über den Begriff der Zahl.* Breslau, 1884.

[19] G. Frege, *Grundgesetze der Arithmetik, begriffsschriftlich abgeleitet.* I./II. Jena, 1893/1903.

[20] Z. Galil, On the complexity of regular resolution and the Davis–Putnam procedure. *Theor. Comp. Sci.* **4** (1977), 23–46.

[21] M. Gardner, *Logic, Machines, and Diagrams, 2nd ed.* U. Chicago Press, 1982.

[22] K. Gödel, Die Vollständigkeit der Axiome des logisch Funktionenkalküls. *Monatshefte für Mathematik und Physik*, **37** (1930), 349–360. [Translated in *From Frege to Gödel*, van Heijenoort, Harvard Univ. Press, 1971.]

[23] A. Haken, The intractability of resolution. *Theoretical Computer Science* **39** (1985), 297–308.

[24] L. Henkin, The discovery of my completeness proofs. *Bull. Symbolic Logic*, **2** (1996), 127–158.

[25] J. Herbrand, Recherches sur la théorie de la demonstration. Ph. D. thesis, 1930, Paris. [Translated in *From Frege to Gödel*, van Heijenoort, Harvard Univ. Press, 1971.]

[26] D. Hilbert and W. Ackermann, *Grundzügen der theoretischen Logik.* Heidelberg: Springer–Verlag, 1928.

[27] G. Huet, A complete proof of the correctness of the Knuth–Bendix completion algorithm. *J. Computer and Systems Sciences* **23** (1981), 11–21.

[28] W. S. Jevons, *Pure Logic.* E. Stanford, 1864.

[29] S. C. Kleene, *Mathematical Logic.* New York: J. Wiley & Sons, 1967.

[30] W. and M. Kneale, *The Development of Logic.* Oxford Univ. Press, 1962.

[31] D. E. Knuth and P. B. Bendix, Simple word problems in universal algebras. in *Computational Problems in Abstract Algebra.* Edited by J. Leech, Oxford: Pergamon Press, 1970, 263–297.

[32] L. Löwenheim, Über Möglichkeiten im Relativkalkül. *Math. Ann.* **68** (1915), 169–207. [Translated in *From Frege to Gödel*, van Heijenoort, Harvard Univ. Press, 1971.]

[33] E. Mendelson, *Introduction to Mathematical Logic.* 3d ed., Monterey: Wadsworth and Brooks/Cole, 1987.

[34] G. Peano, Principles of arithmetic, presented by a new method. 1889. [Translated in *From Frege to Gödel*, van Heijenoort, Harvard Univ. Press, 1971.]

[35] E. Post, Introduction to a general theory of elementary propositions of logic. *Amer. J. Math.* **43** (1921), 163–185.

[36] J.A. Robinson, Logic and logic programming. *Commun. ACM* **35** (1992), 40–65.

[37] J. B. Rosser, Extensions of some theorems of Gödel and Church. *J. Symbolic Logic* **1** (1936), 87–91.

[38] U. Schöning, *Logic for Computer Scientists.* Boston: Birkhäuser, 1989.

[39] E. Schröder, *Algebra der Logik*, vols. 1–3. 1890–1910. New York: Chelsea reprint 1966.

[40] A. Urquhart, Hard examples for resolution. *J. Assoc. Comput. Mach.* **34** (1987), 209–219.

[41] J. van Heijenoort, ed., *A Source Book in Mathematical Logic.* Harvard Univ. Press, 1967.

[42] A. N. Whitehead and B. Russell, *Principia Mathematica, vols. 1–3.* Cambridge Univ. Press, 1910–1913.

INDEX